AKADEMIE DER WISSENSCHAFTEN DER DDR

Zentralinstitut für Physik der Erde (ZIPE)

Seismological Bulletin 1971
Station Moxa (MOX)

By

Johannes Stelzner, Dorothea Güth,
and Joachim Weyrauch

With 2 Figures

AKADEMIE-VERLAG · BERLIN

1976

Erschienen im Akademie-Verlag, 108 Berlin, Leipziger Straße 3—4

Lizenznummer: 202 · 100/448/76
Gesamtherstellung:
VEB Druckerei „Thomas Müntzer", DDR-582 Bad Langensalza
Bestellnummer: 762 401 3 (2004/B/1971) · LSV 1437
Printed in GDR
DDR 42,-- M

PREFACE

The annual Seismological Bulletin 1971 for the Station Moxa (MOX) continues the series of publications about seismological records at the Central Station Moxa edited by the Central Earth Physics Institute of the Academy of Sciences of the German Democratic Republic.

The provisional analysis of the records of station Moxa was performed in the Seismological Service under the direction of JOHANNES STELZNER by JOACHIM WEYRAUCH and BRIGITTE HÄNSCH.

The annual Bulletin 1971 was prepared by JOHANNES STELZNER and DOROTHEA GÜTH with the technical assistance of URSULA DÖRING.

Control of the instruments of the station Moxa was carried out under the direction of CHRISTIAN TEUPSER.

H. KAUTZLEBEN
Director

TABLE OF CONTENTS

Preliminary notes for the interpretation of seismograms

In the Bulletin the international code is used:

1. Phase interpretation

Pg — direct longitudinal wave in near epicentral distances ($D < 10°$)

Pb, Pn — guided longitudinal head waves along the CONRAD- or MOHOROVIČIĆ-discontinuity ($D < 10°$)

P — direct longitudinal wave travelled through the earth mantle

P diff — direct longitudinal wave diffracted around the core boundary

PKIKP — direct longitudinal wave travelled through the inner core (travel-time branch DF)

PKHKP — direct longitudinal wave refracted in the intermediary zone between inner and outer core. Phase symbol according to BOLT [1] (travel-time branch GH)

PKP2 — direct longitudinal wave travelled through the — outer core only (travel-time branch AB)

PKP — first noticeable onset of longitudinal core phase not identified

PP, PPP — waves reflected on the earth surface with permanent longitudinal character

PKKP — core phase reflected once within the core at the outer core boundary

PKPPKP — longitudinal core phase reflected at the earth surface

Sg — direct transversal wave in near epicentral distances ($D < 10°$)

Sb, Sn — guided transversal head waves along the CONRAD- or MOHOROVIČIĆ-discontinuity ($D < 10°$)

S — direct transversal wave travelled through the earth mantle

SKS — direct wave travelled transversal through the mantle and longitudinal through the core

SS, SSS — waves reflected on the earth surface with permanent transversal character

SKKS — wave travelled transversal through the mantle, longitudinal through the core and reflected within the core at the outer core boundary

PcP, ScS, PcS, ScP — longitudinal and transversal waves with steady or changing character reflected at the outer core

PS, SP, PPS — longitudinal and transversal waves with changing character reflected at the surface of the earth

pP, sP, pPP, sPP, pPKIKP, sPKP2, pS — phases of deep-focus earthquakes of longitudinal or transversal waves with steady or changing character. p;s — reflected near the epicentre

pPKP, sPKP — phases of deep focus earthquakes of longitudinal core waves not exactly to be coordinated

SKP, PKS — core phases with different character before and after the direct transit of the core

SKSP — SKS wave with longitudinal character after the reflection at the surface of the earth

P1, P2, P3,...S1, S2,... — multiple onsets of body waves

Pn, Sn — teleseismic Pn and Sn waves in the epicentral distances $23^\circ < D < 40^\circ$ after BÅTH [2]

Pa, Sa — waves probably guided in the astenosphere channel or higher modes of surface waves

PL — leaking modes, normal dispersed train of waves of periods greater than about 10 s, beginning at or near the time of initial P-wave

X, Y, Z — remarkable phases of body waves, not to be identified

LmV, LmH — maximum of the vertical and horizontal component respectively of longperiodical surface waves. If there are several maxima with comparable proportions in A/T, the numeration was carried out in a temporal sequence e.g. Lm1H, Lm2H

The phase symbol is followed by the designation of the type of seismometer from which the time of onsets is taken.

A — seismograph with amplitude characteristic of type A (short-period)
B — seismograph with amplitude characteristic of type B (middle-period)
C — seismograph with amplitude characteristic of type C (long-period)

2. Measurement of amplitudes and calculation of magnitudes

All data of amplitudes and periods printed in the column "remarks" are always taken from the records of the same instruments, from which are taken the onset-times of the corresponding phases. The symbol of phase and component is followed by the symbol of the type of instruments e.g.: PV A, PV B, LmH B, LmV C.

Data of amplitudes obtained from records of instruments of type A are given in units of length of nm (1 nm = 1 nanometre = 10^{-6} millimetre). Data of amplitudes obtained from instruments of type B and such obtained from instruments of type C are given in units of length μm (1 μm = 1 mikrometre = 10^{-3} millimetre) e. g.: PV A 1.3 s 38.6 nm, SH B 10 s 3.2 μm, LmH B 22 s 15 μm.

Magnitudes are determined from all those phases, for which calibrating functions are known and internationally used, i.e.

for maxima of body waves P(PH, PV), PP(PPH, PPV), and S(SH)-Q-functions from Gutenberg and Richter [3] — and

for maxima of surface waves (h < 100 km) LmH, LmV — calibrating functions from Prague σ [4] —.

The station correction S was not yet taken into consideration.

MB — magnitude of vertical component V of the first onset of P-waves given by Neic
MS — magnitude of horizontal component H of the maximum surface wave given by Neic
M — magnitude calculated from given data of station Moxa. Notice the wave type and the type of instruments written on the same line

3. Direction of body-wave onsets

If the direction of motion at the beginning of a wave onset is clearly to be recognized, the sign + or — is placed before the phase symbol. It means:

in the Z component	+	ground motion upwards, compression
	—	ground motion downwards, dilatation
in the N component	+	ground motion to the north
	—	ground motion to the south
in the E component	+	ground motion to the east
	—	ground motion to the west

4. *Further abbreviations*

i	— sharp beginning of phase motion (impetus)
e	— gradual beginning of phase motion (emersio)
D	— epicentral distances in degree (°), calculated according to geocentric coordinates, the maximum error of the own calculations amounts to ± 0,1°
Az	— azimuth: clockwise measured angle between north direction in epicentre and the connecting line from epicentre to station Moxa
h	— depth of focus in km, our data for depth of focus are based on travel-time curves for deep focus earthquakes after GUTENBERG and RICHTER [5]
H	— origin time in GMT (Greenwich Mean Time)
NEIC	— National Earthquake Information Center Boulder, Colorado, USA
BCIS	— Bureau Central International de Séismologie, Strasbourg France,
ANUSSR	— Academia Nauk USSR, Moscow
USAEC	— United States Atomic Energy Commission, Washington, D.C.
ISC	— International Seismological Centre, Edinburgh
HAN-	— Niedersächsisches Landesamt für Bodenforschung Hannover-Buchholz, FRG

For abbreviations of seismological stations and other agencies in the international three letter code see the introductions to the Regional Catalogue of Earthquakes, Edinburgh and the Bulletins of the International Seismological Centre, Edinburgh.

Round brackets indicate uncertainties in interpretation of phase, time, depth of focus or epicentral distances respectively.

[1] BOLT, A.,The velocity of seismic waves near the earths center. Bull. Seism. Soc. Am. **54** (1964) 1, 191—208.

[2] BÅTH, M., Propagation of Sn and Pn teleseismic distances. Pure and Applied Geophysics **64** (1966/II) 19—30.

[3] GUTENBERG, B. and RICHTER, C. F., Magnitude and energy of earthquakes. Annali di Geofisica **9** (1956) 1, 1—15.

[4] KÁRNÍK, V., KONDORSKAJA, N. V. u. a., Standardization of the earthquake magnitude scale. Stud. Geophys. et Geodet., Prague **6** (1962) 41—48.

[5] GUTENBERG, B. and RICHTER, C. F., Materials for the study of deep-focus earthquakes. Bull. Seism. Soc. Am. **26** (1936) 4, 341—390.

Seismological Station Moxa (MOX)
of the Central Earth Physics Institute

Elevation above
mean sea level: 455 m

Bedrock: clay slate of the lower carboniferous formation

Geographic
coordinates: $\varphi = 50°38'46''$N $\lambda = 11°36'58''$ E

Address: Central Earth Physics Institute
Seismological Service
GDR-69, Jena, Burgweg 11
German Democratic Republic
Telex: 058 86275 seis dd

Seismographs and their parameters 1971

T_s — seismometer free period
T_g — galvanometer free period
D_s — seismograph damping
D_g — galvanometer damping
V_0 — magnification factor
N — north-south component
E — east-west component
Z — vertical component
σ^2 — coupling coefficient

SKM — Seismograph Kirnos modified
SSJ — Seismic Station Apparatus Type Jena
VSJ — Vertical Seismograph Type Jena

Type of Seismograph	Comp.	T_s/[s]	T_g/[s]	D_s	D_g	V_0	σ^2
VSJ II	Z	0.23	0.067	0.33	1.2	300000	0.048
VSJ II	Z	1.0	1.0	0.5	0.5	47200	0.56
A SKM III	N	1.64	0.39	0.52	1.98	25700	0.047
	E	1.64	0.39	0.51	1.93	25400	0.046
	Z	1.64	0.39	0.51	1.99	24600	0.050
B SSJ I	N	20	1.13	0.50	8.87	109	0.074
		20	1.14	0.50	8.79	1110	0.074
	E	20	1.13	0.49	8.85	103	0.071
		20	1.16	0.49	8.61	1050	0.071
	Z	20	1.13	0.48	8.82	108	0.048
		20	1.24	0.48	8.05	1070	0.048
C SSJ I/L							
up to Jan., 18.	N	30	74.9	1.25	0.5	1050	0.051
from Feb., 19.	N	30	89.5	1.49	0.5	1000	0.085
	E	30	74.4	1.24	0.5	1010	0.051
up to Jan., 18.	Z	30	85.6	1.43	0.5	955	0.110
from March, 9.	Z	30	91.0	1.52	0.5	1060	0.108
Strain/L	N		85.4		0.70	65*	
	E		86.2		0.70	67*	
(coupled)	N + E		86.2		0.70	42*	
Wiechert 1200 kp	N	8.5		0.32		190	
	E	8.3		0.35		190	
Mainka 150 kp	N	20		0.36		50	
200 kp	E	20		0.47		40	

Since March, 4. the seismographs Wiechert and Mainka are out of operation

*) for wave velocity 5 km · s^{-1}

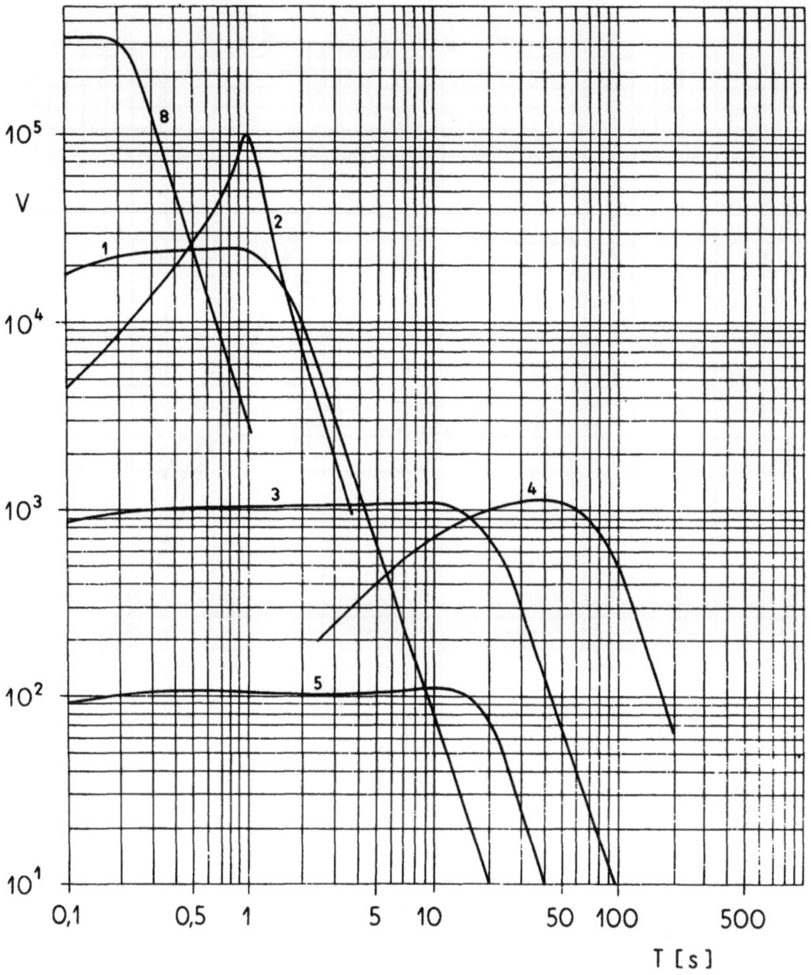

Fig. 1. Station Moxa, mean amplitude characteristics 1971

1 — Seismograph Kirnos Modified-III (SKM-III) (NS-, EW- and Z-component)

2 — Seismograph Type Jena II (Z-component)

3 — Seismic Station Apparatus Type Jena I/1000 (SSJ-I/1000) (NS-, EW- and Z-component)

4 — Seismic Station Apparatus Type Jena I/L (SSJ-I/L) (NS-, EW- and Z-component)

5 — Seismic Station Apparatus Type Jena I/100 (SSJ-I/100) (NS-, EW- and Z-component)

8 — Seismograph Type Jena II (Z-component)

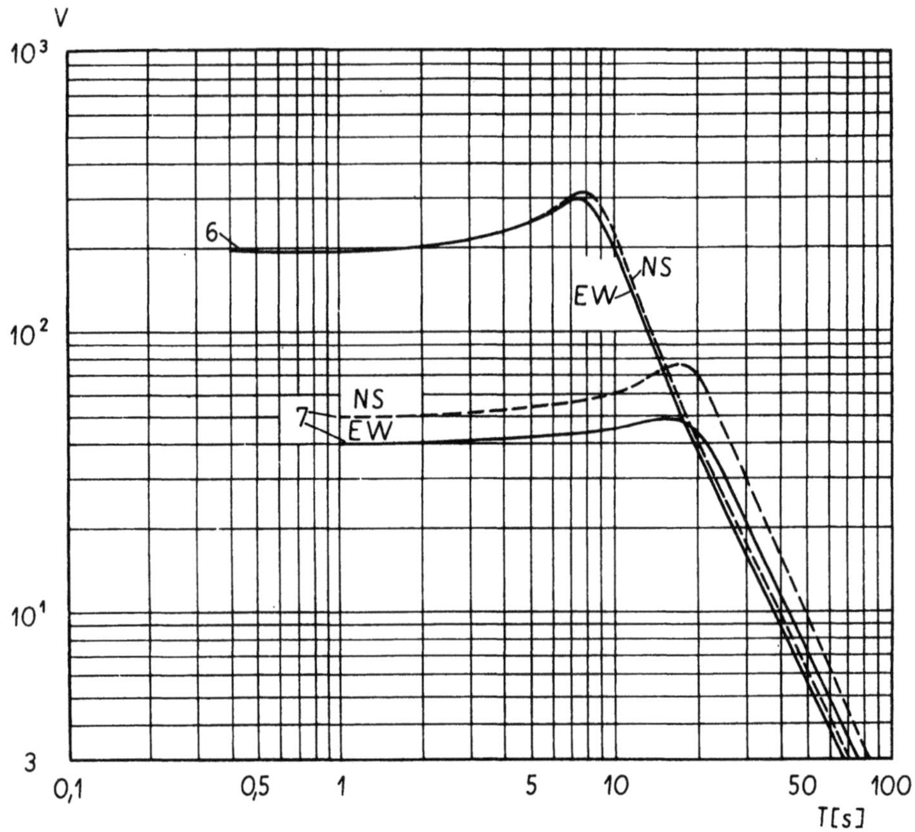

Fig. 2 Station Moxa, mean amplitude characteristic of mechanical seismographs 1971

6 — Wiechert Seismograph (NS- and EW-component)

7 — Mainka Seismograph (NS-and EW-component)

Seismological Recordings at Station Moxa 1971

Day	Phase		h m s	Remarks
1.	iPn	A	01 41 56	**Poland** 50.33 N 18.87 E
	e	A	42 09	H = 01 41 10.0 M = 2.6 (WAR)
	e	AZ	43 41	D = 4.64 Az = 277 (ISC)
1.	e	A	03 16 08	**Greece** 38.0 N 20.5 E
				H = 03 12 49 h = 0 km
				D = 14.08 Az = 336 (ISC)
1.	-eiP	A	04 56 31.5	**Gulf of Alaska** 59.63 N 144.64 W
	epP	A	56 35	H = 04 45 29.0 h = 16 km MB = 5.2
	esP	A	56 39	D = 68.40 Az = 16.0 (NEIC)
	e	A	56 41.5	PV A 1.0s 51.2nm M = 5.7
				e A 57 09
1.	e(PKP)	A	08 17 08.5	**New Guinea** 4.00 S 141.16 E
	ePP	BC	18 06	H = 07 57 58.8 h = 17 km MB = 5.5
	+i	A	18 34.5	D = 117.24 Az = 326.5 (NEIC)
	LmH	B	09 07.6	LmH B 22s 2.6 μm M = 5.8
	LmV	B	09.7	LmV B 20 2.2 μm 5.8
1.	+eiPKHKP	A	08 50 15	**Fiji Islands Region** 19.20 S 178.19 W
				H = 08 31 39.1 h = 638 km MB = 4.6
				D = 147.67 Az = 348.3 (NEIC)
				PKHKP A 1.8s 43.9nm
1.	eP	AZ	12 06 50	**Fox Islands, Aleutian Is.**
				52.98 N 167.04 W
				H = 11 55 03.5 h = 61 km MB = 4.6
				D = 76.75 Az = 0.9 (NEIC)
				PV A 1.0s 9.8nm M = 4.7
2.	eP	A	00 50 20	**Crete** 35.04 N 23.11 E
	epP	A	50 28	H = 00 46 14.95 h = 32 km MB = 4.6
	LmH	B	58.9	D = 17.69 Az = 335.3 (NEIC)
	LmV	B	58.9	PV A 15s 20.1nm M = 4.0
				LmH B 16 0.5 μm 3.8
				LmV B 15 0.6 μm 4.2

Day	Phase		h m s	Remarks
2.	eP	A	03 29 53.5	**Turkey** 36.98 N 28.95 E
				H = 03 25 39.40 h = 40 km MB = 4.3
				D = 18.46 Az = 323.2 (NEIC)
				PV A 1.8s 30.4nm M = 4.2
2.	e(P)	A	07 51 18.5	**Bali Sea** 6.95 S 117.23 E
				H = 07 36 08.02 h = 584 km MB = 5.0 (NEIC)
				D = 105.2
2.	ePKIKP	A	13 57 47.5	**New Ireland Region** 4.57 S 153.29 E
				H = 13 38 55.1 h = 69 km MB = 4.9
				D = 123.98 Az = 331.6 (NEIC)
2.	eP	A	19 20 52.5	**Gulf of Alaska** 59.49 N 144.85 W
	epP	A	20 58	H = 19 09 49.82 h = 24 km MB = 4.9
				D = 68.57 Az = 15.8 (NEIC)
				PV A 1.2s 12.2nm M = 4.9
				pP A 1.1 28.2nm
3.	eiP	A	03 17 40	**Hokkaido, Japan Region**
	e	A	17 48.5	41.34 N 142.50 E
	e	A	17 56.5	H = 03 05 41.97 h = 61 km MB = 5.2
	e	A	18 05	D = 78.84 Az = 330.6 (NEIC)
	LmH	C	53.6	PV A 1.4s 34.9nm M = 5.1
	LmV	C	53.8	
3.	eP	A	10 24 09.5	**South of Honshu, Japan**
	ePcP	A	24 16	33.61 N 138.67 E
	ePP	A	27 24	H = 10 11 39.82 h = 24 km MB = 5.2
	LmH	B	11 02.5	D = 83.99 Az = 329.3 (NEIC)
	LmV	B	07.6	PV A 1.1s 20.2nm M = 5.3
3.	eP	A	14 03 34	**Hokkaido, Japan Region**
	esP	A	03 45	41.36 N 143.44 E
	e	A	03 50	H = 13 51 29.39 h = 25 km MB = 5.3
	ePP	A	06 32	D = 79.17 Az = 331.1 (NEIC)
	LmH	B	37.5	PV A 1.2s 24.4nm M = 5.1
	LmV	B	44.5	LmH B 17 1.4 μm 5.4
				LmV B 16 1.1 μm 5.3

Day	Phase		h m s	Remarks
3.	eP diff	B	17 50 44	<u>South Atlantic Ridge</u> 55.50 S 2.58 W
	ePP	B	54 12	H = 17 35 40.20 h = normal MB = 6.4 (NEIC)
	iSKS	B	18 00 55	D = 106.4
	eS	C	02 05	PPV B 10s 1.03 /um M = 6.5
	e	B	02 20	LmH B 18.5 14.6 /um 6.6
	e	B	03 08	LmV B 17.5 10.3 /um 6.4
	ePS	BC	03 45	e A 17 54 24.5
	ePPS	BC	04 40	
	eiSS	BC	09 25	
	e	B	09 40	
	LmH	B	37.9	
	LmV	B	44.7	
3.	eP1	ABC	23 23 04	<u>Crete</u> 34.67 N 26.34 E
	iP2	A	23 13	H = 23 18 41.29 h = 32 km MB = 5.2
	eS	B	26 38	D = 19.23 Az = 330.6 (NEIC)
	LmH	B	32.5	P1V A 2.6s 243.0nm M = 5.0
	LmV	B	32.5	P2V A 1.3 126.7nm 5.0
				LmH B 13 10.1 /um 5.3
				LmV B 13.5 10.0 /um 5.5
4.	e(PKP)	A	04 18 36.5	<u>Easter Island Region</u> 29.62 S 111.98 W
				H = 03 59 12.61 h = normal MB = 5.2
				D = 133.31 Az = 46.8 (NEIC)
4.	eP	A	10 18 08.5	<u>North Atlantic Ocean</u> 58.37 N 32.27 W
	epP	A	18 15.5	H = 10 12 35.54 h = normal MB = 4.4
	LmH	B	29.1	D = 26.21 Az = 87.9 (NEIC)
	LmV	B	29.1	LmH B 14s 0.8 /um M = 4.4
				LmV B 13 1.1 /um 4.7
4.	eP	A	10 33 06.5	<u>Jan Mayen Island Region</u>
	e	A	33 15	70.36 N 15.10 W
				H = 10 28 01.32 h = normal MB = 4.2
				D = 23.37 Az = 133.8 (NEIC)
4.	ePKIKP	A	11 27 51.5	<u>South of Fiji Islands</u> 23.52 S 179.90 E
	ePKHKP	A	27 59	H = 11 09 08.64 h = 580 km MB = 4.9

Day	Phase		h m s	Remarks
cont. 4.	ePKP2	A	11 28 10	D = 151.43 Az = 344.3 (NEIC)
	epPKP	A	30 08	
4.	+iP	A	19 10 02	<u>West Pakistan</u> 29.20 N 69.24 E
	LmV	B	33.1	H = 19 01 20.35 h = 34 km MB = 4.9
	LmH	B	35.0	D = 47.75 Az = 313.4 (NEIC)
				PV A 0.9s 19.5nm M = 5.1
				LmH B 16 2.0 µm 5.2
				LmV B 16 1.1 µm 5.0
4.	-iP	A	21 21 13.5	<u>Near S. Coast of Honshu, Japan</u>
	LmH	B	22 00.0	34.52 N 137.08 E
	LmV	B	03.2	H = 21 08 53.44 h = 40 km MB = 5.6
				D = 82.53 Az = 328.5 (NEIC)
				PV A 1.1s 40.3nm M = 5.4
				LmH B 13.5 3.6 µm 5.9
				LmV B 13 4.1 µm 6.0
4.	ePb	A	22 10 13	<u>Italy</u> 44.2 N 8.2 E
	ePg	A	10 24	H = 22 08 12 (BCIS)
	e	A	10 57	D = 6.91
	eSn	A	11 11	
	eSg	A	12 08	
5.	e	A	04 21 36.5	<u>Ryukyu Islands</u> 25.45 N 126.30 E
				H = 04 08 19 h = 73 km MB = 4.5 (ISC)
				D = 84.75 traces
5.	eP	A	06 06 22.5	<u>Southern Alaska</u> 61.42 N 147.55 W
				H = 05 55 34.04 h = 46 km MB = 4.5
				D = 67.05 Az = 14.2 (NEIC)
				traces
6.	+iP	A	06 16 41.5	<u>Near East Coast of Honshu, Japan</u>
	epP	A	16 54	36.36 N 141.18 E
	LmH	B	57.6	H = 06 04 21.35 h = 45 km MB = 5.4 (NEIC)

Day	Phase		h m s	Remarks
cont. 6.	LmV	B	06 58.6	D = 82.6 h c.
				PV A 1.8s 60.8nm M = 5.3
				pPV A 2.0 77.0nm
				LmH B 16 1.1/um 5.3
				LmV B 14 1.3/um 5.5
6.	ePg	A	10 57 21.5	Italy 45.1 N 9.4 E
	eSn	A	57 51	H = 10 55 22 (BCIS)
	eSg	AZ	58 33	D = 5.78
6.	ePg	A	11 10 58.5	Italy 45.0 N 9.6 E
	eSn	A	11 34	H = 11 09 02 (BCIS)
	eSg	A	12 15	D = 5.85
6.	iPg	A	11 12 49.5	Italy 45.0 N 9.4 E
	iSn	A	13 25.0	H = 11 10 55 (BCIS)
	iSg	AB	14 06	D = 5.88
	LmH	B	14.3	
	LmV	B	14.4	
6.	ePg	A	11 19 11	Italy 45.1 N 9.3 E
	e	A	19 16	H = 11 17 19 (BCIS)
	eSn	A	19 48	D = 5.79
7.	eP	A	03 01 44	Andreanof Islands, Aleutian Is.
	ePcP	A	01 54	52.44 N 173.32 W
				H = 02 49 57.50 h = 87 km MB = 5.8
				D = 77.21 Az = 356.8 (NEIC)
8.	eP	A	06 13 40	Crete 34.53 N 26.45 E
				H = 06 09 11.5 h = 60 km MB = 4.3
				D = 19.40 Az = 330.6 (NEIC)
8.	eP	A	11 51 19.5	Off East Coast of Kamchatka
	epP	A	51 31	53.70 N 161.53 E
				H = 11 39 50.9 h = 33 km MB = 4.7
				D = 73.00 Az = 340.5 (NEIC) h c. 43 km
				PV A 0.9s 15.5nm M = 5.0

Day	Phase		h m s	Remarks
8.	+iP1	ABC	14 57 22	<u>Kurile Islands</u> 47.37 N 154.35 E
	iP2	A	57 23.5	H = 14 45 29.5 h = 32 km MB = 5.6
	epP	A	57 32.5	D = 77.23 Az = 336.7 (NEIC)
	eS	BC	15 07 00	h = ca. 37 km
	eSS	C	12 15	P1V A 1.0s 43.3nm M = 5.4
	LmH	B	36.8	P2V A 1.5 221.0nm 6.0
	LmV	B	39.5	PV B 8 1.5 μm 6.1
				LmH B 15.5 6.3 μm 6.0
				LmV B 16 5.5 μm 6.0
8.	e	A	22 15 01.5	<u>Dodecanese Islands</u> 35.04 N 26.96 E
				H = 22 10 21.3 h = 67 km MB = 4.2
				D = 19.17 Az = 329.1 (NEIC)
8.	eP1	A	24 00 55.5	<u>West Pakistan</u> 29.15 N 69.14 E
	eP2	A	01 00	H = 23 52 16.3 h = 4 km MB = 5.2
	LmH	B	25.9	D = 47.72 Az = 313.5 (NEIC)
	LmV	B	26.6	P2V A 1.3s 15.3nm M = 5.0
				LmH B 15 2.1 μm 5.2
				LmV B 13 1.2 μm 5.1
10.	ePKP	A	03 56 57	<u>Fiji Islands Region</u> 17.78 S 173.14 E
	LmH	C	04 58.5	H = 03 37 16.9 h = normal MB = 4.9
	LmV	C	59.4	D = 144.03 Az = 339.9 (NEIC)
				PKPV A 2.0s 42.7nm
10.	+iP1 diff	C	07 31 58	<u>West New Guinea</u> 3.13 S 139.70 E
	+iP2 diff	BC	32 03	H = 07 17 03.8 h = normal MB = 7.3
	ePKIKP	A	35 46	D = 115.72 Az = 326.2 (NEIC)
	+ipPKP	B	35 49	P2 diff V B 16s 10.1 μm
	+iPP	B	36 52	PKIKPV B 12 5.7 μm
	-iPPP	B	39 24	PPV B 16 49.9 μm M = 8.0
	iS	B	44 24	LmH B 25.5 839.0 μm 8.2
	iP2KP2	B	46 08	LmV B 25 800.0 μm 8.2
	eiSS	B	52 52	
	eiPSPS	B	53 20	
	eiS3KS2	B	56 54	
	LmV	B	08 24.2	
	LmH	B	25.3	

Day	Phase		h m s	Remarks
10.	ePKP	A	08 02 32	West New Guinea 2.97 S 139.73 E
	e	A	02 40.5	H = 07 43 47.2 h = normal MB = 5.9
	ePP	A	03 30	D = 115.61 Az = 326.3 (NEIC)
				PPV A 1.8s 60.8nm M = 6.0
10.	ePP	A	08 13 40	West New Guinea 3.0 S 139.7 E
				H = 07 53 52.6 h = normal MB = 5.6 (NEIC)
				D = 116.2
10.	ePP	A	09 04 25	West New Guinea 3.2 S 139.7 E
	e	A	04 41	H = 08 44 32.6 h = 21 km MB = 5.5 (NEIC)
10.	ePP	A	11 04 29	West New Guinea 2.98 S 139.62 E
				H = 10 45 06.8 h = 36 km MB = 6.1 (NEIC)
				D = 115.5
				PPV A 2.0s 42.7nm M = 5.8
10.	ePP	A	14 49 45	West New Guinea 3.2 S 139.8 E
	LmH	B	15 43.0	H = 14 29 56.3 h = 28 km
	LmV	B	43.0	MB = 5.8 MS = 5.9 (NEIC)
				D = 115.8
				LmH B 22s 3.1/um M = 5.9
				LmV B 19 3.0/um 5.9
10.	e	A	16 28 01	West New Guinea 3.1 S 139.8 E
	LmH	B	17 20.2	H = 16 07 45.9 h = 33 km
	LmV	B	20.6	MB = 5.7 MS = 5.7 (NEIC)
				D = 115.7
				LmH B 18s 3.5/um M = 6.0
				LmV B 18 4.1/um 6.1
10.	e	A	16 57 48	West New Guinea 3.4 S 140.0 E
	LmH	B	17 50.2	H = 16 38 36.0 h = 27 km
	LmV	B	50.4	MB = 5.5 MS = 6.0 (NEIC)
				D = 116.12
				LmH B 20s 3.0/um M = 5.9
				LmV B 20 3.7/um 6.0

Day	Phase		h m s	Remarks
10.	ePP	A	19 33 37	<u>West New Guinea</u> 3.2 S 139.8 E
				H = 19 13 36.8 h = 33 km
				MB = 5.8 MS = 5.5 (NEIC)
				D = 115.8
10.	e	A	19 53 23	<u>West New Guinea</u> 3.09 S 139.73 E
	ePP	A	54 25.5	H = 19 34 22.3 h = 39 km MB = 5.7
				D = 115.70 Az = 326.2 (NEIC)
10.	eP	A	20 16 03.5	<u>Sea of Okhotsk</u> 54.98 N 142.35 E
	epP	A	16 06.5	H = 20 05 09.0 h = 13 km MB = 5.0
				D = 67.01 Az = 328.4 (NEIC)
				PV A 1.2s 22.4nm M = 5.3
10.	e(PKP)	A	22 33 31	<u>West New Guinea</u> 3.2 S 139.9 E
	ePP	A	34 26	H = 22 14 37.4 h = 30 km
				MB = 6.1 MS = 5.8 (NEIC)
				D = 115.99 Az = 326 (ISC)
				PKPV A 1.8s 40.6nm
10.	e(P)	A	22 44 06	(P)V A 1.1s 20.2nm
10.	LmH	B	23 27.2	LmH B 19s 4.7 µm
	LmV	B	27.1	LmV B 19 4.6 µm
11.	LmH	B	00 37.5	LmH B 20s 1.8 µm
	LmV	B	37.0	LmV B 18 1.9 µm
11.	eP diff	A	12 30 16	<u>West Irian</u> 3.0 S 139.6 E
				H = 12 15 09 h = 55 km
				D = 116.08 Az = 326 (ISC)
11.	eP	A	19 19 14.5	<u>Taiwan</u> 23.69 N 121.72 E
	LmH	B	20 01.7	H = 19 06 45.5 h = 20 km MB = 5.4
	LmV	B	01.8	D = 83.83 Az = 323.0 (NEIC)
				PV A 1.5s 25.1nm M = 5.1
				LmH B 16 1.0 µm 5.3
				LmV B 16 1.1 µm 5.4

Day	Phase		h m s	Remarks
11.	ėPKHKP	A	19 50 22	South of Fiji Islands 25.14 S 178.33 E
	ePKP2	A	50 36	H = 19 31 32.7 h = 606 km MB = 4.5
				D = 152.55 Az = 341.5 (NEIC)
11.	iPKHKP	A	21 22 07.5	Fiji Islands Region 20.47 S 178.46 W
	ePKP2	A	22 13	H = 21 03 23.2 h = 584 km MB = 4.7
				D = 148.85 Az = 347.6 (NEIC)
				PKHKPV A 1.5s 55.3nm
11.	eP	A	23 48 17	USSR-Mongolia Border Region
	e	A	48 39	50.13 N 90.93 E
				H = 23 39 34.9 h = normal MB = 4.5
				D = 48.23 Az = 303.0 (NEIC)
12.	e(PP)	A	15 17 10	West New Guinea 3.2 S 140.0 E
	LmH	B	16 09.8	H = 14 57 11.6 h = normal
	LmV	B	10.8	MB = 5.8 MS = 5.6 (NEIC)
				D = 115.8
				LmH B 20s 2.1/um M = 5.8
				LmV B 19 2.1/um 5.8
12.	LmH	B	15 45.5	Western Caroline Islands
	LmV	B	45.4	9.83 N 138.29 E
				H = 14 42 11 h = 4 km MB = 5.5 (ISC)
				D = 104.2
				LmH B 18s 1.4/um M = 5.5
				LmV B 20 1.9/um 5.6
12.	eP diff	A	18 40 53	West Irian 3.08 S 139.66 E
				H = 18 25 05 h = 14 km
				D = 115.66 Az = 326 (ISC)
13.	LmH	B	11 35.3	West Irian 3.05 S 139.60 E
	LmV	B	36.5	H = 10 22 47 h = 30 km MB = 5.6 (ISC)
				D = 115.6
				LmH B 19s 0.9/um M = 5.4
				LmV B 18 0.9/um 5.4

Day	Phase		h m s	Remarks
13.	ePKP2	A	13 30 03	**Kermadec Islands Region** 31.21 S 179.56 W H = 13 10 12.1 h = 369 km D = 158.87 Az = 340 (ISC)
13.	ePKIKP ePKHKP	A A	13 59 34.5 59 39.5	**South Pacific Cordillera** 55.43 S 128.39 W H = 13 39 43.1 h = normal MB = 4.8 D = 155.73 Az = 84.6 (NEIC) PKHKPV A 1.9s 45.5nm
13.	e	AZ	22 00 29	**Afghanistan-USSR Border Region** 38.92 N 71.01 E H = 21 52 35.6 h = 72 km MB = 4.8 D = 42.67 Az = 306.0 (NEIC)
14.	ePg eSg	A A	03 08 02 09 25	**France** 45.1 N 5.8 E H = 03 05 44 (BCIS) D = 6.8
14.	eP LmH LmV	A C C	04 32 31 05 03.5 09.2	**Near East Coast of Kamchatka** 56.00 N 163.02 E H = 04 21 13.3 h = 45 km MB = 4.7 D = 71.10 Az = 341.2 (NEIC) PV A 1.4s 14.0nm M = 4.8 LmV C 14 0.5 μm 4.9
14.	ePP	A	09 40 22	**Molucca Passage** 2.07 N 126.64 E H = 09 21 56.1 h = normal MB = 5.4 (NEIC) D = 104.0
14.	eP	A	11 28 15	**North Atlantic Ridge** 11.28 N 43.69 W H = 11 18 12.2 h = normal MB = 5.0 D = 59.65 Az = 37.3 (NEIC)
14.	LmH LmV	B B	11 31 30.5	**Near North Coast of West-Irian** 2.99 S 139.61 E H = 10 23 45 h = 34 km MB = 5.2 (ISC) D = 115.6 LmH B 24s 1.2 μm M = 5.4 LmV B 24 1.1 μm 5.4

Day	Phase		h m s	Remarks
14.	ePP	A	18 27 13	Near North Coast of West Irian
	Lm1H	B	19 14.6	2.98 S 139.74 E
	Lm2H	B	17.7	H = 18 07 18 h = 6 km MB = 5.4 (ISC)
	LmV	B	19.4	D = 115.6
				Lm1H B 22s 1.7 µm M = 5.6
				Lm2H B 19 1.4 µm 5.6
				LmV B 20 1.2 µm 5.5
14.	eP	A	23 54 48.5	Off East Coast of Kamchatka
	esP	A	54 57	53.62 N 161.57 E
				H = 23 43 20.8 h = 44 km MB = 5.0
				D = 73.08 Az = 340.5 (NEIC)
				PV A 0.9s 19.5nm M = 5.1
14.	eP	A	24 02 02	Off East Coast of Kamchatka
				53.63 N 161.73 E
				H = 23 50 32.6 h = normal MB = 4.4
				D = 73.11 Az = 340.6 (NEIC)
15.	iPn	A	02 56 04.5	Federal Rep. of Germany 48.15 N 9.20 E
	eiPg	A	56 13	H = 02 55 15.6 h = 3 km
	iSg	AB	56 50	D = 2.96 Az = 31.4 (NEIC)
15.	ePKP2	A	19 53 25	South of Kermadec Islands
	epPKP2	A	53 36	33.12 S 178.36 W
	LmH	C	21 22.5	H = 19 32 42.8 h = normal MB = 4.7 (NEIC)
	LmV	C	27	D = 161.0
				PKP2V A 1.6s 27.5nm
				LmV C 18 0.3 µm
15.	ePKP2	A	22 06 04	South of Kermadec Islands
				33.33 S 178.20 W
				H = 21 45 24 h = 44 km MB = 4.5
				D = 161.25 Az = 340 (ISC)
16.	eP diff	A	04 51 28	West Irian 3.04 S 139.68 E
				H = 04 36 05 h = 16 km
				D = 115.64 Az = 326 (ISC)
				P diff V A 1.1s 12.1nm

Day	Phase		h m s	Remarks
16.	ePP	A	12 58 55	South Sandwich Islands Region
	LmH	C	13 37	55.93 S 27.64 W
				H = 12 39 51.6 h = 81 km MB = 5.8 (NEIC)
				D = 111.1
				PPV A 1.8s 37.2nm M = 5.7
16.	eP	A	15 13 54	Dodecanese Islands 36.67 N 26.81 E
				H = 15 09 53.5 h = 151 km MB = 4.2
				D = 17.73 Az = 326.8 (NEIC)
				PV A 1.8s 37.2nm M = 4.4
16.	eP	A	21 07 57.5	Near Coast of Northern Peru
				9.09 S 78.83 W
				H = 20 54 30.9 h = 75 km MB = 5.3
				D = 97.23 Az = 39.9 (NEIC)
17.	eP	A	00 29 54	Near East Coast of Kamchatka
				56.18 N 162.43 E
				H = 00 18 32.5 h = normal MB = 4.7
				D = 70.83 Az = 340.8 (NEIC)
17.	ePb	A	04 21 37	Switzerland 46.5 N 7.6 E
	ePg	A	21 46	H = 04 20 10 (BCIS)
	eSg	A	22 47	D = 4.92
17.	eP	A	05 22 07.5	Greece 38.08 N 20.51 E
				H = 05 18 47 h = 29 km MB = 4.5
				D = 14.06 Az = 336 (ISC)
17.	eP	A	06 02 43	Greece 39.06 N 21.86 E
				H = 05 59 24 h = 8 km MB = 4.3
				D = 13.65 Az = 331 (ISC)
18.	e	A	00 17 40.5	Molucca Passage 0.98 N 125.83 E
	ePP	A	22 02.5	H = 00 03 31.6 h = normal MB = 5.4
	eSKS	C	28 15	D = 104.36 Az = 323.2 (NEIC)
	eS	C	31 48	LmH C 24s 2.1 μm M = 5.6
	eSS	C	36 40	LmV C 24 2.4 μm 5.7

Day	Phase		h m s	Remarks
cont. 18.	LmH	C	01 08.5	
	LmV	C	11.3	
18.	eP	A	05 14 07	<u>Hokkaido, Japan Region</u> 41.61 N 142.56 E
	epP	A	14 17.5	H = 05 02 08.8 h = 47 km MB = 5.4
	LmV	B	52.0	D = 78.62 Az = 330.6 (NEIC)
	LmH	B	52.5	h = 38 km
				PV A 1.9s 87.0nm M = 5.4
				LmH B 18 2.2 µm 5.5
				LmV B 18 1.6 µm 5.4
18.	eP	A	06 06 26	<u>Molucca Passage</u> 1.06 N 125.91 E
	ePP	A	10 46	H = 05 52 21.48 h = normal MB = 5.7
				D = 104.35 Az = 323.2 (NEIC)
18.	eiP	A	10 39 20	<u>Ionian Sea</u> 37.58 N 20.22 E
				H = 10 35 47.8 h = normal MB = 4.6
				D = 14.44 Az = 337.5 (NEIC)
				PV A 1.0s 19.7nm M = 4.8
19.	eP1	A	03 29 47	<u>Gulf of California</u> 23.84 N 108.58 W
	eP2	A	29 51.5	H = 03 16 53.6 h = normal MB = 5.5
	eP3	A	30 09.5	D = 89.04 Az = 33.4 (NEIC)
	eS	BC	40 44	P1V A 1.3s 13.1nm M = 5.0
	eSS	BC	46 24	P2V A 1.5 30.2nm 5.3
	LQ	BC	53 20	P3V A 2.0 81.2nm 5.6
	LR	B	04 00 28	LmH B 17 8.7 µm 6.2
	LmH	B	07.6	LmV B 16 8.7 µm 6.3
	LmV	B	11.0	
19.	e	A	05 50 39	
19.	eP	A	23 38 14	<u>Crete</u> 34.40 N 24.01 E
	LmH	B	46.0	H = 23 33 56.8 h = normal MB = 4.8
	LmV	B	46.7	D = 18.58 Az = 334.6 (NEIC)
				PnV A 0.9s 23.3nm M = 5.4
				LmH B 12 3.4 µm 4.9
				LmV B 10 2.9 µm 5.0

Day	Phase		h m s	Remarks
20.	+iP1	A	04 57 28	Panama 8.78 N 79.23 W
	eP2	A	57 30.5	H = 04 45 00.1 h = 33 km MB = 5.6
	LmH	B	05 32	D = 83.82 Az = 39.8 (NEIC)
	LmV	B	32	P1V A 1.3s 91.7nm M = 5.6
				P2V A 2.5 384.0nm 5.9
				LmH B 18 2.5 µm 5.6
				LmV B 18 3.1 µm 5.8
20.	e(P)	A	21 38 12	Iran-Iraq Border Region
	e(sP)	A	38 26	35.06 N 46.92 E
	e	A	38 46.5	H = 21 32 03.0 h = 57 km MB = 5.1
	e	A	39 43.5	D = 29.86 Az = 312.4 (NEIC)
	LmH	B	53.3	
	LmV	B	53.3	
21.	ePKIKP	A	06 36 23	East New Guinea Region
				6.82 S 147.51 E
				H = 06 17 30.8 h = 55 km MB = 5.6
				D = 123.02 Az = 328.1 (NEIC)
				PKIKPV A 1.2s 24.4nm
21.	iPKHKP	A	12 43 01.5	Fiji Islands Region 21.50 S 176.67 W
	ePKP2	A	43 07	H = 12 23 32.0 h = 193 km MB = 5.1
				D = 150.19 Az = 349.4 (NEIC)
				PKHKPV A 1.3s 127.0nm
21.	ePKIKP	A	17 37 44	Flores Sea 7.80 S 122.67 E
				H = 17 19 35.3 h = 196 km MB = 5.7
				D = 109.30 Az = 321.0 (NEIC)
				traces
21.	eP	A	21 43 02	North of Ascension Island
	epP	A	43 09.5	1.52 S 12.74 W
				H = 21 33 24.3 h = normal MB = 4.8
				D = 55.98 Az = 18.5 (NEIC)
22.	eP	A	00 53 30	PV A 1.5s 25.2nm

Day	Phase		h m s	Remarks
22.	ePKP	A	02 54 47	Fiji Islands Region 17.97 S 178.39 W H = 02 36 10.9 h = 597 km MB = 4.7 D = 146.43 Az = 348.5 (NEIC) PKPV A 1.2s 16.3nm
22.	ePKHKP	A	04 15 46.5	Fiji Islands Region 21.44 S 179.07 W H = 03 56 59.4 h = 570 km MB = 4.4 D = 149.66 Az = 346.5 (NEIC) PKHKPV A 1.2s 12.2nm
22.	ePKIKP	A	20 18 40	New Britain Region 5.76 S 151.29 E H = 19 59 46.1 h = 57 km MB = 5.3 D = 124.04 Az = 330.2 (NEIC)
24.	+eP	A	13 33 01.8	Kurile Islands 49.16 N 156.29 E H = 13 21 15.7 h = normal MB = 4.9 D = 76.07 Az = 337.7 (NEIC) PV A 1.0s 23.6nm M = 5.3
24.	eP LmH LmV	A B B	13 40 29 14 18.4 20.3	Kurile Islands 49.22 N 156.13 E H = 13 28 43.6 h = 36 km MB = 5.2 D = 75.98 Az = 337.6 (NEIC) PV A 1.0s 27.6nm M = 5.2 LmH B 16 2.3/um 5.6
24.	eP	A	13 59 22	Kurile Islands 49.05 N 156.30 E H = 13 47 36.2 h = 34 km MB = 4.6 D = 76.17 Az = 337.7 (NEIC) PV A 0.8s 9.6nm M = 4.8
24.	eP	A	14 25 03	Kurile Islands 49.11 N 156.32 E H = 14 13 18.5 h = 40 km D = 76.12 (ISC) PV A 1.0s 7.9nm
24.	ePKHKP ePKP2	A A	19 00 13.5 00 20	Fiji Islands Region 20.49 S 178.42 W H = 18 41 28.0 h = 557 km MB = 4.9 D = 148.88 Az = 347.6 (NEIC) PKHKPV A 0.8s 30.8nm PKP2V A 1.4 32.6nm

Day	Phase		h m s	Remarks
24.	eP	A	23 27 56	Kodiak Island Region 56.41 N 152.99 W
				H = 23 16 31.1 h = 38 km MB = 5.2
				D = 72.55 Az = 10.2 (NEIC)
25.	eP	A	00 23 12	Off East Coast of Kamchatka
				53.87 N 162.43 E
				H = 00 11 44.3 h = normal MB = 4.7
				D = 73.01 Az = 341.1 (NEIC)
25.	ePKIKP	AB	00. 37 29	Dentrecasteaux Islands Region
	ipPKIKP	A	37 39	9.60 S 151.35 E
	e	A	39 14	H = 00 18 26.1 h = 38 km MB = 5.8
	ePP	A	39 33	D = 127.36 Az = 328.8 (NEIC)
	iPP	B	39 46	h = 36 km
	ePS	BC	49 38	PKIKPV A 2.2s 165.3nm
	eSS	C	56 25	pPKIKPV A 1.8 2.4nm
	e	C	57 40	PPV B 7.5 3.9 µm M = 6.6
	eSSS	C	01 01 20	LmH B 20 5.8 µm 6.3
	LmV	B	34.2	LmV B 20 2.5 µm 5.9
	LmH	B	34.6	
25.	iP	A	14 59 07.5	Svalbard Region 76.7 N 6.8 E
				H = 14 53 33.1 h = 33 km
				D = 26.13 Az = 173 (ISC)
				PV A 0.6s 11.5nm M = 4.7
25.	-iP	AB	16 20 11	Andreanof Islands, Aleutian Is.
	epP	A	20 26	51.47 N 177.69 W
	eS	BC	30 00	H = 16 08 15.1 h = 38 km MB = 5.9
	eSS	C	34 58	D = 77.96 Az = 354.0 (NEIC)
	LQ	C	41 05	h = 54 km
	eP'P'	A	47 27	PV A 1.7s 167.0nm M = 5.7
	LmH	B	17 57.7	LmH B 19 14.5 µm 6.3
	LmV	B	59	LmV B 14.5 4.7 µm 6.0

Day	Phase		h m s	Remarks
25.	eP	A	16 34 22	**Svalbard Region** 76.86 N 8.12 E H = 16 28 46.8 h = normal MB = 4.7 D = 26.35 Az = 175.0 (NEIC) PV A 1.8s 54.0nm M = 4.9
25.	eP	A	18 54 11	**Kurile Islands** 49.19 N 156.38 E H = 18 42 25.4 h = normal MB = 4.7 D = 76.07 Az = 337.8 (NEIC) PV A 0.8s 13.5nm M = 5.1
25.	eP	A	21 38 57	**Svalbard Region** 76.9 N 9.4 E H = 21 33 18.2 h = 0 km D = 26.37 Az = 177 (ISC) PV A 1.7s 36.4nm M = 4.8
25.	eP	A	23 02 44	**Peru-Ecuador Border Region** 4.06 S 80.94 W H = 22 49 26.2 h = 51 km MB = 4.9 D = 94.73 Az = 39.7 (NEIC)
26.	eP	A	01 07 23.5	**Svalbard Region** 76.92 N 10.57 E H = 01 01 48.7 h = normal MB = 4.6 D = 26.38 Az = 178.5 (NEIC) PV A 1.5s 30.2nm M = 4.7
26.	eP1 eP2	A A	06 29 45 29 50.5	**Svalbard Region** 76.63 N 7.61 E H = 06 24 12.1 h = normal MB = 4.9 D = 26.14 Az = 174.2 (NEIC) P2V A 1.1s 44.3nm M = 5.0
26.	+eP	A	08 14 21.5	**Svalbard Region** 77.11 N 8.49 E H = 08 08 44.1 h = normal MB = 4.6 D = 26.60 Az = 175.6 (NEIC) PV A 1.6s 35.8nm M = 4.8
26.	eP	A	15 26 57	**Svalbard Region** 77.2 N 12 E H = 15 21 13 h = 0 km MB = 4.3 D = 26.68 Az = 181 (ISC) PV A 0.7s 15.3nm M = 4.9

Day	Phase		h m s	Remarks
26.	eP1	A	19 44 00.5	Andreanof Islands, Aleutian Is.
	eP2	A	44 02.5	51.67 N 174.92 W
	LmH	B	20 25.1	H = 19 32 04.9 h = 36 km MB = 5.4
				D = 77.92 Az = 355.8 (NEIC)
				P1V A 1.0s 19.7nm M = 4.9
				P2V A 1.6 46.7nm 5.2
				LmH B 19 2.4/um 5.5
26.	eP	A	22 53 02	Western Caucasus 43.82 N 39.19 E
				H = 22 48 31.1 h = 28 km MB = 4.8
				D = 19.86 Az = 299.8 (NEIC)
				PV A 2.0s 51.3nm M = 4.5
27.	e	A	02 50 29	Svalbard Region 76.7 N 7.7 E
				H = 02 44 49.9 h = 0 km
				D = 26.16 Az = 174 (ISC)
27.	iPn	A	09 17 21	Federal Republic of Germany
	ePg	A	17 29.5	48.69 N 9.00 E
	eSn	A	17 56.5	H = 09 16 35.9 h = 0 km
	eSg	A	18 09	D = 2.59 Az = 40 (ISC)
27.	eP1	A	16 08 35.5	Greece-Albania Border Region
	eP2	A	08 37.5	39.92 N 20.08 E
	e	A	09 08	H = 16 05 41.0 h = 26 km MB = 4.4
	LmH	C	14.4	D = 12.26 Az = 333.8 (NEIC)
				P2V A 0.8s 30.8nm M = 5.6
27.	eP diff	A	16 19 47.5	Molucca Passage 1.66 N 126.38 E
				H = 16 05 41.2 h = normal MB = 5.7
				D = 104.14 Az = 323.4 (NEIC)
27.	iPn	A	17 14 29.5	Austria 47.8 N 11.1 E
	iPb	A	14 37	H = 17 13 45 (BCIS)
	iPg	A	14 42	D = 2.87 Az = 7 (ISC)
	iSn	A	15 04.5	
	iSg	A	15 17	

Day	Phase		h m s	Remarks
27.	eP	A	20 51 15	<u>Svalbard Region</u> 76.69 N 7.05 E
	LmH	BC	21 00	H = 20 46 42.8 h = 31 km MB = 4.9
				D = 26.21 Az = 173.4 (NEIC)
				PV A 1.2s 20.4nm M = 4.5
27.	eP	A	22 40 54.5	<u>Northern Sumatra</u> 1.88 N 99.16 E
	e	A	41 12	H = 22 28 20.8 h = 116 km MB = 5.4
	epP	A	41 24	D = 86.99 Az = 320.4 (NEIC)
				PV A 1.6s 33.0nm M = 5.1
27.	eP	A	23 36 07	<u>Northern Peru</u> 4.30 S 75.23 W
				H = 23 22 54.9 h = 24 km MB = 4.8
				D = 91.29 Az = 39.5 (NEIC)
				PV A 1.9s 30.3nm M = 5.3
28.	ePKIKP	A	00 07 34	<u>New Hebrides Islands</u> 15.03 S 167.29 E
	epPKIKP	A	08 12	H = 23 48 22.2 h = 133 km MB = 5.2
				D = 139.40 Az = 336.2 (NEIC)
				D = 134 km
28.	eP	A	01 50 05	<u>Jan Mayen Islands Region</u>
				70.04 N 16.47 W
				H = 01 44 56.7 h = normal MB = 4.4
				D = 23.49 Az = 131.2 (NEIC)
28.	ePKIKP	A	06 47 32	<u>Fiji Islands Region</u> 17.92 S 178.48 W
	iPKHKP	A	47 35	H = 06 29 01.0 h = 616 km MB = 5.0
	ePKP2	A	47 37.5	D = 146.36 Az = 348.4 (NEIC)
	e	A	48 32.5	PKIKPV A 1.0s 15.8nm
				PKHKPV A 1.2 44.7nm
				PKP2V A 1.1 32.3nm
28.	eP1	A	11 09 21	<u>Svalbard Region</u> 76.49 N 8.09 E
	eP2	A	09 26	H = 11 03 49.5 h = normal MB = 4.5
				D = 25.98 Az = 174.9 (NEIC)
				P1V A 1.6s 22.0nm M = 4.5
				P2V A 1.6 49.5nm 4.8

Day	Phase		h m s	Remarks
28.	e	A	13 09 57	Near S. Coast of Honshu, Japan 35.27 N 139.34 E H = 12 57 43.0 h = 125 km MB = 4.8 D = 82.85 Az = 329.5 (NEIC)
28.	e(P)	A	15 20 28	
29.	eP	A	08 39 43	Svalbard Region 78.78 N 8.02 E H = 08 33 49.7 h = normal MB = 4.8 D = 28.28 Az = 175.2 (NEIC)
29.	eP	A	16 54 56.5	Near East Coast of Honshu, Japan 40.31 N 141.57 E H = 16 42 56.5 h = 67 km MB = 5.3 D = 79.38 Az = 330.2 (NEIC) PV A 1.3s 17.5nm M = 4.8
29.	eP	A	21 35 02	Off East Coast of Kamchatka
	esP	A	35 11	55.66 N 163.46 E H = 21 23 42.7 h = normal MB = 4.6 D = 71.50 Az = 341.5 (NEIC)
29.	-iP1	AB	22 08 37.0	Sea of Okhotsk 51.72 N 150.95 E
	iP2	A	08 38.3	H = 21 58 05.4 h = 544 km MB = 6.1
	+ipP	B	10 28	D = 72.34 Az = 334.2 (NEIC)
	+isP	B	11 17	i 10 30.8 i 13 58 i 14 34 e 15 40
	iPP	B	11 30	P1V A 1.0s 323.0nm M = 5.8
	-iPPP	B	13 00	P2V A 1.8 3082.0nm 6.5
	iS	B	17 19	P1H A 1.1 136.0nm 5.7
	esS	C	20 25	P2H A 1.8 1640.0nm 6.6
	iSS	B	22 10	PV B 8 17.7 μm 6.6
	eP'P'	A	36 10	SH B 10 29.4 μm 6.8
29.	eiP	A	22 26 40	Sea of Okhotsk 51.88 N 151.21 E H = 22 16 06.7 h = 512 km MB = 4.9 D = 72.26 Az = 334.3 (NEIC) PV A 1.4s 74.4nm M = 5.0

Day	Phase		h m s	Remarks
29.	eP	A	23 08 56.5	Sea of Okhotsk 51.91 N 151.16 E
	ePcP	A	09 15.5	H = 22 58 23.2 h = 522 km MB = 4.6
				D = 72.22 Az = 334.3 (NEIC)
				PcPV A 0.7s 15.3nm M = 4.6
29.	ePKIKP	A	23 57 13	South of Fiji Islands 23.94 S 175.95 W
	ePKHKP	A	57 21	H = 23 37 24.5 h = 22 km MB = 5.1
	ePKP2	A	57 31	D = 152.70 Az = 349.5 (NEIC)
				PKIKPV A (1.6)s 16.5nm
30.	eP	A	11 28 39	South of Honshu, Japan 32.79 N 138.21 E
				H = 11 16 08.2 h = 33 km MB = 4.5
				D = 84.49 Az = 329.1 (NEIC)
30.	ePKP	A	13 25 14.5	New Hebrides Islands 19.07 S 169.34 E
				H = 13 06 12.2 h = 264 km MB = 5.2 (NEIC)
				D = 143.8
				PKPV A 1.2s 26.4nm
30.	ePKP2	A	18 49 35.5	South of Fiji Islands 24.09 S 175.98 W
				H = 18 29 31.9 h = 48 km MB = 5.0
				D = 152.85 Az = 349.4 (NEIC)
30.	eP	A	20 24 55.5	Tibet India Border Region
				30.49 N 79.06 E
				H = 20 15 40.8 h = 56 km MB = 4.6
				D = 53.12 Az = 312.7 (NEIC)
31.	eP1	A	04 38 59.5	Svalbard Region 76.61 N 7.29 E
	eP2	A	39 06	H = 04 33 27.4 h = normal MB = 4.7
	eP3	A	39 17	D = 26.13 Az = 173.7 (NEIC)
				P2V A 1.6s 57.7nm M = 5.0
				P3V A 1.4 23.3nm 4.6
31.	eP	A	05 41 55	Turkey 37.16 N 30.58 E
	LmH	B	48.8	H = 05 37 31.1 h = 37 km MB = 4.6
	LmV	B	50.4	D = 19.12 Az = 320.8 (NEIC)
				LmH B 12s 0.7/um M = 4.2
				LmV B 12 0.7/um 4.3

Day	Phase		h m s	Remarks
1.	eP	A	01 16 47	Turkey 37.23 N 30.18 E
	LmH	B	23.5	H = 01 12 26.8 h = 35 km MB = 4.5
	LmV	B	25.2	D = 18.87 Az = 321.2 (NEIC)
				PV A 1.0s 19.7nm M = 4.3
				LmH B 12 0.7 µm 4.2
				LmV B 10 1.3 µm 4.7
1.	+eP	AB	05 31 19	Andreanof Islands, Aleutian Is.
	e	B	31 55	51.67 N 172.95 W
	e	B	39 54	H = 05 19 23.4 h = 40 km MB = 5.5
	eS	BC	41 16	D = 77.99 Az = 357.0 (NEIC)
	eSS	C	46 44	PV A 1.5s 70.4nm M = 5.5
	eSS	B	46 49	LmH B 18.5 4.1 µm 5.8
	LmH	B	06 10.9	LmV B 17 3.1 µm 5.7
	LmV	B	18.1	
1.	ePKIKP	A	06 34 37.5	South of Fiji Islands 25.54 S 176.77 W
	ePKHKP	A	34 46	H = 06 14 50.2 h = 44 km MB = 5.4
				D = 154.12 Az = 347.7 (NEIC)
1.	eP	A	07 53 03	Sicily 38.74 N 14.06 E
				H = 07 50 10.7 h = normal MB = 4.3
				D = 12.03 Az = 352.5 (NEIC)
1.	e(Pn)	A	12 28 41	Northern Italy 44.57 N 7.50 E
	ePg	A	29 09.5	H = 12 26 55.63 h = 18 km MB = 4.4
	eSn	A	29 46.5	D = 6.68 Az = 23.1 (NEIC)
	eSg	A	30 40	
1.	LmH	C	13 50.5	Probably South Sandwich Islands Region
				56.2 S 27.4 W
				H = 12 48 32.8 h = 17 km MB = 5.1 (NEIC)
				D = 111.2
				LmH C 21s 0.4 µm M = 5.0
2.	+eiP	A	01 00 51	Andreanof Islands, Aleutian Is.
	e	A	00 57	51.76 N 173.17 W
	e	A	01 05	H = 00 48 55.3 h = 40 km MB = 5.2
				D = 77.90 Az = 356.9 (NEIC)
				PV A 1.2s 26.2nm M = 5.1

Day	Phase		h m s	Remarks
2.	+iP	A	08 10 39.5	India-East Pakistan Border Region
	ipP	A	10 52.5	23.76 N 91.76 E
				H = 07 59 57.1 h = 48 km MB = 5.4
				D = 65.87 Az = 316.6 (NEIC) h = 50 km
				PV A 1.1s 48.3nm M = 5.4
2.	+iPKHKP	A	15 38 03	Tonga Islands 21.06 S 175.66 W
	epPKP	A	38 26	H = 15 18 21.5 h = 87 km MB = 5.4
				D = 149.92 Az = 350.7 (NEIC)
				PKHKPV A 1.3s 52.3nm
3.	eP	A	04 56 26	Sicily 38.70 N 14.15 E
	e	A	56 36	H = 04 53 34.6 h = normal MB = 4.5
				D = 12.08 Az = 352.3 (NEIC)
3.	ePKP	A	06 26 08.5	Fiji Islands Region 17.91 S 178.80 W
				H = 06 07 34.0 h = 598 km MB = 5.1
				D = 146.30 Az = 348.0 (NEIC)
				PKPV A 1.4s 27.9nm
3.	e	A	08 45 28	
3.	eP	A	20 01 57	Mariana Islands 17.98 N 146.87 E
	+iPP	AB	06 07	H = 19 48 08.7 h = 40 km MB = 5.3
	eSS	C	20 35	D = 101.18 Az = 332.8 (NEIC)
	LmH	B	50.5	PV A 1.8s 43.9nm M = 5.9
	LmV	B	51.5	PPV A 1.8 189.0nm 6.3
				LmH B 13 1.8 µm 5.8
				LmV B 16 1.5 µm 5.6
4.	ePP	A	02 33 00	Mariana Islands Region 17.94 N 147.04 E
				H = 02 15 02.6 h = 46 km MB = 5.0 (NEIC)
				D = 101.3
				PPV A 2.0s 51.3nm M = 5.7
4.	e(PKP2)	A	14 00 35	South of Fiji Islands 26.09 S 177.33 W
				H = 13 40 24.9 h = 93 km MB = 4.3
				D = 154.53 Az = 346.7 (NEIC)

Day	Phase		h m s	Remarks
4.	eP1	A	14 18 32	<u>Sicily</u> 38.77 N 14.34 E
	eP2	A	18 35	H = 14 15 41.0 h = normal MB = 4.4
				D = 12.03 Az = 351.7 (NEIC)
				P1V A 0.8s 11.5nm
				P2V A 1.2 24.2nm
4.	-iP1	AB	15 46 15.4	<u>Northern Sumatra</u> 0.65 N 98.84 E
	+iP2	A	46 19.5	H = 15 33 28.6 h = normal MB = 6.3
	iP3	AB	46 24	D = 87.74 Az = 320.5 (NEIC)
	+iP4	B	46 43	P1V A 2.4s 925.0nm M = 6.7
	e	B	49 17	P1V B 5.5 9.0μm 7.3
	ePP	B	49 36	P2V A 2.0 1150.0nm 6.8
	eiSKS	B	56 34	P3V A 1.9 1440.0nm 7.0
	eS	B	56 52	P3V B 9.5 13.5μm 7.2
	eSP	B	57 54	P4V B 15 17.1μm 7.1
	ePS	B	58 00	SKSH B 7.0 11.8μm
	eP'P'	A	16 12 20	SH B 21 61.4μm 7.3
	LmH	B	24.4	PSH B 15 28.6μm
	LmV	B	29.5	P'P'V A 2.2 207.0nm
				LmH B 22 59.6μm 7.0
				LmV B 20 31.0μm 6.7
4.	eP	A	19 39 46	<u>Turkey</u> 36.83 N 29.33 E
				H = 19 35 26 h = 18 km
				D = 18.76 Az = 323 (ISC)
				PV A 1.5s 20.1nm
4.	ePKP	A	23 45 30	<u>New Hebrides Islands</u> 18.90 S 169.37 E
				H = 23 26 26.6 h = 257 km MB = 4.3
				D = 143.73 Az = 336.0 (NEIC)
5.	eX	A	02 24 38.5	XV A 1.2s 12.2nm
5.	eP1	A	07 44 57.5	<u>Queen Charlotte Islands Region</u>
	eP2	A	45 10	51.77 N 130.78 W
	eS	C	54 34	H = 07 33 29.1 h = normal MB = 5.1
	LmH	B	08 16.5	D = 73.10 Az = 24.0 (NEIC)
	LmV	B	20.4	P2V A 1.6s 27.5nm M = 5.0
				LmH B 17 4.0μm 5.8
				LmV B 17 4.1μm 5.8

Day	Phase		h m s	Remarks
5.	eP	A	09 21 43.5	Yunnan Province, China 25.23 N 99.41 E
	eSSS	C	38 52	H = 09 10 35.70 h = normal MB = 5.3
	LmH	B	51.1	D = 69.57 Az = 317.2 (NEIC)
	LmV	B	54.0	PV A 1.8s 43.9nm M = 5.1
				LmH B 19 3.2/um 5.6
				LmV B 16 1.5/um 5.4
5.	LmH	B	21 54.8	Probably Central Chile 28.11 S 70.71 W
	LmV	B	54.9	H = 20 52 32.8 h = 49 km MB = 5.6 (ISC)
				D = 106.6
				LmH B 18.5s 1.4/um M = 5.5
				LmV B 22 1.8/um 5.6
5.	ePg	A	23 39 08	Poland 50.37 N 18.83 E
	eSg	A	40 03	H = 23 37 38.0 M = 3.3 (WAR)
				D = 4.61 Az = 276 (ISC)
6.	eP	A	01 10 33.5	Kurile Islands 44.02 N 147.85 E
	LmV	B	02 49.1	H = 00 58 42.7 h = 100 km MB = 4.5
	LmH	B	51.0	D = 78.33 Az = 333.3 (NEIC)
				LmH B 18s 0.9/um M = 5.1
				LmV B 20 1.8/um 5.4
6.	ePKP	A	01 50 14	Easter Island Cordillera
				49.63 S 115.97
				H = 01 30 34.5 h = normal MB = 4.8
				D = 146.97 Az = 67.8 (NEIC)
6.	eP	A	08 31 48.5	Molucca Passage 2.19 N 126.89 E
	e	A	36 28	H = 08 17 48.6 h = normal MB = 5.6
	eSS	C	50 50	D = 104.03 Az = 323.6 (NEIC)
6.	+iP	A	10 54 55.5	Kurile Islands 47.04 N 154.14 E
	Pm	A	54 58	H = 10 43 05.6 h = 60 km MB = 5.6
	e	A	55 05	D = 77.47 Az = 336.6 (NEIC)
	e	A	55 14.5	PV A 1.1s 46.4nm M = 5.3
	LmH	C	11 24	PmV A 1.1 129.0nm 5.8

Day	Phase		h m s	Remarks
6.	eP	A	16 42 08	Tadzhik-Sinkiang Border Region
				38.12 N 73.35 E
				H = 16 34 10.7 h = 171 km MB = 4.5
				D = 44.61 Az = 307.0 (NEIC)
6.	eP	A	18 11 06.5	Central Italy 42.49 N 11.85 E
	e	A	11 50.5	H = 18 09 09.1 h = normal MB = 4.6
	LmH	B	14.2	D = 8.16 Az = 359.0 (NEIC)
	LmV	B	15.0	PV A 1.5s 32.6nm
				LmH B 18 1.8 μm M = 3.8
				LmV B 12 1.3 μm
6.	ePKIKP	A	18 26 50	Kermadec Islands Region 27.07 S 177.00 W
	ePKP2	A	27 16	H = 18 07 10.0 h = 129 km MB = 5.0
	e	A	27 26	D = 155.56 Az = 346.7 (NEIC)
	e	A	27 35.5	LmH B 22s 0.6 μm M = 5.3
	LmV	B	19 35.5	LmV B 20 0.7 μm 5.5
	LmH	B	38.0	
6.	eP	AB	22 20 41	Hindu Kush Region 36.00 N 69.88 E
	e	B	23 06	H = 22 12 44.8 h = 109 km MB = 5.0
	LmH	B	44.0	D = 43.71 Az = 308.4 (NEIC)
	LmV	B	44.0	
7.	eP1	A	02 41 23	Andreanof Islands, Aleutian Is.
	eP2	AB	41 27	51.36 N 176.72 W
	iP3	AB	41 34.5	H = 02 29 28.2 h = 50 km MB = 6.0
	iPP	B	44 44	D = 78.13 Az = 354.6 (NEIC)
	eS	BC	51 18	P1V A 1.7s 54.5nm M = 5.2
	ePS	B	51 52	P2V A 1.7 170.0nm 5.7
	ePPS	B	52 20	P2V B 5 1.1 μm 6.1
	ei	BZ	52 24	P3V A 1.8 608.0nm 6.3
	eSS	BC	56 40	P3V B 16 14.1 μm 6.7
	LmH	B	24.3	PPV B 13 8.9 μm 6.7
	LmV	B	28.3	LmH B 16 35.2 μm 6.8
				LmV B 16 27.0 μm 6.7
				ei A 41 58.5 e B 43 00
				ei A 42 17 e A 51 23

Day	Phase		h m s	Remarks
7.	eiP	A	02 54 00.5	**Andreanof Islands, Aleutian Is.** 51.16 N 177.08 W H = 02 42 04.5 h = 49 km MB = 5.8 D = 78.30 Az = 354.4 (NEIC) PV A 1.9s 186.0nm M = 5.7
7.	eP	A	03 02 47.5	**Andreanof Islands, Aleutian Is.** 51.13 N 177.05 W H = 02 50 49.3 h = normal MB = 4.9 D = 78.34 Az = 354.4 (NEIC) PV A 1.0s 15.8nm M = 5.1
7.	+eP	A	03 05 07	**Andreanof Islands, Aleutian Is.** 51.20 N 176.92 W H = 02 53 09.1 h = normal MB = 5.4 D = 78.27 Az = 354.5 (NEIC) PV A 1.8s 94.5nm M = 5.6
7.	eP	A	03 12 34.5	**Andreanof Islands, Aleutian Is.** 51.36 N 177.12 W H = 03 00 39.2 h = 51 km MB = 4.5 D = 78.10 Az = 354.3 (NEIC) traces
7.	eP	A	03 31 12.5	**Andreanof Islands, Aleutian Is.** 51.21 N 176.96 W H = 03 19 12.2 h = 21 km MB = 5.2 D = 78.26 Az = 354.4 (NEIC) PV A (1.0)s 23.6nm M = 5.2
7.	eP	A	03 32 57	**Andreanof Islands, Aleutian Is.** 51.11 N 177.00 W H = 03 20 59.9 h = 43 km MB = 5.4 D = 78.36 Az = 354.4 (NEIC) PV A 1.5s 62.8nm M = 5.4

Day	Phase		h m s	Remarks
7.	eP	A	04 13 54	Andreanof Islands, Aleutian Is.
				51.19 N 177.03 W
				H = 04 01 58.5 h = 54 km MB = 4.6
				D = 78.28 Az = 354.4 (NEIC)
7.	eP1	A	05 04 06	Dodecanese Islands 36.07 N 28.11 E
	eP2	A	04 09	H = 04 59 44.5 h = 22 km MB = 4.2
				D = 18.80 Az = 325.9 (NEIC)
				P2V A 0.8s 15.4nm M = 4.2
7.	ePKHKP	A	17 10 06	Kermadec Islands Region 27.54 S 175.21 W
	e	A	10 15.5	H = 16 50 01.4 h = normal MB = 5.2
	ePKP2	A	10 24	D = 156.34 Az = 349.1 (NEIC)
	LmH	B	18 31.7	LmH B 16s 0.3/um M = 5.2
	LmV	B	37.7	LmV B 16 0.2/um 5.1
8.	eP	A	02 41 07	Andreanof Islands, Aleutian Is.
				51.29 N 178.83 W
				H = 02 29 11.4 h = 44 km MB = 5.2
				D = 78.06 Az = 353.2 (NEIC)
				PV A 1.1s 28.2nm M = 5.2
8.	ePKIKP	A	20 56 50	South Shetland Islands 63.30 S 61.64 W
				H = 20 37 48.7 h = normal MB = 5.3
				D = 127.24 Az = 50.0 (NEIC)
				PKIKPV A 1.6s 19.2nm
8.	eP diff	B	21 20 14	South Shetland Islands 63.45 S 61.22 W
	ePKIKP1	AB	23 23	H = 21 04 21.83 h = normal MB = 6.3
	eiPKIKP2	A	23 27	MS = 7.0 (NEIC)
	ePKIKP3	A	23 32	D = 127.2
	ePP	AB	25 21	P diff V B 15s 0.9/um
	e	BC	26 42	PKIKP1V A 1.8 60.8nm
	ePcPPKP	A	36 47	PKIKP2V A 2.0 265.0nm
	eiSS	C	43 19	PKIKP3V A 2.0 410.0nm
	LmH	B	22 18.3	PPV B 10 5.3/um M = 6.9
	LmV	B	18.4	PcPPKPV A 1.6 38.5nm
				LmH B 18.5 19.6/um 6.8
				LmV B 18.5 19.7/um 6.8

Day	Phase		h m s	Remarks
9.	eiP	AB	14 13 15.5	Southern California 34.40 N 118.40 W
	i	A	13 25	H = 14 00 41.6 h = 13 km MB = 6.2
	ePP	B	16 44	D = 84.54 Az = 29.3 (NEIC)
	ePPP	B	18 40	PV A 1.5s 643.0nm M = 6.6
	eS	B	23 44	PV B 7 3.9 µm 6.8
	ePS	B	24 48	LmV B 15 48.5 µm 7.0
	eSS	B	29 20	LmH B 16 86.0 µm 7.3
	ePKKP	A	31 24	
	LmV	B	53.6	
	LmH	B	54.6	
9.	e(P)	A	14 37 53	Southern California 34.60 N 118.45 W
				H = 14 25 15.6 h = 7 km
				D = 84.39 Az = 29.3 (NEIC)
9.	eP	A	16 10 55.5	Southern California 34.34 N 118.33 W
				H = 15 58 20.7 h = 14 km MB = 5.1
				D = 84.57 Az = 29.4 (NEIC)
				PV A 1.1s 16.1nm M = 5.2
9.	e(P)	A	21 24 07	Greece 38.09 N 22.76 E
				H = 21 20 33.2 h = 29 km MB = 4.4
				D = 14.85 Az = 331.3 (NEIC)
10.	eP	A	05 30 41	Southern California 34.41 N 118.42 W
				H = 05 18 07.0 h = 8 km MB = 4.7
				D = 84.54 Az = 29.3 (NEIC)
10.	eP	A	23 39 38	Afghanistan-USSR Border Region
	ePP	A	41 20	38.70 N 70.59 E
				H = 23 31 42.9 h = 30 km MB = 4.8
				D = 42.53 Az = 306.2 (NEIC)
11.	e(P)	A	00 25 02	Southern Alaska 60.10 N 147.25 W
				H = 00 13 55.92 h = 24 km MB = 4.2
				D = 68.30 Az = 14.3 (NEIC)
11.				In the intervall 07h 40m - 12h 43m the short-period seismographs (type A) have

Day	Phase		h m s	Remarks
cont. 11.				been out of operation.
11.	eiP	A	13 07 49	Andreanof Islands, Aleutian Is.
	LmH	B	44.3	51.22 N 177.22 W
	LmV	B	55.5	H = 12 55 53.8 h = 50 km MB = 5.5
				D = 78.24 Az = 354.3 (NEIC)
				PV A 1.2s 81.3nm M = 5.6
				LmH B 18 0.8 μm 5.1
				LmV B 16 0.7 μm 5.1
11.	eP	A	14 09 48	Kurile Islands 43.37 N 146.48 E
				H = 13 57 49.8 h = 40 km MB = 4.5
				D = 78.45 Az = 332.6 (NEIC)
11.	e	A	17 00 28	Greece-Albania Border Region
				39.82 N 20.92 E
				H = 16 57 09 h = 22 km MB = 5.4
				D = 12.64 Az = 332 (ISC)
11.	eiP	AB	18 51 11	Central Italy 43.24 N 12.58 E
	iSn	AB	52 35	H = 18 49 20.2 h = 4 km MB = 4.9
	e	BE	53 02	D = 7.44 Az = 355.3 (NEIC)
	eSg	B	53 26	PV A 1.2s 61.0nm
	LmH	B	54.2	LmH B 10 2.4 μm M = 4.1
	LmV	B	54.9	LmV B 13 1.6 μm
12.	ePn	A	03 31 31.5	Yugoslavia-Albania Border Region
	e	A	31 37	42.36 N 20.2 E
	eSn	A	33 18	H = 03 29 03 h = 28 km
	LmH	B	34.8	D = 10.17 Az = 328 (ISC)
	LmV	B	35.7	LmH B 14s 1.0 μm M = 3.8
				LmV B 13 0.7 μm
12.	ePn1	A	04 56 43.5	Central Italy 43.32 N 12.52 E
	iPn2	A	56 46.8	H = 04 54 56.9 h = normal MB = 4.7
	eSn	A	58 05	D = 7.36 Az = 355.5 (NEIC)
	eSg	A	59 03	Pn1V A 1.4s 11.6nm

Day	Phase		h m s	Remarks
cont. 12.	LmH	B	05 00.0	Pn2V A 1.0s 35.4nm
	LmV	B	00.4	LmH B 14 1.8 µm M = 3.8
				LmV B 13 1.1 µm
12.	ePn	A	05 33 03.5	Central Italy 43.34 N 12.58 E
	ePg	A	33 42.5	H = 05 31 17.4 h = normal MB = 4.2
	eiSn	A	34 25.5	D = 7.34 Az = 355.2 (NEIC)
12.	iPn	A	07 49 52.3	Yugoslavia 44.20 N 16.89 E
	iSn	A	51 15	H = 07 48 04.9 h = normal MB = 4.8
	eiSg	A	52 06	D = 7.38 Az = 332.9 (NEIC)
12.	ePn	A	10 23 18	Central Italy 43.22 N 12.86 E
	e	A	23 33	H = 10 21 23.8 h = 0 km
	eSn	A	24 40	D = 7.48 Az = 354 (ISC)
12.	ePn	A	11 41 51	Central Italy 43.23 N 12.8 E
	ePg	A	42 29.5	H = 11 39 58.3 h = 0 km
	eSn	A	43 14	D = 7.46 Az = 354 (ISC)
12.	ePKIKP	A	19 25 36.5	East New Guinea Region 6.25 S 146.48 E
	epPKIKP	A	26 11.5	H = 19 06 53.6 h = 113 km MB = 5.7
	e	A	27 48.5	D = 122.00 Az = 327.9 (NEIC) h = 133 km
	ePSPS	B	44 30	PKIKPV A 1.5s 30.2nm
	LmH	B	20 09.5	LmH B 22.5 2.8 µm
	LmV	B	17.6	LmV B 18 1.1 µm
13.	+iPKP	A	11 22 45.0	New Hebrides Islands 19.96 S 169.79 E
				H = 11 03 35.7 h = 239 km MB = 4.2
				D = 144.85 Az = 335.7 (NEIC)
				PKPV A 1.0s 31.5nm
13.	e(PP)	A	22 30 28.5	Eastern New Guinea Region
				6.06 S 146.25 E
				H = 22 09 40.7 h = 114 km MB = 5.4 (ISC)
				D = 121.75

Day	Phase	h m s	Remarks
13.	ePKHKP A	23 54 53	<u>Loyalty Islands Region</u> 22.78 S 170.92 E
	e A	55 06	H = 23 35 09.1 h = 30 km MB = 5.0
			D = 147.84 Az = 335.0 (NEIC)
			PKHKPV A 1.0s 23.6nm
14.	ePn A	00 06 48.5	<u>Central Italy</u> 43.20 N 12.50 E
	eSn A	08 09	H = 00 04 57.9 h = normal MB = 4.2
			D = 7.48 Az = 355.7 (NEIC)
14.	ePKP A	06 21 17	<u>Fiji Islands Region</u> 15.78 S 177.50 W
			H = 06 02 29.0 h = 429 km MB = 4.7
			D = 144.46 Az = 350.0 (NEIC)
14.	eP1 A	16 34 21	<u>Iran</u> 36.56 N 55.63 E
	eiP2 AB	34 27	H = 16 27 36.1 h = 39 km MB = 5.2
	eS C	39 40	D = 34.31 Az = 308.3 (NEIC)
	LmH B	55.2	P1V A 1.0s 17.7nm M = 5.0
	LmV B	57.0	P2V A 1.8 182.0nm 5.7
			P2V B 4 0.9µm 6.1
			LmH B 9.5 3.2µm 5.4
			LmV B 10 2.8µm 5.4
15.	ePKP A	04 03 56	<u>Fiji Islands Region</u> 17.68 S 178.47 W
			H = 03 45 18.4 h = 579 km MB = 4.5
			D = 146.14 Az = 348.5 (NEIC)
			PKPV A 1.5s 65.3nm
15.	ePKIKP AB	08 09 46	<u>South of Fiji Islands</u> 25.22 S 178.35 E
	ePKHKP AB	09 55	H = 07 51 02.6 h = 584 km MB = 5.7
	ePKP2 AB	10 08	D = 152.63 Az = 341.5 (NEIC)
	e(sPKP2) B	13 23	PKIKPV A 1.0s 94.5nm
			PKIKPV B 6 1.6µm
15.	eP A	08 23 56.5	<u>Turkey</u> 39.17 N 29.40 E
	LmH B	31.4	H = 08 19 56.2 h = 28 km MB = 4.9
			D = 16.99 Az = 318.3 (NEIC)
			PV A 1.9s 91.0nm M = 4.6
			LmH B 16 1.9µm 4.4

Day	Phase		h m s	Remarks
15.	e	A	15 30 13	
	e	A	30 32.5	
15.	e	A	22 18 08	Yugoslavia 42.41 N 20.10 E
				H = 22 13 19.5 h = 0 km (ISC)
				D = 10.1
16.	+eiP	A	14 38 36.8	Taiwan Region 24.01 N 122.20 E
	LmH	B	20.3	H = 14 26 09.1 h = 32 km MB = 5.8
	LmV	B	20.3	D = 83.84 Az = 323.2 (NEIC)
				PV A 1.5s 121.0nm M = 5.8
				LmH B 17 5.7 µm 6.0
				LmV B 16 6.8 µm 6.2
16.	+iPKHKP	A	14 57 37.0	Fiji Islands Region 20.19 S 177.82 W
				H = 14 38 46.4 h = 515 km MB = 4.7
				D = 148.71 Az = 348.4 (NEIC)
				PKHKPV A 1.8s 47.3nm
17.	eSn	A	03 36 30	Yugoslavia 41.2 N 21.0 E
	eSg	A	37 37	H = 03 31 58 (ATH)
				H = 03 31 44 42.4 N 20.4 E h = 0 km (ISC)
				D = 11.5
17.	eiP	A	05 25 05.5	Near East Coast of Honshu, Japan
	e	A	25 16	40.17 N 142.35 E
				H = 05 13 00.6 h = 48 km MB = 5.3
				D = 79.80 Az = 330.6 (NEIC)
				PV A 1.6s 57.7nm M = 5.3
17.	eP	A	07 17 51	Northern Sumatra 0.51 N 98.67 E
				H = 07 05 08.1 h = 66 km MB = 5.2
				D = 87.73 Az = 320.5 (NEIC)
				PV A 1.4s 32.6nm M = 5.3
18.	eP	A	03 11 58.5	North of Svalbard 83.99 N 1.45 W
	e	A	12 02	H = 03 05 19.6 h = normal MB = 4.9
	eS	B	17 25	D = 33.67 Az = 164.9 (NEIC)
	eSS	B	19 40	PV A 1.1s 16.1nm M = 4.9

Day	Phase		h m s	Remarks
ont. 18.	LmH	B	03 27.0	LmH B 14s 0.4 μm M = 4.3
	LmV	B	27.9	LmV B 16 0.5 μm 4.5
18.	ePKIKP	A	07 00 55	<u>Near Coast of Central Chile</u>
	e	A	00 16.5	36.78 S 73.21 W
	LmH	B	54.0	H = 07 47 30.1 h = 31 km MB = 5.6
	LmV	B	54.0	D = 114.43 Az = 44.1 (NEIC)
				PKIKPV A 1.0s 7.9nm
18.	ePn	A	23 42 19	<u>Federal Republic of Germany - The Nether-</u>
	e	A	42 22.5	<u>lands Border Region</u> 50.89 N 5.98 E
	iPg	A	42 31.3	H = 23 41 24.4 h = 27 km MB = 4.2
	ei	A	42 32.5	D = 3.59 Az = 91.8 (NEIC)
	eSn	A	43 00	51°02.3' N 5°56.8' E H = 23 41 23 (BCIS)
	eSg	A	43 12	51.04 N 5.64 E H = 23 41 23
				h = 16 km (ISC)
19.	iPn	A	07 13 21.5	<u>Czechoslovakia</u> 50°34' N 14°16' E
	iPg	A	13 23.5	H = 07 12.01
	iSg	A	13 45.5	explosion yield = 23.6 tons (PRU)
				D = 1.70 H c. 07 12 50
19.	LmH	C	07 18.9	<u>Guatemala</u> 14.2 N 91.7 W
				H = 06 27 58.2 h = 68 km MB = 5.0 (NEIC)
				D = 87.3
				LmH C 20s 0.4 μm
19.	eP	A	10 21 31	<u>Afghanistan-USSR Border Region</u>
				36.56 N 71.50 E
				H = 10 13 29.3 h = 110 km MB = 4.8
				D = 44.39 Az = 308.1 (NEIC)
19.	eP	A	18 53 19	<u>Sicily</u> 37.86 N 15.51 E
	e	A	53 24	H = 18 50 15 h = 33 km
				D = 13.08 Az = 349 (ISC)
19.	LmV	B	21 42.5	Probably <u>West New Guinea</u> 3.5 S 140.0 E
				H = 20 29 20.2 h = 40 km MB = 5.2 (NEIC)
				D = 116.2

Day	Phase		h m s	Remarks
20.	e	A	01 09 57.5	
	e	A	11 04.5	
20.	eP	A	07 19 33	Turkey 37.84 N 29.47 E
	e	A	19 37.5	H = 07 15 20.4 h = 19 km MB = 4.6
	eS	BZ	23 03	D = 18.03 Az = 320.9 (NEIC)
	LmH	B	26.7	P1V A 1.5s 32.7nm M = 4.2
	LmV	B	28.5	P2V A 1.6 55.0nm
20.	eP	A	09 59 20.5	North of Svalbard 83.79 N 1.99 W
				H = 09 52 41.7 h = normal MB = 4.2
				D = 33.50 Az = 164.3 (NEIC)
20.	eP	A	12 47 30.5	North of Svalbard 83.92 N 1.61 W
	e	A	47 35.5	H = 12 40 41.7 h = normal MB = 4.7
				D = 33.61 Az = 164.7 (NEIC)
				P1V A 1.0s 9.8nm M = 4.7
				P2V A 1.2 16.3nm
21.	ePKP2	A	03 24 31	Tonga Islands 20.47 S 173.43 W
	eX1	A	24 34	H = 03 04 39.3 h = normal MB = 5.0
	eX2	A	24 43	D = 149.62 Az = 353.6 (NEIC)
	LmH	C	50	X1V A 1.6s 27.5nm
	LmV	B	50.5	X2V A 1.4 41.8nm
21.	ePKHKP	A	03 39 16	South of Fiji Islands 23.79 S 179.97 W
	eX	A	40 35	H = 03 20 20.0 h = 512 km MB = 5.2
				D = 151.72 Az = 344.3 (NEIC)
				XV A 1.5s 20.1nm
21.	e	A	06 03 31.5	Southern California 34.37 N 118.47 W
				H = 05 50 52.3 h = 13 km
				D = 84.59 Az = 29.3 (NEIC)
21.	ePKHKP	A	08 48 42	Fiji Islands Region 20.92 S 179.53 W
	e	A	50 47	H = 08 29 58.2 h = 597 km MB = 4.3
				D = 149.06 Az = 346.2 (NEIC)
				PKHKPV A 1.0s 11.8nm

Day	Phase		h m s	Remarks
21.	+eP	A	10 48 53	<u>Chile-Argentina Border Region</u>
	ipP	AB	49 40	23.85 S 67.16 W
	ePP	B	53 00	H = 10 35 20.1 h = 169 km MB = 6.3
	epPP	B	53 44	D = 102.33 Az = 39.6 (NEIC)
	eSKS	B	59 16	PV A 1.7s 84.9nm M = 6.1
	iS	B	11 00 15	SH B 15 7.2 µm 6.8
	iSP	B	01 48	LmH B 19 13.8 µm
	eiPKKP	A	05 04	LmV B 20 19.5 µm
	e	A	05 24	e B 10 57 02 ei A 11 06 09
	iPKKP	A	05 29	
	e	B	06 48	
	iSS	BC	07 18	
	isSS	BC	08 50	
	eP'P'	A	13 19	
	LmV	B	34.9	
	LmH	B	35.2	
21.	e	A	14 54 30	<u>Philippine Islands Region</u>
				19.38 N 121.12 E
				H = 14 41 39.5 h = normal MB = 4.8
				D = 86.92 Az = 323.1 (NEIC)
				traces
21.	e	A	14 55 50	<u>South Atlantic Ridge</u> 19.94 S 11.78 W
				H = 14 44 03.8 h = normal MB = 4.7
				D = 73.24 Az = 15.3 (NEIC)
21.	eP	A	18 43 14	<u>Hokkaido, Japan Region</u> 41.57 N 142.81 E
	e	A	43 17	H = 18 31 20.4 h = 27 km MB = 4.9
	e	A	43 22	D = 78.75 Az = 330.8 (NEIC)
	e	A	43 28	LmH B 14s 0.7 µm M = 5.1
	LmH	B	19 18.7	LmV B 16 0.6 µm 5.0
	LmV	B	21.3	
22.	eP	AB	14 32 04	<u>Turkey</u> 37.24 N 30.27 E
	e(S)	BZ	35 44	H = 14 27 44.7 h = 45 km MB = 5.0
	LmH	B	38.7	D = 18.90 Az = 321.1 (NEIC)

Day	Phase		h m s	Remarks
cont. 22.	LmV	B	14 40.5	PV A 1.7s 84.9nm M = 4.7
				PV B 5 0.9μm 5.3
				LmH B 14 2.9μm 5.0
				LmV B 14.5 3.0μm 5.2
22.	LmH	C	23 12.0	<u>Molucca Passage</u> 2.22 N 126.93 E
				H = 22 08 27.81 h = normal
				MB = 5.1 MS = 4.9 (NEIC)
				D = 104.2
				LmH C 25s 0.3μm M = 4.8
23.	eP	A	09 14 28	<u>Jan Mayen Island Region</u> 72.25 N 5.87 W
	e	A	14 36	H = 09 09 24.4 h = normal MB = 4.4
				D = 23.04 Az = 150.7 (NEIC)
23.	+eP	A	16 15 34.5	<u>South of Honshu, Japan</u> 33.59 N 138.35 E
				H = 16 03 34.6 h = 276 km MB = 4.9
				D = 83.87 Az = 329.1 (NEIC)
				PV A 1.3s 19.7nm M = 4.8
23.	eP1	ABC	19 45 03	<u>Turkey</u> 39.60 N 27.37 E
	eP2	A	45 05	H = 19 41 23.0 h = 15 km MB = 5.0
	-eiP3	AB	45 08.5	D = 15.64 Az = 320.1 (NEIC)
	eS	B	48 14	P1V A 1.0s 7.9nm M = 3.9
	LmH	B	50.5	P2V A 1.5 45.2nm 4.4
	LmV	B	54.5	P3V A 1.5 181.0nm 5.0
				LmH B 15 22.3μm 5.4
				LmV B 12 15.2μm 5.6
24.	eP	A	17 37 49	<u>Ryukyu Islands</u> 26.63 N 128.76 E
	e	A	38 17	H = 17 25 09.96 h = 19 km MB = 4.6 (NEIC)
	LmH	B	18 21.0	D = 85.2
	LmV	B	21.1	LmH B 15s 1.7μm M = 5.6
				LmV B 14 1.8μm 5.6
25.	eP	A	04 28 19	<u>Near Coast of Nicaragua</u> 12.16 N 87.49 W
	eS	BC	38 40	H = 04 15 41.5 h = 52 km MB = 5.3
	eSS	C	44.5	D = 86.40 Az = 39.0 (NEIC)

Day	Phase		h m s	Remarks
cont. 25.	eSSS	C	04 48.5	PV A 1.5s 17.6nm M = 5.0
	LmH	B	05 01.6	LmH B 22 2.4 µm
	LmV	B	04.8	LmV B 19 2.2 µm
25.	e(P)	A	04 51 16	Turkey 37.05 N 28.94 E
				H = 04 46 55.0 h = 13 km MB = 4.5
				D = 18.40 Az = 323.1 (NEIC)
				PV A 1.6s 22.0nm M = 4.0
25.	eP	A	06 52 39	Fox Islands, Aleutian Is.
				52.13 N 169.49 W
				H = 06 40 44.3 h = 32 km MB = 5.3
				D = 77.59 Az = 359.3 (NEIC)
				PV A 1.1s 16.1nm M = 5.0
25.	ePKP	A	10 46 58.5	Tonga Islands 15.32 S 174.93 W
				H = 10 27 53.5 h = 260 km MB = 4.4
				D = 144.37 Az = 352.8 (NEIC)
25.	+iP	A	14 49 44.5	Mindanao, Philippine Islands
				5.87 N 126.70 E
				H = 14 36 02.2 h = 88 km MB = 5.6
				D = 100.97 Az = 324.0 (NEIC)
				PV A 1.2s 36.6nm M = 5.9
25.	eP	AB	19 40 04	Near West Coast of Honshu, Japan
	eX	A	40 10	37.11 N 138.34 E
	e	A	40 47	H = 19 27 53.1 h = 40 km MB = 5.5
	eS	C	50 15	D = 80.86 Az = 328.9 (NEIC)
	LmV	B	21.0	PV A 1.8s 81.0nm M = 5.4
	LmH	B	21.8	XV A 1.7 151.0nm
				LmH B 11.5 2.6 µm 5.8
				LmV B 15 3.1 µm 5.8
25.	eP	A	22 26 46	Off W. Coast of Northern Sumatra
	eX	A	26 54.5	3.24 N 95.93 E
	eS	C	37 03	H = 22 14 17.7 h = normal MB = 5.4

Day	Phase		h m s	Remarks
cont. 25.	LmH	B	23 10.7	D = 83.90 Az = 320.4 (NEIC)
	LmV	B	10.8	PV A 1.5s 27.6nm M = 5.2
				XV A 2.0 89.7nm
				LmH B 18 0.6 µm 5.0
				LmV B 18 0.9 µm 5.2
26.	iPn	A	02 11 33.5	<u>Switzerland</u> 46.7 N 10.2 E
	i	A	11 37.5	H = 02 10 32 (BCIS)
	ePg	A	11 51	Northern Italy 46.56 N 10.2 E
	iSn	A	12 23.5	H = 02 10 29.9 h = 0 km
	eSg	A	12 45	D = 4.20 Az = 12 (ISC)
26.	ePKIKP	A	05 14 56	<u>Solomon Islands</u> 10.43 S 161.26 E
	e	A	15 23	H = 04 55 50.0 h = 90 km MB = 5.9
	ePP	A	17 10	D = 132.77 Az = 334.0 (NEIC)
26.	iPg	A	10 05 23	<u>Böhmischbruck, Federal Rep. Germany</u>
	iSg	A	05 37.5	49°34.1'N 12°21.4'E
				explosion
				H = 10 05 00 yield 6.7 t (MUN)
				D c. 1.2
26.	eP1	A	11 58 59.5	<u>Turkey</u> 37.48 N 29.83 E
	eP2	A	59 03	H = 11 54 42.1 h = 33 km MB = 4.8
	LmH	C	12 06.0	D = 18.50 Az = 321.2 (NEIC)
				P2V A 2.0s 42.7nm M = 4.2
26.	e	A	19 17 51	<u>New Hebrides Islands</u> 18.00 S 167.08 E
	LmH	B	20 23.0	H = 18 58 04.6 h = 11 km MB = 5.6
	LmV	B	24.0	D = 142.01 Az = 334.6 (NEIC)
				LmH B 20s 0.4 µm M = 5.1
				LmV B 20 0.5 µm 5.3
26.	e	A	21 44 11	X3V A 1.4s 25.6nm
	e	A	44 13	
	eX3	A	44 25	

Day	Phase		h m s	Remarks
27.	ePKHKP	A	00 24 58.5	South of Fiji Islands 22.16 S 179.63 W
				H = 00 06 11.9 h = 587 km MB = 5.0
				D = 150.23 Az = 345.5 (NEIC)
				PKHKPV A 1.1s 18.1nm
27.	eP	A	00 44 06.5	Near Coast of Northern California
	LmH	B	01 20.3	40.36 N 124.80 W
	LmV	B	21.3	H = 00 31 39.9 h = normal MB = 5.3 (NEIC)
				D = 81.7
				PV A 1.8s 54.0nm M = 5.3
				LmH B 16 0.9/um 5.2
				LmV B 16 0.8/um 5.2
27.	eP	A	02 38 37	Kurile Islands 45.56 N 150.80 E
				H = 02 26 41.8 h = normal MB = 5.1
				D = 77.86 Az = 334.8 (NEIC)
				PV A 1.2s 12.2nm M = 4.9
27.	ePKIKP	A	05 03 19	Fiji Islands Region 20.24 S 177.79 W
	eiPKHKP	A	03 24	H = 04 44 35.0 h = 524 km MB = 5.0
	ePKP2	A	03 29.5	D = 148.76 Az = 348.4 (NEIC) h = 538 km
	epPKHKP	A	05 27	PKIKPV A 1.6s 22.0nm
				PKHKPV A 1.2 71.1nm
				PKP2V A 1.2 22.4nm
				pPKHKPV A 1.8 20.3nm
27.	eP	A	05 22 47	Kurile Islands 46.23 N 152.16 E
				H = 05 10 55.4 h = 60 km MB = 4.8
				D = 77.65 Az = 335.5 (NEIC)
				PV A 1.0s 9.8nm M = 4.7
27.	eP	A	09 08 46	Near East Coast of Honshu, Japan
	epP	A	09 02	38.27 N 141.92 E
				H = 08 56 33.5 h = 46 km MB = 5.1
				D = 81.29 Az = 330.6 (NEIC) h = 60 km
				PV A 1.2s 28.4nm M = 5.1
27.	ePKIKP	A	13 45 26	Fiji Islands Region 21.38 S 177.88 W
	ePKHKP	A	45 33	H = 13 26 27.1 h = 390 km MB = 5.0

Day	Phase		h m s	Remarks
cont. 27.	ePKP2	A	13 45 40	D = 149.85 Az = 347.9 (NEIC)
	epPKP	A	47 08	
	e	A	47 11.5	
27.	eP	A	16 44 55.5	Off East Coast of Honshu, Japan
	epP	A	45 06.5	40.65 N 143.50 E
	ePP	A	47 56	H = 16 32 49.4 h = 38 km MB = 5.2
	eS	BC	55 00	D = 79.80 Az = 331.2 (NEIC) h = 40 km
	LmH	B	17 18.8	PV A 1.4s 23.3nm M = 5.0
	LmV	B	23.4	LmH B 17 3.2/um 5.7
				LmV B 17 1.4/um
28.	eP	A	08 39 02	Near Islands, Aleutian Is.
				52.82 N 171.45 E
				H = 08 27 17.5 h = 24 km MB = 4.6
				D = 75.51 Az = 346.9 (NEIC)
28.	ePKHKP	A	11 30 24	Fiji Islands Region 19.34 S 176.49 W
				H = 11 10 48.5 h = 109 km MB = 4.8
				D = 148.11 Az = 350.2 (NEIC)

Day	Phase		h m s	Remarks
1.	ePKHKP	A	12 29 27	West of Macquarie Island
	e	A	29 53	52.83 S 140.17 E
	e	A	29 58	H = 12 09 43.6 h = normal MB = 5.2
	LmH	C	13 36.0	D = 148.62 Az = 287.0 (NEIC)
	LmV	B	40.5	PKHKPV A 1.3s 17.5nm
				LmH C 20 0.7/um M = 5.4
				LmV B 20 0.4/um 5.2
1.	ePKHKP	A	13 21 07.5	West of Macquarie Island
				52.80 S 139.99 E
				H = 13 01 24.3 h = normal MB = 4.8
				D = 148.51 Az = 287.2 (NEIC)
1.	eP	A	13 41 35	Kansu Province 36.2 N 108.6 E
				H = 13 30 36 h = 0 km
				D = 67.04 Az = 317 (ISC)
2.	LmH	B	01 54.0	Probably : South Atlantic Ocean
	LmV	B	57.0	54.4 S 54.1 W
				H = 00 45 42.6 h = normal MB = 4.9 (NEIC)
				D = 118.2
				LmH B 18s 0.4/um M = 5.1
				LmV B 18 0.4/um 5.1
2.	eP	A	03 23 00.5	Iceland Region 64.5 N 26.6 W
				H = 03 17 43 h = 0 km
				D = 24.25 Az = 107 (ISC)
2.	ePn	A	18 54 07	Central Italy 42.70 N 12.17 E
	ePg	A	54 48	H = 18 52 12.0 h = 22 km MB = 4.4
	eSn	A	55 34.5	D = 7.96 Az = 357.5 (NEIC)
	eSg	A	56 41	PgV A 1.1s 20.2nm
				SgV A 1.5 47.7nm
3.	eP	A	02 24 25	Eastern Arabian Peninsula
	e	A	24 33	22.10 N 59.24 E
				H = 02 15 58.1 h = 35 km MB = 5.1
				D = 46.65 Az = 319.7 (NEIC)
				PV A 1.0s 19.7nm M = 5.0

Day	Phase		h m s	Remarks
3.	e	A	02 37 02.5	<u>Poland</u> 50.6 N 18.9 E
	eSg	A	37 06	H = 02 34 32 h = 0 km (ISC)
				D = 4.65
				SgV A 0.8s 17.3nm
3.	eP	A	04 13 56.5	<u>Northeast of Taiwan</u> 26.78 N 125.87 E
				H = 04 01 45.9 h = 167 km MB = 4.5
				D = 83.58 Az = 324.3 (NEIC)
				PV A 0.8s 13.4nm M = 4.7
3.	eP	A	14 59 20	<u>Gulf of California</u> 25.91 N 109.93 W
	LmH	B	15 39.6	H = 14 46 30.7 h = 34 km MB = 5.3
	LmV	B	39.6	D = 87.98 Az = 32.9 (NEIC)
				PV A 2.0s 34.2nm M = 5.3
				LmH B 17 4.6/um 6.0
				LmV B 16.5 5.1/um 6.0
3.	eP	A	19 04 44	<u>Southern Greece</u> 36.39 N 22.30 E
	e	A	04 48.5	H = 19 00 59.6 h = 55 km MB = 4.5
	e	A	08 10.5	D = 16.20 Az = 335.0 (NEIC)
3.	-iP	AB	22 05 44.5	<u>Kurile Islands</u> 48.28 N 153.01 E
	e	A	05 49	H = 21 54 09.1 h = 120 km MB = 5.8
	epP	AB	06 23	D = 76.03 Az = 335.8 (NEIC) h c. 165 km
	esP	A	06 44	PV A 1.5s 186.0nm M = 5.6
				PV B 3 0.9/um 6.0
3.	eP	A	23 16 20.5	<u>Kurile Islands</u> 46.86 N 152.53 E
				H = 23 04 28.1 h = normal MB = 4.8
				D = 77.19 Az = 335.7 (NEIC)
				PV A 1.0s 13.8nm M = 4.9
4.	eiP	A	00 40 33.5	<u>South of Honshu, Japan</u>
	e	A	42 12	30.40 N 138.32 E
	e	A	42 16.5	H = 00 28 36.3 h = 428 km MB = 5.7
	e	A	43 07	D = 86.59 Az = 329.2 (NEIC)
	ePP	A	44 02	PV A 1.5s 75.4nm M = 5.2
	eS	B	50 29	PPV A 1.7 54.6nm 5.4

Day	Phase		h m s	Remarks
cont. 4.	esS	B	00 53 29	SH B 7s 1.8µm M = 5.8
	LmH	B	01 24.3	
	LmV	B	25.9	
4.	e	A	03 29 35.5	<u>West of Macquarie Island</u>
				52.69 S 139.75 E
				H = 03 09 34.9 h = normal MB = 4.7
				D = 148.34 Az = 287.5 (NEIC)
4.	eP	A	08 00 45	<u>Off East Coast of Honshu, Japan</u>
	epP	A	00 58	40.66 N 143.51 E
	LmH	C	34.5	H = 07 48 39.0 h = 37 km MB = 5.2
				D = 79.80 Az = 331.2 (NEIC)
				h = 47 km
				PV A 1.3s 21.8nm M = 5.0
				LmH C 19 1.4µm 5.3
4.	eP	A	10 42 17	<u>Northern Sumatra</u> 0.63 N 98.70 E
	e	A	42 50.5	H = 10 29 29.7 h = normal MB = 5.2
				D = 87.66 Az = 320.5 (NEIC)
				PV A 1.5s 27.6nm M = 5.3
4.	ePKHKP	A	17 36 01.5	<u>Tonga Islands</u> 17.92 S 175.61 W
				H = 17 16 51.5 h = 290 km MB = 4.9
				D = 146.84 Az = 351.6 (NEIC)
				PKHKPV A 1.2s 16.3nm
6.	ePKHKP	A	04 59 41	<u>Fiji Islands Region</u> 18.76 S 177.63 W
	ePKP2	A	59 44.5	H = 04 40 57.5 h = 543 km MB = 4.6
				D = 147.34 Az = 349.1 (NEIC)
				PKHKPV A 1.1s 18.1nm
6.	epP	A	08 13 10	<u>Taiwan Region</u> 21.47 N 121.89 E
	LmH	B	51.7	H = 08 00 19.1 h = 41 km MB = 4.9
	LmV	B	56.3	D = 85.69 Az = 323.2 (NEIC)
				LmH B 14s 0.3µm M = 4.9
				LmV B 16 0.4µm 5.0

Day	Phase		h m s	Remarks
7.	LmH	B	12 33.6	LmH B 20s 0.4 um
	LmV	B	33.6	LmV B 20 0.6 um
7.	e	A	18 01 43.5	PV A 1.2s 14.2nm
8.	LmH	B	15 26.0	Probably <u>Near N. Coast of West New Guinea</u>
	LmV	B	26.3	2.9 S 139.4 E
				H = 14 11 46.2 h = 15 km
				MB = 5.2 MS = 5.4 (NEIC)
				D = 115.3
				LmH B 20s 0.7 um M = 5.3
				LmV B 18 0.8 um 5.4
8.	eP	A	16 40 54.5	<u>Panama</u> 7.24 N 81.57 W
				H = 16 28 09.4 h = 17 km MB = 5.0
				D = 86.48 Az = 39.6 (NEIC)
				PV A 1.3s 21.8nm M = 5.2
8.	+iPn	A	17 44 18.2	<u>Federal Rep. of Germany</u> 49.5 N 8.5 E
	eiSn	A	44 47.5	H = 17 43 35 h = 0 km
				D = 2.33 Az = 59 (ISC)
8.	ePKIKP	A	21 47 02.5	<u>Fiji Islands Region</u> 17.78 S 178.70 W
	ePKHKP	A	47 05	H = 21 28 29.7 h = 607 km MB = 4.7
				D = 146.19 Az = 348.2 (NEIC)
				PKHKPV A 0.6s 24.9nm
8.	eP	A	22 49 07	<u>Turkey</u> 37.52 N 29.74 E
	LmH	B	55.7	H = 22 44 47.33 h = 21 km MB = 4.7
				D = 18.42 Az = 321.2 (NEIC)
				PV A 1.2s 14.2nm M = 4.0
9.	eP	A	05 01 50	<u>Greece</u> 38.71 N 20.33 E
	LmH	B	06.8	H = 04 58 38.9 h = 16 km MB = 4.8
	LmV	B	08.1	D = 13.43 Az = 335.5 (NEIC)
				PV A 1.0s 23.6nm
				LmV B 10 2.6 um
				EW-component of seismograph type B out of operation.

Day	Phase		h m s	Remarks
9.	ePKP2	A	08 30 59	South of Fiji Islands 23.39 S 180.00 E H = 08 11 52.8 h = 511 km MB = 5.0 D = 151.33 Az = 344.5 (NEIC) PKP2V A 0.9s 13.6nm
9.	eP	A	13 30 46	Panama 7.28 N 81.54 W H = 13 18 02.1 h = 24 km MB = 4.8 D = 86.43 Az = 39.6 (NEIC) PV A 0.8s 9.6nm M = 5.1
9.	eP	A	15 26 24.5	Kurile Islands 45.61 N 150.86 E H = 15 14 28.5 h = normal MB = 4.9 D = 77.84 Az = 334.8 (NEIC) PV A 1.2s 16.3nm M = 4.9
9.	ePKP	A	18 26 15	Tonga Islands 15.18 S 173.67 W H = 18 06 41.4 h = normal MB = 5.0 D = 144.37 Az = 354.2 (NEIC)
9.	LmH	C	22 08.0	Off East Coast of Honshu, Japan 36.6 N 142.1 E H = 21 19 52.3 h = 64 km MB = 4.4 (NEIC) D = 82.8
10.	eP e	A A	15 50 05 50 12	Vancouver Island Region 49.32 N 127.39 W H = 15 38 28.7 h = normal MB = 5.0 D = 74.41 Az = 25.7 (NEIC)
11.	eP LmH LmV	A B C	15 36 51 16 10.4 13.2	Gulf of Alaska 59.33 N 146.65 W H = 15 25 45.9 h = 18 km MB = 5.0 D = 68.97 Az = 14.6 (NEIC) PV A 1.6s 24.7nm M = 5.1 LmV C 16 0.6 um 5.0
12.	e e	A A	12 49 06 49 45	

Day	Phase		h m s	Remarks
12.	ePKHKP	A	14 47 41.5	<u>Tonga Islands</u> 18.34 S 175.68 W
				H = 14 28 28.8 h = 280 km MB = 4.7
				D = 147.25 Az = 351.4 (NEIC)
12.	eP	AC	22 45 33	<u>Philippine Islands Region</u>
	eSKS	BC	56 00	18.81 N 119.81 E
	eSP	BC	57 00	H = 22 32 51.0 h = 44 km MB = 5.0
	eSS	C	23 01 45	D = 86.63 Az = 322.7 (NEIC)
	LmH	B	21.7	PV A 0.8s 15.4nm M = 5.3
	LmV	B	30.1	LmH B 18.5 1.9/um 5.5
				LmV B 13.5 1.6/um 5.6
13.	eP	A	03 11 42.5	<u>Near East Coast of Honshu, Japan</u>
	LmH	B	48.0	40.10 N 142.36 E
	LmV	B	51.7	H = 02 59 38.5 h = 57 km MB = 5.1
				D = 79.87 Az = 330.7 (NEIC)
				PV A 1.5s 27.6nm M = 5.0
				LmH B 18 1.2/um 5.3
				LmV B 16 0.7/um 5.1
13.	LmH	C	03 26.5	LmH C 17s 0.4/um
	LmV	C	33.5	LmV C 14 0.7/um
13.	LmH	C	04 21.5	LmV C 16s 0.6/um
	LmV	C	28.3	
13.	LmH	B	04 28.5	LmH B 16s 0.3/um
	LmV	B	28.5	LmV B 14 0.4/um
13.	ePKP	AB	16 07 33	<u>Fiji Islands Region</u> 18.37 S 173.64 E
	e	A	07 36	H = 15 47 57.1 h = 20 km MB = 5.6
	eiX	A	07 38	D = 144.75 Az = 340.1 (NEIC)
	eSS	C	29 30	PKPV A 1.5s 25.1nm
	eSSS	C	35 10	XV A 1.4 79.0nm
	LmV	B	17 10.0	LmH B 21 3.8/um M = 6.1
	LmH	B	10.5	LmV B 22 4.9/um 6.2

Day	Phase		h m s	Remarks
13.	e(P diff)	A	19 27 48	<u>East New Guinea Region</u> 5.72 S 145.37 E
	eiPKIKP	AB	31 05	H = 19 12 25.0 h = 118 km MB = 6.3
	ePP	AB	32 36	D = 120.96 Az = 327.6 (NEIC)
	ePKKP	A	41 10	PKIKPV A 1.4s 172.0nm
	ePS	C	42 25	PPV B 8 1.1 µm M = 5.9
	e(SS)	C	49 55	PKKPV A 1.7 30.3nm
	e(SSS)	C	54 15	LmH B 20 1.4 µm
	LmH	B	20 19.8	LmV B 20 1.5 µm
	LmV	B	25.7	e A 33 08 e B 33 27
13.	-eP	A	24 03 09	<u>Vancouver Islands Region</u>
	e	B	03 15	50.60 N 129.95 W
	eS	BC	12 44	H = 23 51 35.5 h = normal MB = 5.7
	eSKS	BC	13 16	D = 73.95 Az = 24.3 (NEIC)
	eSS	C	17 35	PV A 1.2s 42.7nm M = 5.3
	eSSS	C	21 20	LmH B 15.5 5.3 µm 5.9
	LmH	B	38.8	LmV B 14 5.4 µm 6.0
	LmV	B	40.8	
14.	ePKP	AB	06 28 55.5	<u>Loyalty Islands Region</u>
	e	B	29 12	21.34 S 169.17 E
	e	B	29 40	H = 06 09 14.5 h = 11 km MB = 5.1
	LmH	B	36.9	D = 145.85 Az = 334.3 (NEIC)
	LmV	B	37.3	PKPV A 3.0s 263.0nm
				LmH B 22 2.2 µm M = 5.8
				LmV B 20 1.1 µm 5.6
14.	+eP	A	12 26 39.5	<u>Near East Coast of Kamchatka</u>
				53.86 N 160.53 E
				H = 12 15 14.1 MB = 5.3 h = 35 km
				D = 72.65 Az = 339.9 (NEIC)
				PV A 1.2s 34.6nm M = 5.2
14.	eP	A	12 35 30.5	<u>Near East Coast of Kamchatka</u>
				53.96 N 160.60 E
				H = 12 24 04.5 h = normal MB = 4.8
				D = 72.57 Az = 339.9 (NEIC)

Day	Phase		h m s	Remarks
14.	LmH	C	12 57.5	LmH C 22s 0.4/um
14.	eP	A	20 52 08	Straitsof Gibraltar 35.26 N 5.81 W
				H = 20 47 37.1 h = 28 km
				D = 19.90 Az = 34.1 (NEIC)
15.	+iP	A	05 47 44.2	Hokkaido, Japan Region
	eipP	A	47 54.5	41.74 N 143.73 E
	LmH	B	06 22.2	H = 05 35 44.63 h = 48 km MB = 5.4
	LmV	B	25.9	D = 78.93 Az = 331.2 (NEIC)
				h = 36 km
				PV A 1.4s 60.5nm M = 5.4
				LmH B 17 1.8/um 5.5
				LmV B 18 1.3/um
15.	ePKP	AB	11 07 51.5	Tonga Island 17.65 S 173.03 W
	LmH	B	12 16.9	H = 10 48 10.7 h = normal MB = 5.2
	LmV	B	20.8	D = 146.87 Az = 354.6 (NEIC)
				PKPV A 1.8s 186.0nm
				LmH B 18 0.5/um M = 5.3
				LmV B 18 0.6/um 5.4
15.	LmH	B	15 09.6	South of Honshu, Japan
	LmV	B	10.6	31.1 N 141.8 E
				H = 14 12 10.4 h = normal MB = 4.8 (NEIC)
				D = 87.4
				LmH B 14s 0.5/um M = 5.1
15.	eP	A	15 27 07	Southern Greece 37.19 N 23.97 E
	LmH	B	34.1	H = 15 23 17.9 h = 32 km MB = 4.8
	LmV	B	34.1	D = 16.10 Az = 330.6 (NEIC)
				PV A 1.0s 19.7nm M = 4.2
				LmH B 10 0.6/um 4.1
				LmV B 10 0.6/um 4.3
15.	ePKJKP	A	21 10 56	New Hebrides Islands 15.51 S 167.57 E
	ePKHKP	A	11 03.5	H = 20 51 47.8 h = 118 km MB = 5.4

Day	Phase		h m s	Remarks
cont. 15.	ePKIKP	A	21 11 06	D = 139.94 Az = 336.2 (NEIC)
				PKJKPV A 0.8s 13.5nm
				PKHKPV A 1.3 56.7nm
				PKIKPV A 1.3 70.0nm
16.	ePKIKP	AB	12 51 56.5	Solomon Islands 10.92 S 163.69 E
	ePP	AB	54 28	H = 12 32 30.6 h = 42 km MB = 5.8
	ePPS	BC	13 06 40	D = 134.23 Az = 335.4 (NEIC)
	e	B	11 35	PKIKPV A 1.3s 32.8nm
	eSS	B	12 14	PPV A 2.3 73.1nm
	LmH	B	44.5	LmH B 20 7.6/um M = 6.4
	LmV	B	53.7	LmV B 18 5.2/um 6.3
				e B 12 55 00 e B 55 52 e B 13 14 20
16.	eP	AB	20 48 57	Philippine Islands Region
	e	AB	48 22	8.30 N 127.12 E
	i	A	48 26.5	H = 20 35 18.8 h = 52 km MB = 5.7
	ePP	B	53 00	D = 99.25 Az = 324.4 (NEIC)
	eSKS	B	59 32	PV A 1.4s 34.9nm M = 5.7
	ePS	B	21 02 18	LmH B 16 3.9/um
	LmH	B	39.1	LmV B 17 6.0/um
	LmV	B	39.2	
17.	LmH	C	04 21.0	West New Guinea 3.2 S 139.8 E
	LmV	C	22.0	H = 03 13 59.7 h = normal MB = 5.6 (NEIC)
				D = 115.8
				LmH C 27s 0.5/um M = 5.0
				LmV C 27 0.3/um 4.8
18.	ePKHKP	A	02 53 41.5	South of Fiji Islands 23.79 S 179.95 E
	ePKP2	A	53 53.5	H = 02 34 48.7 h = 542 km MB = 5.0
				D = 151.70 Az = 344.2 (NEIC)
18.	iP	A	03 25 01.3	Kurile Islands 49.25 N 156.26 E
	e	A	25 11.5	H = 03 13 15.4 h = normal MB = 5.1
	e	A	25 17.5	D = 75.98 Az = 337.7 (NEIC)
	LmH	B	58.0	PV A 1.1s 48.4nm M = 5.4

Day	Phase		h m s	Remarks
cont. 18.	LmV	B	04 04.0	LmH B 16s 0.5/um M = 4.9 LmV B 16 0.6/um 5.0
18.	ePKHKP	A	05 39 53	Tonga Islands 20.90 S 174.38 W H = 05 20 03.7 h = normal MB = 5.2 D = 149.94 Az = 352.4 (NEIC) PKHKPV A 1.1s 32.3nm
18.	-iP	AB	16 12 06.3	Dodecanese Islands 36.51 N 26.96 E H = 16 08 04.1 h = 154 km MB = 4.7 D = 17.93 Az = 326.8 (NEIC) PV A 1.1s 109nm M = 5.1
18.	+iP	A	19 20 48.8	Tadzhik-Sinkiang Border Region 38.18 N 73.60 E H = 19 12 43.0 h = 145 km MB = 4.9 D = 44.73 Az = 307.0 (NEIC) PV A 1.0s 27.6nm M = 4.9
19.	ePKP e	A A	02 42 06.5 42 14	Loyalty Islands Region 21.30 S 169.23 E H = 02 22 29.6 h = 36 km MB = 5.3 D = 145.83 Az = 334.4 (NEIC) PKPV A 1.6s 38.5nm
19.	ePKP	A	03 28 05	Fiji Islands Region 16.16 S 176.94 W H = 03 09 16.5 h = 429 km MB = 4.5 D = 144.92 Az = 350.5 (NEIC)
19.	eP e	A A	06 25 13 25 33	Oaxaca, Mexico 17.10 N 95.13 W H = 06 12 31.9 h = 83 km MB = 5.5 D = 87.14 Az = 37.6 (NEIC) PV A 1.2s 20.3nm M = 5.0
19.	eP e	A A	13 36 10.5 36 18	Arabian Sea 14.58 N 56.23 E H = 13 27 11.7 h = normal MB = 5.3 D = 50.82 Az = 324.8 (NEIC) P traces

Day	Phase		h m s	Remarks
19.	eP	A	16 59 32	Molucca Passage 2.13 N 126.83 E
	LmH	B	17 50.1	H = 16 45 30.2 h = normal MB = 5.4
	LmV	B	52.6	D = 104.04 Az = 323.6 (NEIC)
				LmH B 19.5s 0.6 µm M = 5.2
				LmV B 17 0.6 µm 5.2
19.	eP1	A	21 27 00	North Atlantic Ridge 49.47 N 28.49 W
	eP2	A	27 02.5	H = 21 21 31.8 h = normal MB = 4.8
	LmH	B	36.6	D = 25.56 Az = 71.9 (NEIC)
	LmV	B	37.2	P2V A 1.4s 30.2nm M = 4.7
				LmH B 15 1.7 µm 4.7
				LmV B 16 1.5 µm 4.8
20.	eX	A	02 50 19.5	Windward Islands 14.24 N 60.56 W
	LmH	C	03 14.0	H = 02 39 06.8 h = 68 km MB = 5.2
	LmV	C	14.0	D = 67.83 Az = 40.9 (NEIC)
				XV A 2.0s 42.7nm
20.	eP	A	05 41 31	Kurile Islands 48.29 N 154.33 E
				H = 05 29 43.8 h = normal MB = 4.5
				D = 76.37 Az = 336.6 (NEIC)
				PV A 1.3s 13.1nm M = 4.9
20.	eP	AB	07 54 45	North of Ascension Island
	ePa	B	58 00	1.41 S 13.71 W
	eS	B	08 02 28	H = 07 45 05.2 h = normal MB = 5.0
	LmH	B	17.0	D = 56.19 Az = 19.1 (NEIC)
	LmV	B	17.8	PV A 2.5s 138.0nm M = 5.5
				SH B 18 3.5 µm 5.9
				LmH B 17 5.4 µm 5.4
21.	ePKP	A	15 35 25.5	Fiji Islands Region 18.06 S 178.44 W
				H = 15 16 56.9 h = 686 km MB = 4.2
				D = 146.51 Az = 348.4 (NEIC)
				PKPV A 1.1s 18.2nm

Day	Phase		h m s	Remarks
22.	+iP	AB	04 40 46.9	<u>Eastern Kazakh SSR</u> 49.74 N 78.19 E
	+iPn	A	42 19.1	H = 04 32 57.8 h = 0 km MB = 5.8
	iPP	AB	42 29	D = 41.32 Az = 297.8 (NEIC)
				PV A 0.7s 268.0nm M = 6.1
				PH A 0.6 155.0nm 6.1
				underground explosion MB = 6.3 (UPP)
22.	eP	A	10 28 44.5	<u>North of Ascension Island</u>
	eS	B	36 30	0.19 S 17.92 W
	LmH	B	50.0	H = 10 19 03.7 h = normal MB = 5.2
	LmV	B	54.8	D = 56.54 Az = 22.1 (NEIC)
				PV A 1.2s 22.4nm M = 5.1
				LmH B 17 0.8 μm 4.9
				LmV B 18 0.9 μm 4.9
22.	eP	A	10 52 22.5	<u>Off East Coast of Honshu, Japan</u>
	e	A	52 29	37.19 N 142.06 E
	e	A	52 35	H = 10 40 04.1 h = 46 km MB = 5.3
	LmH	B	11 32.0	D = 82.28 Az = 330.7 (NEIC)
	LmV	B	34.2	PV A 1.0s 23.6nm M = 5.2
				LmV B 14.5 0.8 μm 5.2
				EW-component of seismograph type B out of operation.
22.	eP	A	21 22 07.5	<u>Carlsberg Ridge</u> 2.19 S 68.12 E
				H = 21 10 50.1 h = 33 km MB = 5.0
				D = 71.22 Az = 326 (ISC)
23.	ePKIKP	AB	02 35 08	<u>South of Fiji Islands</u> 22.88 S 176.36 W
	iPKHKP	A	35 15	H = 02 15 26.9 h = 76 km MB = 6.0
	ePKJKP	A	35 18.5	D = 151.59 Az = 349.3 (NEIC)
	ePKP2	A	35 23.5	h = 71 km
	epPKIKP	A	35 28	PKIKPV A 1.8s 118.2nm
	epPKHKP	A	35 35	PKHKPV A 1.5 367.0nm
	epPKJKP	A	35 39.5	PKJKPV A 1.5 166.0nm
	epPKP2	A	35 44	PKP2V A 1.2 105.8nm
	eSKSP	B	49 10	pPKIKPV A 1.6 159.3nm
	eiSS	C	58 12	pPKHKPV A 1.3 258.0nm

Day	Phase		h m s	Remarks
cont. 23.	eSSS	C	03 04 00	pPKJKPV A 1.2s 105.8nm
	eSSSS	C	08 15	pPKP2V A 1.5 151.0nm
	LmH	B	40.8	e C 02 48 35 e C 54 00
	LmV	B	40.8	EW-component of seismograph type B out of operation.
23.	+iP	A	07 05 37.8	Ural Mountains Region 61.29 N 56.47 E
	ei	A	05 41	H = 06 59 56.0 h = 0 km MB = 5.6
	LmH	B	17.1	D = 26.72 Az = 267.1 (NEIC)
	LmV	B	17.1	underground explosion MB = 5.9 (UPP)
				PV A 0.6s 28.8nm M = 5.3
				LmH B 10 2.9/um 5.1
				LmV B 11 3.0/um 5.2
23.	+iP1	AB	09 31 22.5	Jan Mayen Island Region
	iP2	AB	31 31.0	70.99 N 7.03 W
	PmV	A	31 36	H = 09 26 29.3 h = normal MB = 6.0
	eiS	B	35 20	D = 22.15 Az = 147.3 (NEIC)
	LmV	B	40.7	P1V A 1.1s 262.0nm M = 5.6
	LmH	B	40.9	P2V A 1.5 705.0nm 5.9
				PmV A 1.6 1000.0nm 6.0
				P2V B 16.5 10.3/um 6.0
				P2H B 18 16.2/um 6.3
				SH B 17 31.3/um 6.5
				LmH B 15 56.0/um 6.1
				LmV B 15 40.7/um 6.1
23.	+iP1	A	10 00 36.4	Kirgiz-Sinkiang Border Region
	-iP2	A	00 37.8	41.49 N 79.26 E
	eiP3	AB	00 42.4	H = 09 52 12.3 h = normal MB = 5.7
	ePP	A	02 27	D = 46.30 Az = 305.5 (NEIC)
	LmH	B	21.4	P1V A 1s 23.6nm M = 5.1
	LmV	B	21.4	P2V A 1.3 200.0nm 6.0
				P3V A 1.5 427.0nm 6.2
				PPV A 2.0 172.0nm 5.6
				LmH B 12 14.5/um 6.1
				LmV B 12 15.9/um 6.3

Day	Phase		h m s	Remarks
23.	+eP1	AB	20 55 42	Kirgiz-Sinkiang Border Region
	-iP2	AB	55 45	41.45 N 79.26 E
	+iP3	AB	55 51.5	H = 20 47 17.4 h = normal MB = 6.0
	ePcP	AB	57 26.5	D = 46.32 Az = 305.5 (NEIC)
	ePP	AB	57 31	P1V A 1.5s 125.8nm M = 5.6
	eS	C	21 02 28	P2V A 1.4 376.0nm 6.1
	eScS	BC	05 32	P3V A 1.4 604.0nm 6.3
	eSS	C	05 50	LmH B 12 15.3 µm 6.2
	LmH	B	16.5	LmV B 11 12.9 µm 6.2
	LmV	B	17.9	ei C 20 57 55 e C 21 02 50 e A 12 13
24.	ePKIKP	A	02 45 05.5	New Britain Region 5.28 S 151.52 E
	epPKIKP	A	45 20	H = 02 26 14.7 h = 71 km MB = 5.6
	eSS	C	03 04.4	D = 123.75 Az = 330.5 (NEIC)
	eSSS	C	08.5	h = 50 km
	LmH	B	39.7	PKIKPV A 1.4s 32.6nm
	LmV	B	42.5	LmH B 24 1.1 µm
				LmV B 20 0.9 µm
24.	eP1	A	05 14 30	Greece 38.03 N 20.23 E
	+iP2	A	14 36.7	H = 05 11 12.3 h = 17 km MB = 4.7
	LmH	C	20.9	D = 14.02 Az = 336.8 (NEIC)
	LmV	C	21.0	P1V A 1.0s 13.8nm
				P2V A 0.9 39.0nm
				LmH C 13.5 1.3 µm
				LmV C 12.5 1.7 µm
24.	eP	A	13 48 13	West Pakistan 30.33 N 67.80 E
				H = 13 39 46.3 h = 12 km MB = 4.9
				D = 46.07 Az = 312.7 (NEIC)
				PV A 1.5s 30.2nm M = 5.0
24.	eP1	AB	14 04 35.5	Tsinghai Province, China
	Pm	A	04 40	35.46 N 98.17 E
	+iP2	BC	04 40	H = 13 54 17.7 h = 13 km MB = 5.8
	ePP	B	06 56	D = 61.53 Az = 313.7 (NEIC)
	e	B	08 35	PmV A 1.9s 288.0nm M = 6.1
	eS	B	13 00	P2V B 7 1.4 µm 6.3

Day	Phase		h m s	Remarks
cont. 24.	eSS	B	14 17 12	LmH B 14.5 14.5/um M = 6.3
	eSSS	B	19 25	LmV B 16.5 18.6/um 6.4
	LmH	B	31.8	
	LmV	B	32.8	
24.	eP	A	18 37 10.5	<u>Kurile Islands</u> 45.00 N 148.85 E
				H = 18 25 17.9 h = 56 km MB = 5.0
				D = 77.77 Az = 333.7 (NEIC)
				PV A 1.0s 11.8nm M = 4.8
24.	eP1	A	21 02 55.5	<u>Kirgiz-Sinkiang Border Region</u>
	eP2	A	02 59	41.47 N 79.48 E
				H = 20 54 28.6 h = 18 km MB = 5.3
				D = 46.45 Az = 305.5 (NEIC)
				P1V A 1.0s 17.7nm M = 5.0
				P2V A 1.1 46.8nm 5.4
24.	eiP1	A	21 10 21	<u>Kirgiz-Sinkiang Border Region</u>
	eiP2	A	10 25.5	41.39 N 79.41 E
	ePP	A	12 15	H = 21 01 54.9 h = 25 km MB = 5.3
	eSS	C	19 48	D = 46.45 Az = 305.6 (NEIC)
	LmV	B	31.0	P1V A 1.3s 17.1nm M = 4.9
	LmH	B	32.3	P2V A 1.3 72.0nm 5.6
				PPV A 1.7 60.6nm 5.3
				LmH B 10 1.0/um 5.1
				LmV B 11 0.9/um 5.0
25.	eP	A	00 59 55	<u>Greece</u> 38.03 N 20.07 E
				H = 00 56 37.0 h = 46 km MB = 4.5
				D = 13.98 Az = 337.2 (NEIC)
25.	-eP	A	03 43 58.5	<u>Andreanof Islands, Aleutian Is.</u>
	e	A	44 11	50.52 N 176.76 W
				H = 03 31 53.7 h = 11 km MB = 5.3
				D = 78.96 Az = 354.6 (NEIC)
				PV A 1.4s 51.2nm M = 5.4

Day	Phase		h m s	Remarks
25.	-eP	A	03 43 58.5	<u>**Andreanof Islands, Aleutian Is.**</u>
	e	A	44 11	50.52 N 176.76 W
				H = 03 31 53.7 h = 11 km MB = 5.3
				D = 78.96 Az = 354.6 (NEIC)
				PV A 1.4s 51.2nm M = 5.4
25.	eP	A	11 38 34.5	<u>**Ascension Island Region**</u> 7.62 S 13.46 W
	e	A	38 40.5	H = 11 28 17.1 h = normal MB = 5.0
				D = 61.96 Az = 17.8 (NEIC)
25.	ePKP	A	14 08 38.5	<u>**Tonga Islands**</u> 15.31 S 173.15 W
	e	A	08 41	H = 13 49 03.7 h = normal MB = 5.2
	LmH	B	15 16.3	D = 144.54 Az = 354.8 (NEIC)
	LmV	B	18.3	LmH B 16s 0.4 μm M = 5.2
				LmV B 19 0.8 μm 5.5
25.	eP	A	15 30 49.5	<u>**Crete**</u> 34.29 N 23.95 E
	e	A	30 52	H = 15 26 33.0 h = 42 km MB = 4.6
	e	A	30 55	D = 18.66 Az = 334.9 (NEIC)
	LmH	C	39.3	LmH C 15s 0.7 μm M = 4.1
	LmV	B	39.3	
25.	+iP	A	16 32 03.2	<u>**Near East Coast of Honshu, Japan**</u>
	e	A	32 13	38.45 N 142.06 E
	eipP	A	32 15	H = 16 19 51.8 h = 53 km MB = 5.5
	e	A	32 37	D = 81.18 Az = 330.6 (NEIC)
	LmH	B	17 11.0	h = 48 km
	LmV	B	11.0	PV A 1.3s 87.4nm M = 5.6
				pPV A 1.6 68.6nm
				LmH B 16 0.6 μm
				LmV B (16) 0.7 μm
26.	eP	A	01 49 44	<u>**Kurile Islands**</u> 45.1 N 149.6 E
				H = 01 37 48 MB = 5.2
				D = 77.9 (ANUSSR)
26.	e	A	07 03 50.5	
	e	A	04 12	
	e(Sg)	A	04 38	

Day	Phase		h m s	Remarks
26.	ePKP	A	09 28 02	South Pacific Cordillera
	eSS	B	52 08	55.44 S 129.10 W
	LmV	B	10 33.0	H = 09 08 06.6 h = normal MB = 5.7
	LmH	B	36.5	D = 156.14 Az = 85.1 (NEIC)
				LmV B 20s 1.9/um M = 5.9
				EW-component of seismograph type P out of operation.
26.	ePKP2	A	15 58 15	Kermadec Islands Region
				27.99 S 176.88 W
				H = 15 37 55.4 h = normal MB = 4.9
				D = 156.47 Az = 346.4 (NEIC)
				PKP2V A 1.6s 16.5nm
26.	+iP	AC	17 46 14.5	Southeastern Alaska 60.34 N 140.99 W
	ei	A	46 21	H = 17 35 18.0 h = 7 km MB = 5.5
	ePP	C	48 44	D = 67.18 Az = 18.5 (NEIC)
	eS	C	55 16	PV A 1.3s 170.0nm M = 6.1
	eSS	C	59 42	PV C 12 1.4/um 6.1
	LmH	C	18 17.9	LmH C 17 2.9/um 5.6
	LmV	B	18.0	LmV B 18 3.3/um 5.7
26.	eP	A	17 53 05.5	South Eastern Alaska 60.19 N 140.8 W
				H = 17 42 06.6
				D = 67.29 Az = 19 (ISC)
				PV A 0.8s 19.2nm M = 5.4
27.	ePKP	A	01 38 28	Tonga Islands 20.32 S 173.85 W
				H = 01 18 38.7 h = 40 km MB = 4.9
				D = 149.43 Az = 353.2 (NEIC)
27.	LmH	C	01 43.0	Off Coast of Northern Peru
	LmV	C	43.0	7.5 S 81.6 W
				H = 00 53 25.0 h = 29 km MB = 5.4
				MS = 4.8 (NEIC)
				D = 97.8
				LmH C 30s 0.5/um M = 4.8
				LmV C 30 0.7/um 5.0

Day	Phase		h m s	Remarks
27.	-eP1	A	17 21 31	Andreanof Islands, Aleutian Is.
	-iP2	AB	21 31.8	52.55 N 174.53 W
	epP	A	22 04	H = 17 09 52.3 h = 138 km MB = 5.6
	ePS	C	32 00	D = 77.05 Az = 356.0 (NEIC)
	LQ	C	42 00	h = 136 km
	LR	C	46.0	P1V A 0.9s 11.7nm M = 4.6
	LmH	C	45	P2V A 1.0 315.0nm 6.0
	LmV	C	50	LmH C 40 0.7/um
				LmV C (30) 0.7/um
27.	eP	A	18 18 19	North of Svalbard 84.5 N 1.4 E
				H = 18 11 30 h = 0 km
				D = 34.05 Az = 169 (ISC)
28.	eP	A	01 29 26	Andaman Islands Region 13.4 N 94.6 E
				H = 01 17 44 h = 33 km
				D = 75.29 Az = 319 (ISC)
				PV A 1.2s 8.1nm M = 4.7
28.	eP	A	03 05 39	Burma 19.10 N 96.34 E
	LmH	B	36.0	H = 02 54 12.2 h = 34 km MB = 4.4
	LmV	B	40.0	D = 72.17 Az = 318.2 (NEIC)
				PV A 0.9s 9.7nm M = 4.8
				LmH B 18 0.7/um 5.0
				LmV B 20 0.4/um 4.8
28.	eiP	A	08 28 11	Andaman Islands Region 12.8 N 95.4 E
				H = 08 16 23
				D = 76.3 (ANUSSR)
				PV A 1.4s 32.6nm M = 5.3
28.	eP1	A	08 35 11	Andaman Islands Region 11.77 N 95.05 E
	eP2	A	35 14	H = 08 23 19.9 h = normal MB = 5.5
	eP3	AB	35 17	D = 76.85 Az = 319.5 (NEIC)
	iP4	A	35 23.5	P2V A 0.7s 13.4nm M = 5.1
	e(PP)	B	38 24	P3V A 1.7 115.0nm 5.6
	eS	B	44 56	P4V A 1.5 148.0nm 5.8
	e	B	51 10	LmH B 20 10.7/um 6.2

Day	Phase		h m s	Remarks
cont. 28.	eSSS	B	08 53 20	LmV B 21s 10.3/um M = 6.2
	LmH	B	09 13.8	
	LmV	B	13.8	
28.	eP	A	08 53 26	<u>Andaman Islands Region</u> 11.1 N 94.9 E H = 08 41 26.4 h = 0 km MB = 5.1 D = 77.19 Az = 320 (ISC) PV A 1.2s 7.9nm M = 4.9
28.	eP	A	14 32 48.5	<u>Andaman Islands Region</u> 11.83 N 94.82 E H = 14 20 59.0 h = normal MB = 5.0 D = 76.66 Az = 319.5 (NEIC) PV A 1.0s 7.9nm M = 4.7
28.	eP	A	14 51 33	<u>Andaman Islands Region</u> 11.77 N 95.09 E
	e	A	51 41	H = 14 39 42.4 h = normal MB = 4.8 D = 76.87 Az = 319.5 (NEIC) PV A 1.5s 15.1nm M = 4.8
28.	e(PKP)	A	18 15 43	<u>Timor</u> 9.55 S 124.37 E
	ePP	A	16 17	H = 17 56 56.1 h = normal MB = 5.7
	LmH	C	51.0	D = 111.71 Az = 320.8 (NEIC) (PKP)V A 1.0s 7.9nm PPV A 1.5 20.1nm M = 5.6 LmH C 50 1.0/um 5.0
29.	LmH	B	04 37.4	Probably <u>Near Coast of Central Chile</u>
	LmV	B	37.7	33.3 S 71.9 W H = 03 31 51.4 h = normal MB = 4.9 MS = 4.9 (NEIC) D = 111.2 LmH B 18s 0.6/um M = 5.2 LmV B 18 0.8/um 5.3

Day	Phase		h m s	Remarks
29.	LmV	C	06 42.0	Probably Near Coast of Central Chile
	LmH	C	46.0	33.3 S 71.9 W
				H = 05 38 26.5 h = 26 km
				MB = 4.7 MS = 5.0 (NEIC)
				D = 111.2
				LmH C 18s 0.4/um M = 5.1
				LmV C 20 0.6/um 5.2
29.	LmV	C	07 21.0	Northern Easter I. Cordillera
	LmH	C	27.5	9.0 S 109.8 W
				H = 06 19 06.6 h = normal
				MB = 4.9 MS = 5.3 (NEIC)
				D = 116.7
				LmH C 18s 0.4/um M = 5.0
				LmV C 23 0.7/um 5.2
29.	ePKP2	A	12 11(25)	Loyalty Islands Region 22.10 S 170.25 E
				H = 11 51 38.0 h = 14 km
				D = 146.96 Az = 334.8 (NEIC)
				traces
29.	eP	A	18 07 38	Andaman Islands Region 11.10 N 95.06 E
	epP	A	07 45.5	H = 17 55 44.0 h = 25 km MB = 5.2
				D = 77.36 Az = 319.6 (NEIC)
				h = 28 km
				PV A 0.7s 13.4nm M = 5.1
29.	eP	A	19 32 43.5	Eastern Sea of Japan 41.24 N 137.55 E
	epP	A	33 54	H = 19 21 20.3 h = 289 km MB = 4.9
	ePP	A	35 37	D = 77.02 Az = 328.1 (NEIC)
				h = 310 km
				PV A 1.2s 24.4nm M = 4.8
30.	eP	A	00 33 23	Greece 38.88 N 20.58 E
	LmH	C	38.3	H = 00 30 14.1 h = 48 km MB = 4.5
				D = 13.36 Az = 334.6 (NEIC)
				PV A 9s 11.7nm M = 4.6
				LmH C 14 0.5/um 3.7

Day	Phase		h m s	Remarks
30.	eP	A	03 49 03.5	Northern Sumatra 3.11 N 96.31 E
	epP	A	49 10	H = 03 36 34.4 h = 38 km MB = 5.1
				D = 84.24 Az = 320.4 (NEIC)
				h = 24 km
				PV A 1.8s 27.0nm M = 5.3
30.	eP	A	04 11 59	Kurile Islands 43.68 N 147.83 E
	epP	A	12 10.5	H = 03 59 59.5 h = 42 km MB = 4.2
	LmH	C	44.4	D = 78.63 Az = 333.3 (NEIC)
				h = 42 km
30.	ePKP	A	06 16 37	Loyalty Islands Region
	e	A	17 00	21.36 S 169.17 E
				H = 05 57 00.7 h = 43 km MB = 4.9
				D = 145.87 Az = 334.3 (NEIC)
30.	eP	A	11 42 38.5	Andreanof Islands, Aleutian Is.
	-iX	A	42 49	51.19 N 177.49 W
	eS	C	52 34	H = 11 30 38.9 h = 20 km MB = 5.7 MS = 5.4
	LmH	B	12 23.4	D = 78.25 Az = 354.1 (NEIC)
	LmV	B	27.8	PV A 1.0s 21.6nm M = 5.2
				XV A 1.3 153.0nm
				LmH B 17 1.6 µm 5.4
				LmV B 16 1.1 µm 5.3
30.	eP	A	19 43 20	Greece-Albania Border Region
	e	A	43 57	39.01 N 20.62 E
	LmH	B	48.0	H = 19 40 11.6 h = normal MB = 4.8
	LmV	B	49.6	D = 13.26 Az = 334.3 (NEIC)
				PV A 1.0s 15.8nm
				LmH B 13 1.8 µm M = 4.3
				LmV B 10 1.1 µm
30.	ePKHKP	A	22 59 22	South of Fiji Islands 24.82 S 175.95 W
	ePKP2	A	59 35	H = 22 39 25.0 h = normal MB = 5.0
	LmH	C	24 28.0	D = 153.56 Az = 349.1 (NEIC)
	LmV	B	28.0	PKHKPV A 1.5s 20.1nm
				LmH C 16 0.4 µm M = 5.3
				LmV B 16 0.4 µm 5.3

Day	Phase		h m s	Remarks
31.	eP	A	08 27 14	Burma 26.18 N 96.64 E
				H = 08 16 19.6 h = 22 km MB = 5.0
				D = 67.17 Az = 316.5 (NEIC)
31.	eP	A	09 27 35.5	Mindanao, Philippine Islands
	e	A	27 41	9.01 N 125.71 E
	LmH	B	10 15.7	H = 09 14 03.8 h = 49 km MB = 5.1
	LmV	B	17.4	D = 97.86 Az = 324.1 (NEIC)
				PV A 2.0s 34.2nm M = 5.5
				LmH B 16 0.9/um 5.3
				LmV B 16 0.6/um 5.2
31.	eP	A	10 37 35.5	Mindanao, Philippine Islands
	e	A	37 40	8.96 N 125.68 E
	ePP	A	41 40	H = 10 23 57.3 h = normal MB = 5.4
	LmH	C	11 28.9	D = 97.89 Az = 324.1 (NEIC)
	LmV	B	29.1	PV A 1.4s 27.3nm M = 5.6
				LmH C 16 1.0/um 5.4
				LmV B 16 0.9/um 5.4
31.	ePKIKP	A	11 53 02.5	Kermadec Islands 29.10 S 177.50 W
	ePKHKP	A	53 14.5	H = 11 33 14.1 h = 67 km MB = 5.2
	ePKP2	A	53 34	D = 157.41 Az = 344.8 (NEIC)
	LmV	C	12 56.0	PKP2V A 1.2s 28.4nm
	LmH	C	56.3	LmH C 28 0.5/um M = 5.1
				LmV C 26 1.2/um 5.6
31.	LmV	C	15 45.4	Southern California 34.3 N 118.5 W
	LmH	C	46.4	H = 14 52 22.5 h = 8 km MB = 4.8 (NEIC)
				D = 84.7
				LmH C 14s 0.2/um M = 4.7
				LmV C 14 0.4/um 5.0
31.	eP	A	20 08 42.5	Southern Sinkiang Prov., China
				39.58 N 74.82 E
				H = 20 00 31.5 h = 38 km MB = 4.8
				D = 44.67 Az = 306.1 (NEIC)

Day	Phase		h m s	Remarks
1.	ePKIKP	A	00 30 12	New Hebrides Islands 16.86 S 167.27 E
	e	A	30 16	H = 00 10 43.2 h = 31 km MB=4.9 MS=5.4
	e	A	32 46	D = 141.05 Az = 335.3 (NEIC)
	ePP	A	33 13	LmH C 22s 1.2/um M = 5.5
	ePS	C	45 40	LmV B 20 1.0/um 5.6
	eSS	C	51 40	
	LmH	C	01 34.9	
	LmV	B	35.0	
1.	ePKIKP	A	05 55 12	New Hebrides Islands 16.51 S 167.26 E
	ePP	A	58 16.5	H = 05 35 44.8 h = 24 km MB = 5.2
	eSS	B	06 16 45	D = 140.73 Az = 335.5 (NEIC)
	LmH	B	59.7	PPV A 2.0s 34.2nm M = 5.3
	LmV	B	07 06.5	LmH B 16 0.7/um 5.4
				LmV B 16 0.7/um 5.5
1.	ePKIKP	A	06 21 45	Samoa Islands Region 16.62 S 172.70 W
	e	A	21 49	H = 06 02 06.5 h = normal MB = 4.5
	e	A	21 56	D = 145.88 Az = 355.1 (NEIC)
1.	eP	A	12 45 52	Mascarene Islands Region
	e	A	45 08	17.91 S 66.04 E
	e	A	45 23	H = 12 33 24.5 h = normal MB = 5.0
	eS	B	56 16	D = 83.28 Az = 328.6 (NEIC)
	eSS	B	13 01 10	PV A 1.6s 27.5nm M = 5.1
	LmH	B	24.5	LmH B 18 0.7/um 5.1
	LmV	B	24.5	LmV B 20 0.6/um 5.0
1.	ePKIKP	A	19 57 48	Kermadec Islands Region
	ePKHKP	A	58 00	29.84 S 179.26 W
	eiPKP2	AB	58 23.5	H = 19 38 18.8 h = 214 km MB = 5.3
	ePP	B	20 02 05	D = 157.67 Az = 341.6 (NEIC)
	eSSS	C	30 00	PKIKPV A 2.0s 85.5nm
				PKHKPV A 1.5 35.2nm
				PKP2V A 1.4 209.0nm
2.	eP	A	01 28 36.5	Northern Sumatra 3.11 N 96.26 E
	epP	A	28 44	H = 01 16 06.4 h = normal MB = 5.1

Day	Phase		h m s	Remarks
cont. 2.				D = 84.21 Az = 320.4 (NEIC) h = 27 km PV A 1.4s 20.9nm M = 5.1
2.	ePn eiSn LmH LmV	A A B B	01 45 48.5 47 15 49.0 49.8	Central Italy 43.06 N 13.09 E H = 01 43 56.6 h = 39 km MB = 4.5 D = 7.66 Az = 353.0 (NEIC) PnV A 0.9s 21.4nm LmH B 16 0.5 µm M = 3.3 LmV B 12 0.4 µm
2.	ePKP2	A	03 00 08	South of Fiji Islands 25.82 S 177.41 W H = 02 40 14.9 h = 193 km MB = 4.6 D = 154.26 Az = 346.7 (NEIC) PKPV A 1.3s 21.8nm
2.	ePn eSn LmH	A A B	04 35 24.5 36 52.5 38.5	Central Italy 43.04 N 12.97 E H = 04 33 35.7 h = normal MB = 4.5 D = 7.67 Az = 353.5 (NEIC) PnV A 0.9s 11.7nm LmH B 18 0.4 µm
2.	LmH LmV	B B	05 16.8 21.3	Mindanao, Philippine Islands 8.9 N 125.7 E H = 04 16 08.8 h = 31 km MB = 5.3 (NEIC) D = 98.0 LmH B 18s 0.5 µm M = 5.1 LmV B 16 0.5 µm 5.1
2.	eP	A	12 45 33	Dodekanese Islands 36.2 N 27.4 E H = 12 41 17 h = 0 km D = 18.42 Az = 327 (ISC)
2.	ePn	A	15 36 08	Poland 50.28 N 18.95 E H = 15 34 52.0 M = 3.1 (WAR) D = 4.70 Az = 277 (ISC)

Day	Phase		h m s	Remarks
2.	ePKIKP	A	19 34 22.5	Solomon Islands 6.88 S 154.49 E
	e	A	34 25.5	H = 19 15 21.7 h = 41 km MB = 5.2
	LmH	C	20 34.0	D = 126.56 Az = 331.4 (NEIC)
	LmV	C	34.5	PKIKPV A 1.2s 16.3nm
				LmH C 18 0.2/um M = 4.8
				LmV C 20 0.5/um 5.2
3.	eX	A	02 08 05	XV A 0.7s 7.7nm
3.	eP	A	04 06 41.5	Tyrrhenian Sea 38.98 N 14.79 E
				H = 04 03 58.6 h = 297 km MB = 4.4
				D = 11.88 Az = 350.1 (NEIC)
				PV A 1.2s 32.5nm
3.	eP1	AB	04 59 22	Tibet 32.26 N 95.06 E
	eP2	AB	59 24	H = 04 49 03.4 h = normal MB = 5.7
	e	B	05 04 00	D = 61.89 Az = 314.2 (NEIC)
	eS	BC	07 45	P1V A 1.6s 27.5nm M = 5.1
	eSSS	B	09 35	P2V A 1.6 168.0nm 5.9
	LmH	B	25.1	LmH B 15.5 11.1/um 6.1
	LmV	B	31.1	LmV B 13.5 12.7/um 6.3
3.	eP1	AB	05 01 05	Tibet 32.32 N 95.44 E
	eP2	A	01 10	H = 04 50 45.6 h = normal MB = 5.8 (NEIC)
	eS	B	09 35	D = 62.1
	eSSS	BC	16 14	P2V A 1.9s 102.0nm M = 5.6
3.	eP	A	09 18 32	Kurile Islands 45.52 N 151.93 E
				H = 09 06 33.0 h = normal MB = 4.6
				D = 78.23 Az = 335.5 (NEIC)
3.	ePKIKP	A	13 57 10	West Chile Rise 42.63 S 83.33 W
				H = 13 38 15.3 h = normal MB = 5.0
				D = 124.12 Az = 50.0 (NEIC)
				PKIKPV A 1.2s 10.2nm
3.	ePn	A	19 04 50.5	Yugoslavia 43.0 N 17.8 E
	eSn	A	06 25	H = 19 02 46 (BCIS)

Day	Phase		h m s	Remarks
cont. 3.	LmH	C	19 07.7	D = 8.78
				PV A 1.1s 14.1nm
				LmH C 12 0.7/um M = 3.6
4.	-ePg	AB	05 01 17	Schierstedt at Aschersleben, German Dem.Rep.
	iSg	B	01 33	Rockburst 51.75 N 11.52 E
				H = 05 00 53.0 h = 0 km MB = 4.6
				D = 1.107 Az = 176.7 (NEIC)
				PV B 1s 3.7/um
4.	eP	A	06 20 52.5	West Pakistan 30.45 N 67.75 E
				H = 06 12 27.5 h = 15 km
				D = 45.95 Az = 312.7 (NEIC)
				PV A 0.8s 9.6nm M = 4.8
4.	ePKIKP	A	10 35(23)	Easter Island Cordillera
	ePKHKP	A	35 31	56.25 S 122.46 W
	ePKP2	A	35 43	H = 10 15 37.2 h = normal MB = 6.2
	ePP	B	39 14	D = 152.50 Az = 82.2 (NEIC)
	eiSS	BC	58 53	PKHKPV A 1.9s 152.0nm
	LmH	B	11 53.3	PKP2V A 1.3 161.0nm
	LmV	B	55.7	LmV B 19 7.6/um M = 6.5
				e ABC 10 35 27.5 e A 35 38
				e BC 35 46
				Type B : EW-Component out of operation
4.	eP	A	12 27 24	Azores Islands 37.98 N 30.90 W
	e	A	27 29	H = 12 20 54.0 h = normal MB = 4.8
	e	A	27 39	D = 32.50 Az = 53.2 (NEIC)
4.	eP	A	18 30 51	Kurile Islands 47.14 N 154.03 E
				H = 18 18 56.1 h = normal MB = 3.9
				D = 77.35 Az = 336.5 (NEIC)
4.	+iP	AB	18 51 51.5	Near East Coast of Honshu, Japan
	e	A	54 45	38.37 N 142.09 E
	ePP	BC	55 00	H = 18 39 39.2 h = 52 km MB = 5.8
	ePPP	BC	56 48	D = 81.27 Az = 330.7 (NEIC)

Day	Phase		h m s	Remarks
cont. 4.	eS	BC	19 02 00	PV A 1.3s 153.0nm M = 5.8
	eSS	B	07 12	PV B 8 0.9/um 5.8
	LmH	C	25.8	LmH C 24 7.8/um
	LmV	B	31.1	LmV B 17 5.5/um
5.	+eP	A	05 16 17	<u>Molucca Sea</u> 0.31 S 124.73 E
	ePP	A	20 39	H = 05 02 11.8 h = normal MB = 5.8
	LmH	C	56.0	D = 104.72 Az = 322.7 (NEIC)
	LmV	C	56.5	PV A 1.4s 60.5nm M = 6.3
				PPV A 1.9 94.7nm 6.0
				LmH C 40 1.4/um 5.2
				LmV C 48 2.3/um 5.4
5.	-iP	AB	09 16 16.8	<u>Fox Islands, Aleutian Is.</u>
	epP	BC	16 51	53.36 N 170.55 W
	ipP	A	16 56.5	H = 09 04 42.8 h = 153 km MB = 5.8
	esP	BC	17 00	D = 76.35 Az = 358.6 (NEIC)
	e	BZ	21 36	h = 160 km
	ePS	BC	26 35	PV A 1.5s 337.0nm M = 5.9
	e	C	41 32	PV B 5 2.1/um 6.1
	e(pP'P')	A	44 00	pP A 1.9 352.0nm
	LmH	B	54.8	LmH B 17 0.8/um
	LmV	B	10 00.7	LmV B 18 0.5/um
5.	ePKIKP	A	14 45 31	<u>South of Fiji Islands</u> 25.90 S 179.78 W
	ePKHKP	A	45 40	H = 14 26 30.7 h = 432 km MB = 5.0
	ePKP2	A	45 55	D = 153.78 Az = 343.5 (NEIC)
	epPKIKP	A	47 30.5	h = 344 km
	epPKHKP	A	47 41	PKIKPV A 1.2s 12.2nm
	epPKP2	A	47 56	PKP2V A 1.2 20.4nm
				pPKIKPV A 1.1 18.2nm
				pPKHKPV A 1.3 15.3nm
5.	eP	A	17 01 07	<u>Kurile Islands</u> 45.46 N 151.85 E
	e	A	01 08.5	H = 16 49 08.7 h = normal MB = 5.1
				D = 78.27 Az = 335.4 (NEIC)

Day	Phase		h m s	Remarks
5.	eP	A	20 07 21.5	Hokkaido, Japan Region 41.17 N 141.98 E
				H = 19 55 24.8 h = 71 km MB = 4.9
				D = 78.79 Az = 330.4 (NEIC)
				PV A 0.8s 7.7nm M = 4.7
5.	e(PKHKP)	A	20 19 57	Fiji Islands Region 20.14 S 178.31 W
				H = 20 01 09.5 h = 522 km MB = 4.7 (NEIC)
				D = 148.5
				(PKHKP)V A 1.0s 19.7nm
6.	+eP	A	00 29 13.5	Hokkaido, Japan Region
	epP	A	29 23.5	42.32 N 144.99 E
	e	A	30 22	H = 00 17 11.8 h = 34 km MB = 5.3
	eS	C	39 20	D = 78.87 Az = 331.9 (NEIC)
	eSS	C	45.0	h = 37 km
	LmH	B	01 08.0	PV A 1.5s 37.7nm M = 5.2
	LmV	B	08.0	LmH B 16 1.0 µm 5.3
				LmV B 16 1.3 µm 5.4
6.	eP	A	06 57 00	Southern Iran 29.80 N 51.89 E
	+iX	A	57 13	H = 06 49 52.9 h = 10 km MB = 5.2
	ePP	B	58 28	D = 36.53 Az = 316.3 (NEIC)
	eS	B	07 02 40	PV A 1.1s 12.1nm M = 4.6
	LmH	B	13.3	XV A 1.0 25.6nm
	LmV	B	14.0	LmH B 16 2.4 µm 5.1
				LmV B 16 2.0 µm 5.1
6.	eP	A	09 47 37.5	Hokkaido, Japan Region 42.31 N 145.12 E
	eX	A	47 43	H = 09 35 35.8 h = 35 km MB = 5.1
	LmH	B	10 26.5	D = 78.93 Az = 331.9 (NEIC)
	LmV	B	26.6	PV A 1.6s 33.0nm M = 5.1
				XV A 2.4 103.8nm
				LmH B 16 1.5 µm 5.4
				LmV B 16.5 1.8 µm 5.5
6.	+iPKIKKP	B	11 25 09	South of Fiji Islands 22.22 S 179.57 W
	-iPKIKP	A	25 10.5	H = 11 06 30.6 h = 603 km MB = 5.6
	iPKHKP	AB	25 16	D = 150.31 Az = 345.6 (NEIC)

Day	Phase		h m s	Remarks
cont. 6.	iPKP2	AB	11 25 23	h = 615 km
	epPKIKP	B	27 27	PKIKPV A 1.3s 127.0nm
	esPKIKP	B	28 36	PKHKPV A 1.0 620.0nm
	eSS	BC	47 22	PKP2V A 1.4 516.0nm
6.	eP	A	12 05 52.5	Hokkaido, Japan Region 42.25 N 145.04 E
	epP	A	06 02.5	H = 11 53 51.8 h = 43 km MB = 5.2
	LmH	B	44.7	D = 78.95 Az = 331.9 (NEIC)
	LmV	B	44.8	h = 37 km
				PV A 1.3s 17.5nm M = 4.9
				LmH B 16 0.6 μm 5.1
				LmV B 17 0.9 μm 5.2
6.	ePKIKP	A	16 20 12	South of Fiji Islands 23.82 S 179.16 E
	iPKHKP	A	20 19	H = 16 01 25.2 h = 540 km MB = 5.3
	ePKP2	A	20 30	D = 151.52 Az = 343.3 (NEIC)
	epPKP	A	22 26	h = 600 km
				PKIKPV A 1.0s 19.7nm
				PKHKPV A 1.2 77.2nm
				PKP2V A 1.5 81.5nm
7.	-eP	A	05 13 44.5	Halmahera 2.44 N 129.12 E
	ePP	B	17 55	H = 04 59 39.4 h = 47 km MB = 6.5
	ePPS	B	27 40	D = 105.13 Az = 324.2 (NEIC)
	ePKKP	A	29 40	PV A 1.9s 144.0nm M = 6.6
	LmH	B	06 09.0	PV B 7.5 2.5 μm 7.2
	LmV	B	15.2	LmH B 16 16.8 μm 6.7
				LmV B 16 20.3 μm 6.8
7.	e(PP)	A	15 40 14.5	Probably Mendoza Province, Argentina
				32.57 S 69.08 W
				H = 15 10 38.0 h = 122 km MB = 5.7 (NEIC)
				D = 109.0
				(PP)V A 1.2s 12.2nm
8.	+iP	AB	07 59 08.7	Southern Sumatra 4.33 S 102.40 E
	epP	A	59 29.5	H = 07 45 57.8 h = 75 km MB = 6.3
	ePP	B	08 02 15	D = 93.82 Az = 320.4 (NEIC)

Day	Phase		h m s	Remarks
cont. 8.	e	B	08 05 40	h = 75 km
	iSKS	B	09 36	PV A 1.2s 171.0nm M = 6.4
	eS	B	10 08	PV B 9 2.4 µm 6.7
	eisS	BC	10 45	SH B 11 5.0 µm 6.7
	eSP	B	11 20	LmH B 21.5 4.8 µm
	eSPP	B	11 44	LmV B 20 5.0 µm
	eSS	C	16 35	
	eSSS	C	21 00	
	eSSSS	C	23 48	
	eP'P'	A	24 26	
	LmH	B	46.4	
	LmV	B	47.3	
8.	eP	A	18 22 36	<u>Crete</u> 34.3 N 23.8 E
	LmH	B	30.3	H = 18 18 10 (BCIS)
	LmV	B	31.1	D = 18.65
				LmH B 14s 0.9 µm M = 4.2
				LmV B 10 0.5 µm 4.3
8.	eP	A	18 41 38.5	<u>Tadzhik-Sinkiang Border Region</u>
	epP	A	42 06	38.37 N 73.33 E
				H = 18 33 38.3 h = 129 km MB = 5.0
				D = 44.45 Az = 306.8 (NEIC)
				h = 129 km
				pPV A 1.0s 11.8nm
9.	eP	A	03 21 55	<u>Kurile Islands</u> 43.37 N 147.68 E
				H = 03 09 53.6 h = 33 km
				D = 78.86 Az = 333 (ISC)
9.	eP	A	04 08 07.5	<u>Kurile Islands</u> 43.41 N 147.63 E
	LmH	B	40.0	H = 03 56 07.6 h = 50 km MB = 4.8
	LmV	C	45.0	D = 78.81 Az = 333.2 (NEIC)
				PV A 1.2s 16.3nm M = 4.9
				LmH B 20 0.2 µm 4.5
				LmV B 15 0.2 µm 4.7

Day	Phase		h m s	Remarks
9.	ePn	A	06 10 47.5	Yugoslavia 45.8 N 14.2 E
	ePg	A	11 13.5	H = 06 09 30 (BCIS)
	eSn	A	11 44.5	D = 5.15
	eSg	A	12 15.5	
9.	eP	A	06 20 12	Eastern Sea of Japan 44.43 N 139.06 E
	eX	A	20 16.5	H = 06 08 33.0 h = normal MB = 5.6
	LmH	B	51.2	D = 74.90 Az = 328.4 (NEIC)
	LmV	B	55.0	PV A 1.7s 72.7nm M = 5.4
				XV A 1.2 61.0nm
				LmH B 17 2.3 μm 5.5
				LmV B 18 1.0 μm 5.2
9.	eP	A	06 22 22	Mid-Indian Rise 24.07 S 69.35 E
	e	A	22 30	H = 06 09 25.7 h = normal MB = 4.7 (NEIC)
	e	A	22 35	D = 90.1
				PV A 1.7s 32.3nm M = 5.4
9.	eP	A	08 44 35	Near East Coast of Kamchatka
				56.30 N 162.74 E
				H = 08 33 20.2 h = 45 km MB = 4.9
				D = 70.77 Az = 341.0 (NEIC)
9.	ePn	A	09 30 54	Czechoslovakia 50°45.7' N 14°25.5' E
	e(Sg)	A	31 18	explosion yield 23.5 t (PRU)
				D = 1.26
9.	eP	A	14 01 24	Mindanao, Philippine Islands
				5.08 N 123.59 E
				H = 13 48 32.1 h = 493 km MB = 5.1
				D = 99.76 Az = 323.2 (NEIC)
				PV A 1.0s 9.9nm M = 5.2
9.	eiP	A	15 19 55.5	Kurile Islands 44.26 N 147.02 E
				H = 15 08 09.7 h = 126 km MB = 5.1
				D = 77.84 Az = 332.8 (NEIC)
				PV A 1.3s 35.0nm M = 5.0

Day	Phase		h m s	Remarks
9.	eP	A	15 21 51.5	Near East Coast of Kamchatka
				53.60 N 160.7 E
				H = 15 10 18 h = 0 km MB = 5.0
				D = 72.92 Az = 340 (ISC)
				PV A 1.0s 7.9nm
9.	eP	A	22 13 34.5	Crete 34.77 N 24.21 E
	LmH	B	22.0	H = 22 09 21.5 h = 40 km MB = 4.5
	LmV	B	22.2	D = 18.32 Az = 333.8 (NEIC)
				PV A 0.8s 9.6nm M = 4.0
				LmH B 16 0.3 μm 3.6
				LmV B 15 0.5 μm 4.0
10.	eP	A	00 48 37	Fox Islands, Aleutian Is.
	epP	A	48 49.5	52.06 N 169.91 W
	LmH	B	01 33.8	H = 00 36 44.0 h = 51 km MB = 4.8
	LmV	B	36.6	D = 77.66 Az = 359.0 (NEIC)
				h = 46 km
				PV A 1.5s 30.2nm M = 5.0
				LmH B 17.5 0.4 μm 4.8
				LmV B 16 0.4 μm 4.8
10.	+iPKIKP	AB	01 41 01.2	Fiji Islands Region 21.27 S 178.77 W
	eiPKHKP	AB	41 06	H = 01 22 17.2 h = 542 km MB = 5.7
	eiPKP2	A	41 13	D = 149.56 Az = 346.9 (NEIC)
	epPKP	AB	43 14.5	h = 567 km
	e	A	43 19.5	PKIKPV A 2.0s 150.0nm
				PKHKPV A 1.6 550.0nm
				PKP2V A 1.5 156.0nm
10.	eP	A	03 00 39	Yugoslavia 42.50 N 20.07 E
	LmH	B	04.0	H = 02 58 05.9 h = 21 km MB = 4.6
	LmV	B	04.8	D = 10.01 Az = 327.4 (NEIC)
				LmH B 10.5s 2.0 μm M = 4.2
				LmV B 12 0.9 μm
10.	eP	A	06 12 49.5	Norwegian Sea 74.83 N 15.20 E
	LmH	B	51.2	H = 06 07 35.5 h = normal MB = 4.8

Day	Phase		h m s	Remarks
cont. 10.	LmV	B	07 01.3	D = 24.32 Az = 185.5 (NEIC)
				PV A 1.0s 13.8nm M = 4.3
				LmH B 16 2.1/um 4.7
				LmV B 12.5 1.1/um 4.7
10.	ePKHKP	A	13 22 02	Fiji Islands Region 20.77 S 178.68 W
	ePKP2	A	22 09	H = 13 03 17.5 h = 582 km MB = 4.7
				D = 149.10 Az = 347.2 (NEIC)
				PKHKPV A 1.2s 14.2nm
10.	eP1	A	13 25 52	Turkey 38.92 N 29.13 E
	eP2	A	25 55	H = 13 21 55.0 h = 28 km MB = 4.7
	LmH	B	34.0	D = 17.04 Az = 319.2 (NEIC)
	LmV	B	35.1	P2V A 1.9s 30.3nm M = 4.1
				LmH B 16 0.6/um 3.9
				LmV B 12.5 0.5/um 4.1
11.	eP	A	08 05 12	Mid-Indian Rise 12.58 S 67.02 E
				H = 07 53 07.7 h = normal MB = 5.1
				D = 79.28 Az = 327.8 (NEIC)
11.	e(P)	A	08 54 06	Kurile Islands 46.16 N 152.97 E
				H = 08 42 05.7 h = normal MB = 4.5
				D = 77.95 Az = 336.0 (NEIC)
				traces
11.	e	A	09 27 40	Northern Celebes 0.21 S 122.91 E
				H = 09 09 06.8 h = 119 km MB = 5.1 (NEIC)
				D = 113.5
11.	eP	A	18 44 13.5	Kurile Islands 45.25 N 150.14 E
				H = 18 32 17.1 h = normal MB = 4.9
				D = 77.95 Az = 334.5 (NEIC)
				PV A 1.0s 11.8nm M = 4.9
12.	+iP1	AB	19 10 56.8	Southern Iran 28.31 N 55.60 E
	+iP2	A	11 04.2	H = 19 03 25.9 h = 44 km MB = 6.0
	ePP	B	12 30	D = 39.85 Az = 316.4 (NEIC)

Day	Phase		h m s	Remarks
cont. 12.	eX	BC	19 13 18	P1V A 1.4s 195.5nm M = 5.7
	iS	B	17 00	PV B 5.5 1.9/um 6.1
	eSS	B	19 56	P2V A 1.7 327.0nm 5.8
	eSSS	B	20 24	PPV B 9.5 1.9/um 5.8
	LmH	B	27.7	SH B 12 3.9/um 6.2
	LmV	B	30.2	LmH B 19.5 11.3/um 5.7
				LmV B 16 5.1/um 5.6
12.	ePKIKP	A	19 54 19.5	South of Fiji Islands 23.53 S 179.74 W
	ePKHKP	A	54 26.5	H = 19 35 33.2 h = 546 km MB = 4.9
	ePKP2	A	54 37.5	D = 151.52 Az = 344.8 (NEIC)
	epPKIKP	A	56(33)	h = 577 km
	epPKHKP	A	56 38	PKHKPV A 1.2s 24.4nm
	epPKP2	A	56 48.5	PKP2V A 1.2 16.3nm
12.	ePKP	A	21 01 20	New Hebrides Islands 19.11 S 169.61 E
	epPKP	A	02 27.5	H = 20 42 18.1 h = 272 km MB = 5.2
				D = 144.01 Az = 336.1 (NEIC)
				h = 267 km
				PKPV A 1.0s 17.7nm
				pPKPV A 1.2 10.2nm
12.	ePKIKP	AB	21 19 09.5	Fiji Islands Region 17.89 S 178.19 W
	eiPKHKP	A	19 13	H = 21 00 37.4 h = 612 km MB = 5.3
	eiPKP2	A	19 14.3	D = 146.39 Az = 348.7 (NEIC)
	epPKP	A	21 30	h = 628 km
				PKIKPV A 1.5s 57.8nm
				PKHKPV A 1.8 246.0nm
				PKP2V A 1.3 140.0nm
13.	e	AB	05 37 46.5	South of Tonga Islands
	ePKHKP	A	37 53	24.69 S 175.77 W
	ePKP2	AB	38 05	H = 05 17 54.3 h = normal MB = 5.3
	ePP	BC	41 35	D = 153.46 Az = 349.4 (NEIC)
	LmV	B	07 00.0	LmH B 15s 0.5/um M = 5.3
	LmH	B	05.4	LmV B 16 0.6/um 5.5
13.	ePKP	A	06 17 04	Tonga Islands 15.86 S 173.99 W

Day	Phase		h m s	Remarks
cont. 13.	e	A	06 17 17.5	H = 05 57 34.5 h = 73 km MB = 5.5
	e	A	17 31.5	D = 145.01 Az = 353.8 (NEIC)
	ePP	A	20 38	PKPV A 1.2s 46.8nm
				PPV A 1.6 27.5nm M = 5.2
13.	eiP	AB	12 56 40.5	Turkey 39.03 N 29.89 E
	e	B	13 00 00	H = 12 52 34.6 h = 13 km MB = 5.1
	LmH	B	02.7	D = 17.35 Az = 318.0 (NEIC)
	LmV	B	04.5	PV A 1.4s 46.5nm M = 4.4
				LmH B 16 1.7 μm 4.4
				LmV B 12 2.0 μm 4.7
13.	ePKIKP	A	18 06 00.5	Fiji Islands Region 17.72 S 178.79 W
	eiPKHKP	AB	06 03.4	H = 17 47 24.2 h = 559 km MB = 5.3
	ei(PKP2)	A	06 05.7	D = 146.12 Az = 348.1 (NEIC)
				PKIKPV A 1.5s 60.4nm
				PKHKPV A 1.6 292.0nm
				PKHKPV B 4 1.2 μm
				(PKP2)V A 1.5 176.0nm
13.	eP	A	20 50 32	Southern Iran 28.16 N 55.63 E
				H = 20 43 00.3 h = 44 km MB = 4.8
				D = 39.98 Az = 316.5 (NEIC)
				PV A 0.8s 9.6nm M = 4.6
13.	ePKIKP	A	23 51 25	Easter Island Cordillera
	ePKHKP	A	51 29.5	54.60 S 118.8 W
	e	A	51 34.5	H = 23 31 31 h = 33 km
	LmV	C	24 53.0	D = 150.15 Az = 77 (ISC)
				PKHKPV A 1.2s 20.3nm
				PKP2V A 1.5 27.7nm
				LmV C 20 0.3 μm M = 5.0
14.	eP	A	03 21 40	Guatemala 14.13 N 90.13 W
				H = 03 09 09.1 h = 124 km MB = 4.8
				D = 86.49 Az = 38.6 (NEIC)

Day	Phase		h m s	Remarks
14.	eP	A	11 51 28	Baja California 27.60 N 112.36 W
	eSKS	B	12 02 00	H = 11 38 42.1 h = normal MB = 5.4
	eS	B	02 16	D = 87.45 Az = 31.9 (NEIC)
	ePS	B	03 12	PV A 1.7s 27.3nm M = 5.3
	LmV	B	33.0	LmH B 15.5 2.0 µm 5.6
	LmH	B	33.7	LmV B 16 2.5 µm 5.7
14.	eP	A	19 47 56	Baja California 27.68 N 112.20 W
	eSKS	C	58(30)	H = 19 35 05.3 h = normal MB = 5.2
	LmH	B	20 29.3	D = 87.58 Az = 32.0 (NEIC)
	LmV	B	29.5	LmH B 16s 0.4 µm M = 4.9
				LmV B 16 0.6 µm 5.1
15.	epPKP2	A	00 58 49.5	Off E. Coast of North Island, N.Z.
				36.73 S 177.27 E
				H = 00 37 28.63 h = 245 km MB = 5.5 (NEIC)
				D = 162.7
15.	LmH	C	04 36.0	Mindanao, Philippine Islands
	LmV	C	40.0	9.0 N 125.8 E
				H = 03 35 10.4 h = 35 km MB = 4.9 (NEIC)
				D = 97.9
				LmV C 20s 0.3 µm M = 4.7
15.	ePKP	A	06 09 49.5	Fiji Islands Region 15.46 S 177.71 W
				H = 05 51 00.4 h = 395 km MB = 4.7
				D = 144.11 Az = 349.9 (NEIC)
15.	eP	A	19 06 03	Eastern Gulf of Aden 12.86 N 48.53 E
	LmV	C	25.5	H = 18 57 24.3 h = normal MB = 5.0 (NEIC)
	LmH	C	25.7	D = 48.1
				PV A 1.4s 23.3nm M = 5.0
				LmV C 16 0.3 µm 4.5
16.	e	A	13 40 39	Czechoslovakia, explosion
				49.96 N 14.39 E
				H = 13 39 07 h = 0 km (PRU)
				D = 1.91 Az = 292 (ISC)

Day	Phase		h m s	Remarks
16.	eP	A	21 33 02	Jordan-Syria Region 33.70 N 35.48 E
	LmH	C	43.0	H = 21 27 43.0 h = 13 km MB = 4.7
	LmV	C	45.0	D = 24.32 Az = 321.3 (NEIC)
				PV A 1.4s 97.6nm M = 5.2
17.	ePKIKP	A	02 00 16	Solomon Islands 6.29 S 154.74 E
	LmH	C	47.0	H = 01 41 20.1 h = 72 km MB = 5.7
	LmV	B	59.7	D = 126.17 Az = 331.8 (NEIC)
				PKIKPV A 0.9s 11.7nm
				LmH C 24 0.7/um
				LmV B 18 0.6/um
17.	ePKP	A	06 24 22.5	Fiji Islands Region 16.00 S 177.73 W
				H = 06 05 38.6 h = 463 km MB = 4.6
				D = 144.62 Az = 349.7 (NEIC)
17.	eP	A	08 31 47	Dodekanese Islands 36.02 N 27.30 E
	LmH	B	40.4	H = 08 27 30 h = 33 km
				D = 18.49 Az = 327 (ISC)
				PV A 1.2s 12.2nm M = 3.9
17.	-eP	A	16 42 11.5	Turkey 41.01 N 37.02 E
	eS	BC	46 05	H = 16 37 38.4 h = normal MB = 4.8
	LmH	B	51.6	D = 20.06 Az = 307.2 (NEIC)
	LmV	B	54.2	PV A 1.5s 57.7nm M = 4.6
				LmH B 18 0.6/um 4.0
				LmV B 14 0.4/um 4.1
17.	eSKS	C	17 43 35	Galapagos Islands 0.2 S 91.6 W
	eS	C	44 20	H = 17 19 12.8 h = normal
	eSS	C	51 00	MB = 5.4 MS = 5.2 (NEIC)
	LQ	C	18 00 20	D = 98.5
	LmH	B	20.0	LmH B 16s 0.4/um M = 5.0
	LmV	B	20.0	LmV B 18 0.4/um 5.0
17.	ePKP	A	20 21 12	New Hebrides Islands 19.83 S 168.05 E
	eX	A	21 15	H = 20 01 35.8 h = 10 km MB = 5.1

Day	Phase		h m s	Remarks
cont. 17.	eY	A	20 21 24.5	D = 144.04 Az = 334.3 (NEIC) PKPV A 1.0s 33.4nm XV A 1.1 36.2nm YV A 1.6 52.2nm
17.	eP	A	21 09 56.5	Off East Coast of Kamchatka
	LmV	B	48.5	51.65 N 159.34 E
	LmH	C	48.7	H = 20 58 19.1 h = normal MB = 5.0 D = 74.46 Az = 339.3 (NEIC) PV A 1.0s 7.9nm M = 4.7 LmH C 18 0.3/um 4.6
18.	e	A	02 46 17	
18.	eP	A	07 32 13.5	Tadzhik SSR 39.08 N 71.67 E H = 07 24 14.3 h = normal MB = 4.5 D = 42.99 Az = 306.0 (NEIC) PV A 1.1s 12.1nm M = 4.5
18.	eSKS	C	18 05 48	Galapagos Islands 0.2 S 91.4 W
	eS	C	06 35	H = 17 41 27.9 h = normal
	eSS	C	13 20	MB = 5.7 MS = 5.5 (NEIC)
	LmH	B	38.7	D = 98.4
	LmV	B	42.4	SH C 20s 0.75/um M = 5.9 LmH B 16 0.8/um 5.3 LmV B 17.5 1.1/um 5.4
18.	eP	A	19 12 50.5	PV A 1.2s 24.4nm
18.	iP	A	23 22 27	Kirgizya-Sinkiang Border Region 41.01 N 79.71 E H = 23 13 55 h = 24 km MB = 4.6 D = 46.86 Az = 306 (ISC)
19.	eP1	A	02 47 01	Greece-Albania Border Region
	eiP2	A	47 02	38.95 N 20.50 E
	e	A	47 10.5	H = 02 43 52.2 h = 16 km MB = 5.1

Day	Phase		h m s	Remarks
cont. 19.	LmH	B	02 52.0	D = 13.27 Az = 334.7 (NEIC)
	LmV	B	53.3	P1V A 0.9s 11.7nm M = 5.1
				P2V A 0.8 40.4nm
				LmH B 14 10.1 µm 5.0
				LmV B 12.5 4.1 µm
19.	eP	A	05 20 24	<u>Near Coast of Guatemala</u>
				13.91 N 90.54 W
				H = 05 07 48.5 h = 92 km MB = 5.0
				D = 86.91 Az = 38.6 (NEIC)
				PV A 1.0s 7.9nm M = 4.7
19.	eP	A	16 27 03	<u>Oaxaca, Mexico</u> 16.8 N 94.05 W
				H = 16 14 38.6 h = 189 km MB = 4.3
				D = 86.73 Az = 38 (ISC)
				PV A 1.2s 12.2nm
20.	LmH	C	01 10.0	<u>Near North Coast of New Guinea</u>
	LmV	C	18.5	4.8 S 145.8 E
				H = 00 07 07.9 h = 36 km MB = 5.1 (NEIC)
				D = 120.5
				LmV C 20s 0.5 µm M = 5.1
20.	eP	A	03 56 27.5	<u>Tadzhik-Sinkiang Border Region</u>
				38.29 N 73.47 E
				H = 03 48 27.0 h = 130 km MB = 4.9
				D = 44.59 Az = 306.9 (NEIC)
				traces
20.	eP	A	13 52 26	<u>Near East Coast of Kamchatka</u>
				52.91 N 158.81 E
				H = 13 41 02.1 h = 83 km MB = 5.1
				D = 73.17 Az = 338.9 (NEIC)
				PV A 1.0s 19.7nm M = 4.9
20.	eP	A	14 26 29	<u>Northern Chile</u> 21.76 S 68.96 W
	epP	A	26 52	H = 14 12 46.8 h = 93 km MB = 5.7
	eX	A	30 03.5	D = 100.80 Az = 39.7 (NEIC)

Day	Phase		h m s	Remarks
cont. 20.	ePP	A	14 30 31.5	h = 87 km
	epPP	AB	30 56	pPV A 1.5s 22.6nm
	iSKS	BC	37 00	XV A 1.4 11.6nm
	eiS	BC	37 46	PPV A 1.6 35.6nm M = 5.8
	eSP	CZ	39 32	SH B 10 1.75/um 6.3
	ePKKP	A	42 38.5	LmH B 20 0.9/um
	epPKKP	A	43 04	LmV B 20 1.0/um
	eSS	BC	45 00	
	LmH	B	15 08.9	
	LmV	B	12	
20.	eP	A	16 27 07	Talaud Islands 4.75 N 126.50 E
				H = 16 13 21.0 h = 79 km MB = 5.6
				D = 101.75 Az = 323.8 (NEIC)
				PV A 1.2s 22.3nm M = 5.7
21.	+eP	AB	06 54 02	South of Alaska 53.85 N 161.52 W
	eX1	A	54 12.5	H = 06 42 16.3 h = 21 km MB = 5.1
	eX2	A	54 23	D = 75.72 Az = 4.5 (NEIC)
	eS	B	07 03 44	PV A 1.0s 39.4nm M = 5.4
	LmH	B	41.1	PV B 7 0.49/um 5.7
	LmV	B	42.4	X1V A 1.2 40.7nm
				X2V A 1.2 36.6nm
				SH B 13 1.0/um 5.8
				LmH B 16 3.0/um 5.7
				LmV B 14 2.2/um 5.9
21.	e	A	08 55 16	Tonga 24.76 S 175.6 W
	e	A	55 27.5	H = 08 35 04.5 h = 33 km MB = 4.7
				D = 153.56 Az = 350 (ISC)
21.	LmV	B	13 03.9	Halmahera 1.3 N 128.2 E
	LmH	B	04	H = 11 47 38.1 h = 49 km
				MB = 5.1 MS = 4.8 (NEIC)
				D = 105.6
				LmH B 16s 0.2/um M = 4.6
				LmV B 16 0.3/um 4.9

Day	Phase		h m s	Remarks
21.	eP	A	13 39 57.5	North Atlantic Ridge 32.38 N 40.17 W
	eS	B	46 24	H = 13 32 08.4 h = normal MB = 5.0
	eSS	B	49 39	D = 41.91 Az = 48.5 (NEIC)
	e	B	51 40	PV A 2.0s 42.7nm M = 4.8
	LmH	B	54.4	SH B 12 0.57μm 5.3
	LmV	B	54.4	LmH B 20 1.2μm 4.8
				LmV B 20 1.6μm 5.0
21.	eP	A	14 48 16	Kirgiz-Sinkiang Border Region
	epP	A	48 22	41.47 N 79.18 E
				H = 14 39 53.1 h = 40 km MB = 5.1
				D = 46.27 Az = 305.5 (NEIC)
				h = 27 km
				PV A 1.5s 10.0nm M = 4.5
				pPV A 1.5 20.1nm
21.	e(P)	A	16 33 59	Sicily 37.7 N 15.2 E
				H = 16 31.0 (BCIS)
				D = 13.19 Az = 350 (ISC)
				(P)V A 1.0s 11.8nm
21.	LmH	B	16 40.0	LmH B 12s 0.3μm
	LmV	B	40.0	LmV B (12) 0.2μm
21.	ePKHKP	A	19 22 06	Fiji Islands Region 18.86 S 177.74 W
	ePKP2	A	22 10	H = 19 03 13.4 h = 450 km MB = 4.3
				D = 147.42 Az = 348.9 (NEIC)
21.	eP	A	20 10(22)	Crete 34.30 N 24.10 E
	e	A	10 26	H = 20 06 06.1 h = 56 km MB = 4.2
	LmH	B	18.2	D = 18.70 Az = 334.6 (NEIC)
	LmV	B	18.6	LmH B 8s 0.5μm M = 4.2
				LmV B 12 0.3μm 4.0
22.	eP1	A	09 31 00.5	Albania 41.91 N 20.34 E
	eP2	A	31 03.5	H = 09 28 27.3 h = 36 km MB = 4.8
	LmH	B	34.8	D = 10.62 Az = 328.4 (NEIC)

Day	Phase		h m s	Remarks
cont. 22.	LmV	B	09 35.6	P1V A 1.0s 11.8nm
				P2V A 1.0 23.6nm
				LmH B 8 1.0 µm M = 4.1
				LmV B 8 0.8 µm
22.	+iP	AB	22 42 07	Jan Mayen Island Region
	eX	A	42 11.5	71.62 N 3.22 W
	LmH	B	50.7	H = 22 37 12.7 h = normal MB = 4.7
	LmV	B	52.2	D = 22.09 Az = 154.3 (NEIC)
				PV A 1.3s 59.0nm M = 4.9
				XV A 2.0 98.2nm
				LmH B 16 0.3 µm 3.8
				LmV B 14 0.3 µm 4.0
23.	eP	A	10 53 53	Taiwan Region 22.33 N 121.77 E
	e	A	53 06	H = 10 41 20.5 h = 41 km MB = 5.1
	LmH	B	11 36.6	D = 84.94 Az = 323.1 (NEIC)
	LmV	B	36.8	PV A 1.4s 37.2nm M = 5.4
				LmH B 15 0.3 µm 4.9
				LmV B 15 0.6 µm 5.2
23.	eP	A	11 01 37	Off Coast of Oregon 44.60 N 129.11 W
				H = 10 49 29.4 h = normal MB = 4.9
				D = 79.19 Az = 24.2 (NEIC)
23.	ePKP	A	21 25 53	Fiji Islands Region 17.78 S 178.68 W
				H = 21 07 17.1 h = 599 km MB = 5.0
				D = 146.19 Az = 348.2 (NEIC)
24.	e	A	02 55 10	Santa Cruz Islands 10.69 S 166.00 E
				H = 02 35 58.3 h = 122 km MB = 5.3
				D = 134.94 Az = 337.1 (NEIC)
				traces
25.	+iP	AB	03 40 46.5	Eastern Kazakh SSR 49.82 N 78.09 E
	ePn	AB	42 18.5	H = 03 32 58.0 h = 0 km MB = 5.9
	LmV	B	58.8	D = 41.23 Az = 297.7 (NEIC)

Day	Phase		h m s	Remarks
cont. 25.	LmH	B	03 59.3	PV A 0.8s 388.0nm M = 6.2
				PV B 4 0.44 /um 5.5
				LmH B 8 1.0 /um 5.1
				LmV B 9 1.4 /um 5.3
25.	ePKIKP	A	02 54 58	<u>Solomon Islands</u> 7.29 S 155.80 E
				H = 04 03 34.5 h = 63 km MB = 4.7
				D = 127.54 Az = 332.0 (NEIC)
25.	eP1	AB	04 42 09	<u>Southern Italy</u> 39.29 N 15.20 E
	eP2	A	42 11.5	H = 04 39 29.0 h = 297 km MB = 4.5
				D = 11.63 Az = 348.6 (NEIC)
				P1V A 1.1s 36.3nm
				P2V A 1.5 106.0nm
25.	eP	A	17 50 42.5	<u>Gulf of Aden</u> 12.02 N 43.58 E
				H = 17 42 16.5 h = normal MB = 3.9
				D = 46.53 Az = 332.3 (NEIC)
25.	eP1	AB	17 53 05	<u>Iceland Region</u> 68.25 N 18.14 W
	eP2	A	53 11.5	H = 17 48 03.9 h = normal MB = 4.9
	eP3	A	53 14	D = 22.84 Az = 125.5 (NEIC)
	eS	C	57 15	P1V A 1.0s 19.7nm M = 4.6
	eX	BC	57 30	P1V B 5 1.1 /um 5.6
	LmV	B	18 02.8	P2V A 1.2 40.6nm 4.8
	LmH	B	03.0	P3V A 1.5 80.4nm 5.0
				XH B 5 1.0 /um
				LmH B 14.5 1.9 /um 4.7
				LmV B 16 1.5 /um 4.7
26.	ePKHKP	A	04 39 15	<u>South of Fiji Islands</u> 24.55 S 176.24 W
	e	BC	39 22	H = 04 19 18.8 h = normal MB = 5.3
	ePKP2	A	39 27	D = 153.24 Az = 348.9 (NEIC)
	LmH	B	06 01.0	LmH B 18s 0.4 /um M = 5.2
	LmV	B	02.5	LmV B 18 0.6 /um 5.5

Day	Phase		h m s	Remarks
26.	eP	A	10 21 40	<u>Northwest of Kurile Islands</u>
				51.80 N 155.44 E
				H = 10 10 32.3 h = 244 km MB = 5.0
				D = 73.43 Az = 336.9 (NEIC)
				PV A 1.2s 16.3nm M = 4.6
26.	LmH	B	19 23.0	<u>Near Coast of Nicaragua</u> 11.4 N 86.8 W
	LmV	B	23.5	H = 18 32 20.3 h = 69 km MB = 4.7 (NEIC)
				D = 86.8
				LmH B 17s 0.3/um
				LmV B 17 0.4/um
26.	eP	A	22 36 27.5	<u>Southwesten Ryukyu Islands</u>
	LmH	B	23 13.7	25.75 N 125.66 E
	LmV	B	18.9	H = 22 23 55.6 h = 19 km MB = 5.1
				D = 84.31 Az = 324.2 (NEIC)
				LmH B 15s 0.7/um M = 5.2
				LmV B 13 0.4/um 5.1
27.	LmH	C	04 26.5	<u>West New Guinea</u> 3.6 S 140.0 E
	LmV	C	29.5	H = 03 20 30.1 h = 59 km MB = 5.0 (NEIC)
				D = 116.3
				LmV C 22s 0.4/um M = 5.0
				type C : NS-component out of operation
27.	eX	A	05 19 44	XV A 0.6s 11.5nm
27.	ePn	A	06 38 23.5	<u>Federal Republic of Germany</u>
	e	A	38 28	49.91 N 8.3 E
	i	A	38 44.7	H = 06 37 44.5 h = 0 km
	e	A	38 49	D = 2.27 Az = 70 (ISC)
				PnV A 0.4s 39.9nm
27.	eP	A	10 16 52	<u>Philippine Islands Region</u>
	eSKS	B	27 15	19.21 N 121.22 E
	ePS	B	28 20	H = 10 04 06.3 h = 18 km MB = 5.0
	LmH	B	11 01.5	D = 87.12 Az = 323.1 (NEIC)
	LmV	B	01.5	LmH B 17s 0.4/um M = 4.9
				LmV B 18 0.6/um 5.1

Day	Phase		h m s	Remarks
27.	e	A	11 45 37	
27.	LmH	B	12 49.0	LmV B 18s 0.5/um
	LmV	B	49.0	The horizontal components of all instruments are disturbed
27.	LmH	B	13 48.5	LmH B 18s 0.2/um
	LmV	B	48.5	LmV B 18 0.3/um
27.	eP	A	14 55 58	Kirgiz SSR 39.21 N 72.91 E
	LmH	B	15 16.2	H = 14 47 55.2 h = 37 km MB = 4.8
	LmV	B	16.2	D = 43.69 Az = 306.1 (NEIC)
				PV A 1.6s 22.0/um M = 4.6
				LmH B 11 0.6/um 4.8
				LmV B 12 0.8/um 4.9
27.	ePKIKP	AB	20 15 13	Kermadec Islands 30.94 S 176.97 W
	ePKP2	AB	15 51	H = 19 55 18.1 h = 38 km MB = 4.9
	e	A	15 55	D = 159.29 Az = 344.4 (NEIC)
	LmV	B	21 36.0	LmH B 16.5s 0.4/um M = 5.2
	LmH	B	38.0	LmV B 18 0.3/um 5.2
28.	ePKHKP	A	08 09 11.5	Fiji Islands Region 19.95 S 178.00 W
				H = 07 50 30.1 h = 595 km MB = 4.5
				D = 148.43 Az = 348.3 (NEIC)
28.	ePKHKP	A	12 49 59	Fiji Islands Region 20.51 S 178.30 W
	ePKP2	A	50 05	H = 12 31 15.8 h = 580 km MB = 4.5
				D = 148.93 Az = 347.7 (NEIC)
				PKHKPV A 1.3s 17.5nm
28.	eP	A	15 21 11.5	West Pakistan 34.39 N 73.55 E
				H = 15 12 42.2 h = 40 km MB = 4.9
				D = 47.05 Az = 309.9 (NEIC)
				PV A 1.2s 16.3nm M = 4.9
28.	eP1	AB	15 43 28	Burma-China Border Region
	+eiP2	A	43 30.5	22.93 N 101.03 E

Day	Phase		h m s	Remarks
cont. 28.	ePcP	AB	15 43 50	H = 15 32 00.9 h = 15 km MB = 5.6
	eS	B	52 52	D = 72.25 Az = 318.1 (NEIC)
	eSS	B	57.28	P1V A 1.2s 28.5nm M = 5.3
	eSSS	C	16 01 00	P1V B 5 0.55 µm 5.9
	LmH	B	13.4	P2V A 2.6 294.2nm 6.0
	LmV	B	18.1	PcPV B 6 0.87 µm
				SH B 12 2.04 µm 6.2
				LmH B 18 18.5 µm 6.4
				LmV B 13.5 9.8 µm 6.3
29.	eP	A	00 59 31.5	Luzon, Philippine Islands
	e	A	59 34.5	18.25 N 120.72 E
	LmH	B	01 42.0	H = 00 46 45.4 h = normal MB = 5.2
	LmV	B	42.0	D = 87.59 Az = 323.0 (NEIC)
				PV A 0.8s 17.3nm M = 5.4
				LmH B 16 0.2 µm 4.7
				LmV B 16 0.2 µm 4.7
29.	eP	A	04 36 13	Federal Republic of Germany
	e	A	36 19.5	48.50 N 8.89 E
	iPg	A	36 22.4	H = 04 35 29.4 h = normal
	i	A	36 27	D = 2.79 Az = 38.6 (NEIC)
	iSn	A	36 45.6	
	iSg	A	36 59	
	LmH	C	37.2	
	LmV	C	37.3	
29.	eP	A	08 24 42.5	Central Mid-Atlantic Ridge
	e	A	24 47	0.52 S 19.76 W
	e	A	24 51	H = 08 14 53.6 h = normal MB = 5.1
	eS	B	32 41	D = 57.55 Az = 23.1 (NEIC)
	eSS	B	36 40	PV A 1.7s 51.5nm M = 5.3
	LmH	B	47.0	SH B 20 0.76 µm 5.1
	LmV	B	50.3	LmH B 16 0.6 µm 4.8
				LmV B 17 0.7 µm 4.9

Day	Phase		h m s	Remarks
29.	iPg	A	12 00 32	<u>Czechoslovakia</u> 50°25' N 13°50' E
	eiSg	A	00 51	explosion yield 14.5 t (PRU)
				D = 1.44
29.				In the intervall 29d 13h 19m - 14h 00m
				the short-period seismographs (type A)
				had been out of operation.
29.	eiP	A	15 38 16	<u>Near Islands, Aleutian Is.</u>
	LmH	B	16 18.9	53.21 N 170.61 E
	LmV	B	19.8	H = 15 26 35.2 h = 27 km MB = 5.1
				D = 75.01 Az = 346.3 (NEIC)
				PV A 1.4s 39.5nm M = 5.3
				LmH B 14 0.3 μm 4.7
				LmV B 16 0.3 μm 4.7
29.	ePKP2	A	16 35 35	<u>South Pacific Cordillera</u>
	e(SS)	C	59 35	54.46 S 136.70 W
	LmV	B	17 51.0	H = 16 14 57.6 h = normal MB = 5.0
	LmH	B	51.2	D = 160.46 Az = 88.2 (NEIC)
				LmH B 19s 0.7 μm M = 5.4
				LmV B 16 0.8 μm 5.6
29.	eP1	AB	20 09 27.5	<u>Luzon, Philippine Islands</u>
	eP2	A	09 29.5	12.64 N 122.34 E
	eiP3	A	09 31	H = 19 56 11.60 h = 4 km
	ePP	AB	13 12	MB = 5.5 MS = 5.7 (NEIC)
	eSKKS	B	20 20	D = 93.0
	eS	B	20 32	P1V B 5s 0.82 μm M = 6.4
	eSP	B	21 37	P2V A 1.5 32.7nm 5.5
	eSS	C	26 46	P3V A 2.0 188.0nm 6.2
	LmH	B	49.1	PPV B 5 0.55 μm 6.2
	LmV	B	21 00.4	SH B 9.5 1.85 μm 6.2
				LmH B 20 6.2 μm 6.2
				LmV B 15.5 3.5 μm 5.9
30.	eX	A	02 23 13	XV A 0.7s 15.3nm

Day	Phase		h m s	Remarks
30.	eP1	A	03 26 15	Ascension Island Region
	eP2	A	26 17	11.43 S 12.78 W
	eP3	A	26 21	H = 03 15 35.2 h = normal MB = 4.7
	LmH	B	55.5	D = 65.38 Az = 16.8 (NEIC)
	LmV	B	56.0	P2V A 1.0s 13.8nm M = 5.0
				P3V A 1.2 28.5nm 5.2
				LmH B 14 0.2 µm 4.4
				LmV B 16 0.4 µm 4.7
30.	eP	A	06 14 55	Turkey 37.77 N 36.15 E
	LmH	C	25.3	H = 06 09 59.6 h = 30 km MB = 4.3
	LmV	C	25.3	D = 21.67 Az = 314.3 (NEIC)
				LmH C 15s 0.3 µm M = 3.8
				LmV C 15 0.4 µm 4.1
30.	eP	A	09 12 43	Iran 34.61 N 50.31 E
	LmH	B	28.0	H = 09 06 16.5 h = 42 km MB = 4.7
	LmV	B	32.3	D = 32.23 Az = 311.7 (NEIC)
				LmH B 15s 0.2 µm M = 3.9
				LmV B 12 0.3 µm 4.3
30.	eiPg	A	11 42 59.5	Czechoslovakia 50°33.8' N 14°00.4' E
	eiSg	A	43 19.5	explosion yield 20.5 t (PRU)
				D = 1.51
30.	eiPg	A	13 46 58	D c. 1.4
	eiSg	A	47 16.5	
30.	eiP	A	14 17 36	Rat Islands, Aleutian Is.
				51.70 N 179.93 E
				H = 14 05 49.0 h = 93 km MB = 5.2
				D = 77.55 Az = 352.4 (NEIC)
30.	eiPg	A	14 45 29	Adelebsen/Solling, Fed. Rep. of Germany
	eiSg	A	45 50	51° 36.5' N 09° 44.82' E
				H = 14 45 00.53 yield 7.0 t (Hannover)
				D = 1.5

Day	Phase		h m s	Remarks
30.	e	A	15 01 09.5	Eastern Sea of Japan 42.55 N 137.64 E
				H = 14 49 13.7 h = normal MB = 4.7
				D = 75.95 Az = 327.9 (NEIC)
30.	+eiP	A	15 59 50	Near Islands, Aleutian Is.
	eS	C	16 09(28)	52.76 N 172.45 E
	ePS	C	10 20	H = 15 48 06.5 h = 37 km MB = 5.5
	eSS	C	14 30	D = 75.70 Az = 347.5 (NEIC)
	LmH	B	36.1	PV A 1.4s 51.2nm M = 5.4
	LmV	B	36.2	LmH B 18 0.8 μm 5.1
				LmV B 17 0.8 μm 5.1
30.	eP	A	18 02 09	Caribbean Sea 17.07 N 85.18 W
	e	A	18 02 16	H = 17 49 50.8 h = 6 km MB = 4.9
				D = 81.20 Az = 39.8 (NEIC)
				PV A 1.0s 7.9nm M = 4.8

Day	Phase		h m s	Remarks
1.	eP	A	04 15 33	North Atlantic Ocean 18.42 N 36.76 W
				H = 04 06 37.0 h = normal MB = 4.8
				D = 49.93 Az = 38.5 (NEIC)
1.	eP1	AB	13 48 57.5	Turkey 40.88 N 27.85 E
	eP2	AB	49 03	H = 13 45 25.8 h = 13 km MB = 4.6
	eP3	A	49 18	D = 14.93 Az = 316.3 (NEIC)
	e	C	52 05	P1V A 1.5s 20.1nm
	LmV	B	56.2	P2V A 2.2 81.9nm
	LmH	B	56.3	P3V A 1.7 66.6nm
				LmH B 11.5 3.0/um M = 4.7
				LmV B 12 2.7/um 4.8
1.	ePKIKP	A	14 35 01.5	Loyalty Islands Region
	eiPKHKP	AB	35 04	22.29 S 171.61 E
	eiPKP2	AB	35 08	H = 14 15 36.2 h = 147 km MB = 5.8
	ePP	B	38 24	D = 147.66 Az = 336.0 (NEIC)
	ePPP	B	41 49	PKHKPV A 1.5s 108.0nm
	eSKKS	B	45 10	PKP2V A 1.5 121.0nm
	eSKSP	B	48 30	PPV B 6 0.8/um M = 6.1
	ePPS	C	51 05	LmH B 18 0.8/um
	eSS	C	57 20	LmV B 22 1.1/um
	eSSS	C	15 03 05	
	LmV	B	34.1	
	LmH	B	34.4	
1.	eP	A	14 44 45	El Salvador 13.26 N 88.50 W
	e	A	44 55	H = 14 32 12.8 h = 93 km MB = 5.4
	epP	A	45 05	D = 86.17 Az = 38.9 (NEIC)
				h = 92 km
1.	eP	A	22 22 49	Southern Italy 39.68 N 15.20 E
				H = 22 20 10.4 h = 291 km MB = 4.3
				D = 11.25 Az = 348.2 (NEIC)
				PV A 1.2s 26.4nm M = 4.3

Day	Phase		h m s	Remarks
2.	eSS	C	02 09.8	South of Mariana Islands
	LmV	B	42.3	11.9 N 143.2 E
	LmH	B	42.6	H = 01 36 18.0 h = 20 km MB = 5.1 (NEIC)
				D = 104.8
				LmH B 18s 0.9 µm M = 5.4
				LmV B 18 1.0 µm 5.4
2.	+iP1	AB	06 20 22	Andreanof Islands, Aleutian Is.
	eiP2	AB	20 29	51.43 N 177.21 W
	ePP	C	23 20	H = 06 08 27.3 h = 43 km MB = 6.0
	ePPP	C	25 04	D = 78.02 Az = 354.3 (NEIC)
	eiPPPP	C	26 40	P1V A 1.9s 151.1nm M = 5.7
	e	C	29 00	P1V B 6 0.9 µm 5.9
	iS	B	30 14	P2V A 1.6 357.0nm 6.1
	iPS	B	31 11	P2V B 14 16.9 µm 6.8
	eSS	C	35 05	P2H B 16 8.3 µm 6.8
	e(PKKP)	A	39 29	SH B 11 7.55 µm 6.7
	e	A	47 13	LmH B 21 139.0 µm 7.3
	eP'P'	A	47 25	LmV B 20 58.9 µm 7.0
	e	A	47 46	
	LmV	B	55.0	
	LmH	B	57.7	
2.	eP	A	09 20 54	Andreanof Islands, Aleutian Is.
				51.54 N 177.21 W
				H = 09 08 59.2 h = 47 km MB = 5.3
				D = 77.91 Az = 354.3 (NEIC)
2.	e	A	12 35 29	
3.	+eP	AB	00 43 05	Tibet 30.78 N 84.47 E
	ePP	C	45 10	H = 00 33 22.5 h = 16 km MB = 5.4
	eS	B	50 52	D = 56.34 Az = 313.0 (NEIC)
	LmH	B	01 04.9	PV A 1.7s 60.5nm M = 5.4
	LmV	B	09.9	LmH B 25 4.4 µm 5.5
				LmV B 16 3.6 µm 5.6

Day	Phase		h m s	Remarks
3.	ePn	A	05 54 48.5	Switzerland 47.3 N 8.4 E
	e	A	55 06	H = 05 53 51 (BCIS)
	ePg	A	55 10.5	D = 3.92
	eiSn	A	55 32	
	eiSg	A	55 54	
3.	eP	A	06 32 32	Mindanao, Philippine Islands
				8.73 N 124.14 E
				H = 06 19 05.7 h = 68 km
				D = 97.17 Az = 323.6 (NEIC)
3.	ePP	BC	11 11 12	Solomon Islands 11.13 S 161.57 E
	eSKP	BC	12 08	H = 10 49 26.9 h = 45 km MB = 4.6 (NEIC)
	eSS	C	28 52	D = 133.6
	LmH	B	12 09.1	LmH B 19s 2.1 μm M = 5.8
	LmV	B	09.1	LmV B 20 2.1 μm 5.8
3.	ePKHKP	A	11 48 06.5	Fiji Islands Region 17.94 S 178.69 W
	ePKP2	A	48 09	H = 11 29 34.0 h = 638 km MB = 4.4
				D = 146.35 Az = 348.1 (NEIC)
				PKHKPV A 1.0s 15.7nm
4.	eP1	A	02 18 00	Sunda Strait 6.61 S 105.37 E
	eP2	AB	18 03.5	H = 02 04 32.6 h = 69 km MB = 6.1
	e	A	18 07	D = 97.45 Az = 320.2 (NEIC)
	eX	A	18 32	P1V A 1.2s 12.2nm M = 5.4
	eiPP1	AB	21 57.5	P2V A 1.4 65.2nm 6.0
	iPP2	A	22 02	P2V B 8 1.3 μm 6.5
	eSKS	C	28 40	XV A 1.3 39.3nm
	ePS	C	30 46	PP1V A 2.0 231.0nm 6.4
	eP'P'	A	42 58	PP2V A 1.8 480.0nm 6.7
	LmV	B	03 10.1	LmH B 20.5 7.8 μm
	LmH	B	11.9	LmV B 17 3.9 μm
				e A 02 21 45 e A 21 53 e C 29 26
				e A 45 22 e A 45 41.5 e A 46 11.5
4.	e	A	05 05 46	
	e	A	06 51	

Day	Phase		h m s	Remarks
4.	ePn	A	14 05 05.5	Federal Rep. of Germany, rockburst
	ePg	A	05 14.5	51.5 N 7.2 E
	eSn	A	05 37	H = 14 04 18 (BCIS)
	eSn	A	05 37	D = 2.87
	eSg	A	05 52	
	e	A	05 54	
4.	eSg	A	23 20 26	Austria 47.7 N 16.0 E
				H = 23 18 12 (BCIS)
				D = 4.10
5.	ePn	A	01 18 08	Albania 41.89 N 20.24 E
	eSn	A	20 06	H = 01 15 33.0 h = 9 km MB = 4.9
	e(Sg)	B	21 25	D = 10.59 Az = 328.7 (NEIC)
	LmH	C	33.5	PnV A 1.3s 15.3nm
	LmV	C	37.4	LmV C 18 0.3/um
5.	eP	A	01 55(55)	Philippine Islands Region
	LmH	C	02 38.4	19.75 N 121.33 E
	LmV	C	38.4	H = 01 43 17.4 h = 64 km MB = 4.8 (NEIC)
				D = 86.7
				P : traces
				LmH C 20s 0.2/um
				LmV C 20 0.3/um
5.	ePKHKP	A	03 11 09	South of Fiji Islands 26.85 S 176.43 W
	ePKP2	A	11 24	H = 02 51 09.8 h = 50 km MB = 5.0
				D = 155.46 Az = 347.6 (NEIC)
5.	ePn	A	09 44 07.5	Southern Italy 41.20 N 15.29 E
	eiSn	A	45 52.5	H = 09 41 44.1 h = 12 km MB = 4.3
	e	A	46 11.5	D = 9.78 Az = 346.1 (NEIC)
	LmH	C	47.6	LmH C 15s 1.4/um M = 3.9
	LmV	C	49.4	LmV C 9.5 1.5/um
5.	iPg	A	22 32 59	Near Volkenroda, German Democr. Rep.
	i	A	33 00	rockburst
	eiSg	A	33 10.5	51.3 N 10.6 E

Day	Phase		h m s	Remarks
cont. 5.	i	AB	22 33 12	H = 22 32 41 (BCIS)
	i	B	33 14	D = 0.88
	LmH	B	33.5	LmH B 1.7s 0.69 μm
	LmV	B	33.5	LmV B 1.6 0.33 μm
6.	LmH	C	01 19.5	North Atlantic Ridge 28.6 N 43.7 W
	LmV	C	19.8	H = 00 55 19.1 h = normal MB = 4.6 (NEIC)
				D = 46.7
				LmV C 20s 0.4 μm M = 4.5
6.	eP	A	03 11 55	Andreanof Islands, Aleutian Is.
	e	A	12 28	51.51 N 177.28 W
				H = 02 59 58.9 h = 49 km MB = 4.8
				D = 77.95 Az = 354.2 (NEIC)
6.	ePn	A	03 47 34	Southern Italy 41.14 N 15.24 E
	e	A	47 36.5	H = 03 45 13.6 h = normal MB = 4.9
	e	A	47 59.5	D = 9.83 Az = 346.4 (NEIC)
	eiSn	AB	49 17	LmH B 12s 6.3 μm M = 4.7
	LmH	B	51.1	LmV B 10.5 4.5 μm
	LmV	B	52.9	
6.	eP1	A	04 28 37.5	Turkey 39.04 N 29.71 E
	eP2	AB	28 41	H = 04 24 33.9 h = 23 km MB = 4.6
	LmH	B	34.5	D = 17.25 Az = 318.2 (NEIC)
	LmV	B	36.4	P1V A 2.0s 55.6nm M = 4.3
				P2V A 2.2 142.0nm 4.7
				P2V B 4.5 0.55 μm 5.0
				LmH B 16 1.8 μm 4.4
				LmV B 12 1.4 μm 4.6
6.	ePg	A	14 39 33	France 44.8 N 6.8 E
	eSn	A	40(13)	H = 14 37 19 (BCIS)
	e	A	40 30.5	D = 6.73
	eSg	A	40 54.5	

Day	Phase		h m s	Remarks
6.	eP	A	17 38 21.5	Near East Coast of Kamchatka
				54.93 N 162.11 E
				H = 17 26 59.7 h = normal MB = 4.9
				D = 71.95 Az = 340.7 (NEIC)
				PV A 1.0s 23.6nm M = 5.1
7.	ePKHKP	A	00 28 29	South of Fiji Islands 23.45 S 179.95 W
	e	A	28 30.5	H = 00 09 38.5 h = 574 km MB = 5.2
	ePKP2	A	28 40	D = 151.40 Az = 344.5 (NEIC)
	epPKHKP	A	30 40.5	h = 580 km
	epPKP2	A	30 48	pPKP2V A 1.3s 30.6nm
7.	e	A	00 35 23.5	Celebes 2.61 S 119.60 E
	ePP	A	39 31	H = 00 21 13.9 h = 35 km MB = 5.8 MS =5.8
	e	B	39 37	D = 103.37 Az = 321.5 (NEIC)
	eX	AC	39 39	XV A 2.0s 128.0nm
	eSP	C	48 38	LmH B 28 15.0/um M = 6.4
	eSPP	C	49 32	LmV B 20 6.5/um 6.2
	eSS	C	54 15	
	e	A	01 02 09.5	
	LmH	B	19.9	
	LmV	B	29.0	
7.	eP	A	00 43 13.5	Southern Sumatra 4.66 S 102.32 E
				H = 00 29 56.7 h = normal
				D = 94.01 Az = 320.3 (NEIC)
8.	ePKIKP	A	01 08 14	S. Chile-Argentina Border Region
	epPKIKP	A	08 56	42.22 S 71.69 W
	ePP	AB	09 28	H = 00 49 45.6 h = 151 km MB = 5.9
	eiSKP	A	11 34	D = 117.46 Az = 45.4 (NEIC)
	eSKS	B	14 52	PKIKPV A 1.0s 23.6nm
	eSKKS	B	16 05	PPV A 1.8 64.1nm M = 5.6
	e	B	17 08	SKP A 1.5 60.4nm
	e	B	18 18	PKKP A 1.6 27.5nm
	ePKKP	A	18 41	LmH B 17 0.7/um
	eSP	B	19 00	LmV B 18 0.6/um
	LmV	B	58.3	
	LmH	B	59.3	

Day	Phase		h m s	Remarks
8.	eP	A	03 08 27.5	West Pakistan 29.58 N 66.32 E H = 03 00 08.9 h = normal MB = 4.9 D = 45.64 Az = 313.4 (NEIC) PV A 1.2s 38.6nm M = 5.2
8.	LmH LmV	B B	03 29.6 33.5	Either North Atlantic Ridge 22.3 N 45.1 W H = 02 55 01.8 h = normal MB = 4.8 MS = 4.6 (NEIC) D = 52.1 or: West Pakistan 29.6 N 66.3 E H = 03 00 08.9 h = normal MB = 4.9 D = 45.7 LmH B 17s 0.9 µm LmV B 15 0.7 µm
8.	eP ePP	A A	11 07 42 11 03.5	South of Honshu, Japan 32.10 N 139.56 E H = 10 55 17.1 h = 140 km MB = 5.0 D = 85.66 Az = 329.8 (NEIC) PV A 1.4s 14.0nm M = 4.6
8.				In the intervall $12^h\ 00^m$ - $12^h\ 04^m$ the short-period seismographs (type A) had been out of operation.
8.	eP	A	22 50 20	Greenland Sea 73.5 N 8.5 E H = 22 45 13 h = 33 km D = 22.98 Az = 175 (ISC) PV A 1.5s 17.6nm M = 4.4
8.	ePKP	A	22 52 53	Tonga Islands 16.63 S 173.82 W H = 22 32 55.6 h = 56 km MB = 4.3 (NEIC) D = 145.8
8.	eP	A	23 05 27	Greenland Sea 73.4 N 6.8 E H = 23 00 07 h = 0 km D = 22.92 Az = 172 (ISC)

Day	Phase		h m s	Remarks
8.	eP	A	23 39 32.5	Greenland Sea 73.1 N 7.8 E
				H = 23 34 29 h = 0 km
				D = 22.61 Az = 174 (ISC)
				PV A 1.1s 10.1nm
9.	eP	A	03 18 34	Hokkaido, Japan Region 42.55 N 144.96 E
				H = 03 06 37.7 h = 70 km MB = 5.2
				D = 78.66 Az = 331.8 (NEIC)
9.	ePKP	A	07 52 37	Tonga Islands 16.69 S 174.77 W
				H = 07 33 21.8 h = 212 km MB = 4.7
				D = 145.74 Az = 352.8 (NEIC)
				PKPV A 1.6s 38.5nm
9.	ePKIKP	A	08 44 08	Southern Pacific Ocean 39.78 S 104.84 W
	eX1	AB	44 19.5	H = 08 25 01.7 h = normal MB = 6.2
	eiX2	A	44 29.3	D = 135.24 Az = 54.0 (NEIC)
	ePP	C	47 04	PKIKPV A 1.4s 14.0nm
	e	A	47 08	X1V A 1.5 115.6nm
	iPKS	AB	47 55	X2V A 1.3 118.0nm
	eSKKS	C	53 45	LmH B 21 4.0/um M = 6.1
	eSS	B	09 04 44	LmV B 20 4.0/um 6.1
	eiPSPS	C	05 20	
	eSSS	C	09 32	
	LmH	B	42.3	
	LmV	B	42.4	
9.	ePKP2	A	12 59 06.5	Tonga 20.3 S 173.4 W
				H = 12 39 17.2 h = 33 km MB = 4.5
				D = 149.4 Az = 354 (ISC)
				PKP2V A 1.6s 13.7nm
9.	ePKIKP	A	18 20 19	Southern Pacific Ocean 39.84 S 104.89 W
	LmH	C	19 20.0	H = 18 00 59.9 h = normal MB = 5.4
				D = 135.30 Az = 54.1 (NEIC)

Day	Phase		h m s	Remarks
9.	ePKIKP	A	18 54 27	Easter Island Cordillera
	LmH	B	19 55.0	39.72 S 104.98 W
	LmV	B	57.3	H = 18 35 09.8 h = normal MB=5.4 MS=5.4
				D = 135.29 Az = 54.0 (NEIC)
				PKIKPV A 1.4s 14.0nm
				LmH B 13 1.0 μm M = 5.7
				LmV B 16 1.3 μm 5.7
9.	+eiP	AB	19 34 41	West Pakistan 35.59 N 71.13 E
	e	A	34 57.5	H = 19 26 34.6 h = 85 km MB = 5.5
	e	A	35 11.5	D = 44.76 Az = 308.8 (NEIC)
	e	B	36 44	PV A 1.2s 89.5nm M = 5.5
	e	A	36 49	
	e	B	37 40	
9.	ePn	A	23 05 33	Italy 41.0 N 15.2 E
	eSn	A	07 15	H = 23 03 02 (BCIS)
	LmH	C	08.5	D = 10.00
	LmV	C	10.0	LmH C 14s 0.6 μm M = 3.6
10.	eiP	A	00 13 08.5	Fox Islands, Aleutian Is.
	e	A	13 22	52.17 N 171.01 W
	e	A	13 33.5	H = 00 01 16.3 h = 48 km MB = 5.3
	LmH	B	50.7	D = 77.54 Az = 358.3 (NEIC)
	LmV	B	50.9	LmH B 19s 0.5 μm M = 4.9
				LmV B 16 0.4 μm 4.9
10.	eP	A	08 13 21	Mid-Indian Rise 14.50 S 66.16 E
	e	A	13 26.5	H = 08 01 11.53 h = normal MB = 5.0
				D = 80.45 Az = 328.3 (NEIC)
10.	e	A	09 40 13	Tonga Islands 15.07 S 173.79 W
				H = 09 20 25.0 h = normal MB = 4.9
				D = 144.25 Az = 354.1 (NEIC)
10.	eP	A	12 07 33	Mindanao, Philippine Islands
	e(pP)	A	07 44.5	7.96 N 126.84 E
	e(PP)	A	11 30	H = 11 53 53.0 h = normal MB = 5.8

Day	Phase		h m s	Remarks
cont. 10.	eSKS	C	12 18 18	D = 99.37 Az = 324.3 (NEIC)
	eS	C	19 05	PV A 1.5s 57.7nm M = 5.9
	LmH	B	13 09.3	(pP)V A 1.5 50.2nm
	LmV	B	09.5	LmH B 17 1.7/um 5.6
				LmV B 14.5 1.7/um 5.7
10.	+iP1	AB	14 59 26.5	<u>Kirgiz SSR</u> 42.78 N 71.38 E
	iP2	AB	59 31.5	H = 14 51 46.8 h = normal MB = 5.6
	ePP	B	15 00 58	D = 40.73 Az = 302.5 (NEIC)
	e	A	01 07.5	P1V A 1.7s 85.0nm M = 5.2
	eS	B	05 51	P2V A 1.8 108.0nm 5.3
	eSS	C	08 25	MPPV B 5 1.37/um 5.9
	LmH	B	17.7	LmH B 13.5 12.0/um 5.9
	LmV	B	17.7	LmV B 14 14.2/um 6.1
10.	eP	A	23 10 00	<u>Andreanof Islands, Aleutian Is.</u>
	LmV	C	39.0	51.42 N 177.24 W
	LmH	C	39.5	H = 22 58 05.2 h = 50 km MB = 5.3
				D = 78.04 Az = 354.3 (NEIC)
				PV A 1.4s 18.6nm M = 4.9
				LmH C 28 0.3/um
				LmV C 28 0.7/um
10.	e	A	16 38 05	<u>Poland</u> 50.32 N 18.92 E
				H = 16 35 27.0 M = 2.7 (WAR)
				D = 4.15
11.	eP1	A	00 22 05	<u>Taiwan Region</u> 22.30 N 121.12 E
	eP2	A	22 10	H = 00 09 35.0 h = 46 km MB = 5.4
	e	A	25 26	D = 84.60 Az = 322.9 (NEIC)
	LmH	C	01 01.0	P2V A 1.3s 32.8nm M = 5.3
	LmV	C	02.9	LmH C 27 0.6/um 4.9
				LmV C 25 1.1/um 5.2
11.	LmV	B	23 02.7	LmH B 16s 0.3/um
	LmH	B	04.3	LmV B 16 0.7/um

Day	Phase		h m s	Remarks
12.	ePKIKP	A	03 07 41.5	<u>New Britain Region</u> 6.30 S 151.85 E
	LmV	C	04 05.3	H = 02 48 44.4 h = 38 km MB = 5.3
				D = 124.78 Az = 330.3 (NEIC)
12.	eP	A	05 22 29	<u>Turkey</u> 37.42 N 32.1 E
	e	A	22 36	H = 65 17 59 h = 34 km
				D = 19.69 Az = 319 (ISC)
				PV A 1.4s 11.6nm
12.	eP1	AB	06 29 29	<u>Turkey</u> 37.59 N 29.76 E
	iP2	B	29 32	H = 06 25 13.0 h = 23 km MB = 5.5
	iP3	AB	29 35	D = 18.38 Az = 321.0 (NEIC)
	-iP4	A	29 44.5	P1V A 1.8s 108.0nm M = 4.8
	iS	B	32 56	P3V A 2.2 1230.0nm 5.8
	LmH	B	37.6	P3V B 8 15.5nm 6.3
	LmV	B	37.6	P4V A 2.2 2260.0nm 6.0
				SH B 12 16.3/um 6.6
				LmH B 13 54.2/um 6.0
				LmV B 12 52.9/um 6.2
12.	eP	A	06 48 28	<u>Turkey</u> 37.61 N 30.18 E
				H = 06 44 09.7 h = 15 km MB = 4.4
				D = 18.58 Az = 320.5 (NEIC)
12.	eP	A	08 30 49	<u>Turkey</u> 37.60 N 29.81 E
				H = 08 26 33.6 h = normal MB = 4.3
				D = 18.40 Az = 320.9 (NEIC)
				PV A 1.4s 14.0nm M = 4.0
12.	eP1	A	10 14 40	<u>Turkey</u> 37.49 N 29.69 E
	eP2	AB	14 50	H = 10 10 24.9 h = normal MB = 4.8
	iP3	AB	14 56	D = 18.42 Az = 321.3 (NEIC)
	eS	BC	18 20	P1V A 1.4s 14.0nm M = 4.0
	LmH	B	22.9	P2V A 1.7 242.0nm 5.2
	LmV	B	22.9	P3V A 1.7 388.0nm 5.4
				P3V B 6 2.82/um 5.7
				SH B 10 3.74/um 6.1
				LmH B 10.5 10.9/um 5.4
				LmV B 12 8.2/um 5.4

Day	Phase		h m s	Remarks
12.	eP1	AB	13 01 38	<u>Turkey</u> 37.59 N 29.60 E
	iP2	A	01 40.5	H = 12 57 24.8 h = normal MB = 5.4
	iP3	A	01 43.5	D = 18.30 Az = 321.2 (NEIC)
	iP4	A	01 45.3	P1V B 12s 5.74 μm M = 6.7
	iS	BC	05 03	P2V A 1.0 70.8nm 4.8
	LmH	B	09.8	P3V A 1.0 157.5 μm 5.2
	LmV	B	09.8	P4V A 1.8 676.0 μm 5.6
				SH B 10.5 10.0 μm 6.5
				LmH B 11.5 31.6 μm 5.8
				LmV B 12 26.3 μm 5.9
12.	ePKP	A	13 55 27.5	<u>Tonga Islands</u> 15.57 S 175.30 W
				H = 13 35 51.4 h = normal MB = 5.1
				D = 144.57 Az = 352.4 (NEIC)
12.	eP	A	14 27 50.5	<u>Turkey</u> 37.55 N 29.61 E
	e	A	27 54.5	H = 14 23 35.4 h = normal MB = 4.3
				D = 18.33 Az = 321.3 (NEIC)
				PV A 1.5s 20.1nm M = 4.1
12.	eP	A	14 43 58	<u>Turkey</u> 37.46 N 29.76 E
				H = 14 39 35.6 h = 0 km
				D = 18.48 Az = 321 (ISC)
12.	eP	A	15 16 10	<u>Turkey</u> 37.52 N 29.97 E
				H = 15 11 51.7 h = normal MB = 4.3
				D = 18.54 Az = 320.9 (NEIC)
				PV A 0.8s 7.7nm M = 4.0
12.	eP1	A	17 16 41	<u>Turkey</u> 37.58 N 29.96 E
	eP2	AB	16 45	H = 17 12 25.4 h = 28 km MB = 4.7
	eS	B	20 13	D = 18.49 Az = 320.8 (NEIC)
	LmH	B	25.0	P1V A 1.8s 47.4nm M = 4.4
	LmV	B	25.7	P2V A 2.0 150.0nm 4.9
				P2V B 5 0.76 μm 5.2
				SH B 10 1.15 μm 5.1
				LmH B 9.5 3.0 μm 4.9
				LmV B 10.5 1.7 μm 4.8

Day	Phase		h m s	Remarks
12.	eP1	AB	17 52 18	**Turkey** 37.44 N 29.64 E
	eP2	A	52 24	H = 17 48 03.0 h = normal MB = 4.4
	eS	B	55 50	D = 18.43 Az = 321.5 (NEIC)
	LmH	B	18 00.3	P1V A 1.6s 24.7nm M = 4.1
	LmV	B	00.3	P2V A 1.6 44.0nm 4.3
				LmH B 11 1.0/um 4.4
				LmV B 12 1.1/um 4.5
12.	eP1	AB	19 06 41	**Turkey** 37.51 N 29.80 E
	eP2	A	06 47	H = 19 02 25.6 h = normal MB = 4.5
	eS	C	10 15	D = 18.45 Az = 321.1 (NEIC)
	LmV	B	14.7	P1V A 1.4s 14.0nm M = 3.9
	LmH	B	14.8	P2V A 1.8 40.6nm 4.3
				LmH B 10 1.0/um 4.4
				LmV B 12 0.9/um 4.4
12.	eP	AB	20 17 24.5	**Turkey** 37.50 N 29.88 E
	eS	C	20 50	H = 20 13 08.1 h = 36 km MB = 4.8
	LmV	B	25.4	D = 18.51 Az = 321.1 (NEIC)
	LmH	B	26.0	PV A 2.4s 124.0nm M = 4.7
				PV B 7 0.76/um 5.0
				SH C 10 1.09/um 5.0
				LmH B 9 4.3/um 5.1
				LmV B 8 1.7/um 4.9
12.	eP	A	21 52 48.5	**Kurile Islands** 47.53 N 154.44 E
				H = 21 40 56.5 h = normal MB = 5.1
				D = 77.10 Az = 336.8 (NEIC)
				PV A 1.1s 34.3nm M = 5.3
12.	LmH	C	24 23.1	**Kyushu, Japan** 31.0 N 130.3 E
	LmV	C	23.2	H = 23 28 52 MB = 5.1 (ANUSSR)
				D = 82.3
				LmH C 16s 0.2/um M = 4.6
				LmV C 16 0.5/um 5.0
13.	eP1	A	04 11 39.5	**Turkey** 37.58 N 29.97 E
	eP2	A	11 43.5	H = 04 07 21.3 h = 23 km MB = 4.4

Day	Phase		h m s	Remarks
cont. 13.	eS	C	04 15 16	D = 18.48 Az = 320.8 (NEIC)
	LmH	C	17.8	P2 A 1.4s 27.9nm M = 4.3
	LmV	C	19.0	LmH C 19 0.5 µm 3.8
				LmV C 20 0.3 µm 3.8
13.	eP1	AB	04 49 46	<u>Turkey</u> 37.46 N 29.74 E
	eP2	A	49 49	H = 04 45 27.5 h = 14 km MB = 4.8
	eS	B	53 19	D = 18.46 Az = 321.3 (NEIC)
	LmH	B	56.1	P1V A 1.5s 35.2nm M = 4.3
	LmV	B	59.2	P1V B 4.0 0.55 µm 5.1
				P2V A 2.0 128.0nm 4.7
				LmH B 12.5 1.7 µm 4.5
				LmV B 13 0.9 µm 4.4
13.	eP	AB	08 18 51.5	<u>Turkey</u> 37.61 N 30.05 E
	eS	B	22 26	H = 08 14 33.0 h = 14 km MB = 4.6 (NEIC)
	LmH	B	27.1	D = 18.5
	LmV	B	27.8	LmH B 11.5s 0.4 µm M = 3.9
				LmV B 11 0.3 µm 4.0
13.	eP	A	08 34 41	<u>Turkey</u> 37.68 N 30.05 E
	eS	B	38 16	H = 08 30 25.0 h = normal MB = 4.4
	LmH	B	42.9	D = 18.45 Az = 320.5 (NEIC)
	LmV	B	43.0	LmH B 11s 0.8 µm M = 4.2
				LmV B 11 0.6 µm 4.3
13.	eP	AB	13 32 55	<u>Turkey</u> 37.64 N 29.93 E
	e	A	33 00	H = 13 28 39.5 h = normal MB = 4.7
	eS	B	36 28	D = 18.43 Az = 320.7 (NEIC)
	LmH	B	41.1	PV A 2.5s 61.5nm M = 4.3
	LmV	B	41.1	SH B 11 0.58 µm 4.7
				LmH B 11 1.4 µm 4.5
				LmV B 12 1.4 µm 4.6
13.	eP	A	18 40 31	<u>Iceland Region</u> 63.9 N 23.9 W
	LmH	B	49.7	H = 18 35 30.1 h = 33 km MB = 4.4

Day	Phase		h m s	Remarks
cont. 13.	LmV	B	18 51.6	D = 22.92 Az = 108
				PV A 0.8s 11.5nm M = 4.4
				LmH B 16 0.3/um 3.8
				LmV B 14 0.2/um 3.9
13.	eP	A	20 13 46	<u>Iceland Region</u> 63.94 N 23.11 W
	LmH	B	23.0	H = 20 08 46.2 h = normal MB = 4.1
	LmV	B	25.0	D = 22.63 Az = 109.5 (NEIC)
				LmH B 15s 0.4/um M = 4.0
13.	eP1	A	22 51 24.5	<u>Turkey</u> 37.66 N 29.91 E
	eP2	A	51 28	H = 22 47 07.3 h = 17 km MB = 4.4
	LmV	B	59.5	D = 18.39 Az = 320.7 (NEIC)
	LmH	B	59.6	P2V A 2.0s 51.3nm M = 4.3
				LmH B 9 0.3/um 4.0
				LmV B 11 0.3/um 4.0
13.	eP2	A	23 37 13	<u>Turkey</u> 37.59 N 29.92 E
	eS	B	40 48	H = 23 32 54.9 h = normal MB = 4.4
	LmV	B	45.3	D = 18.45 Az = 320.8 (NEIC)
	LmH	B	45.4	PV A 1.9s 15.2nm M = 3.9
				LmH B 13 0.7/um 4.1
				LmV B 12 0.4/um 4.1
14.	eP2	A	04 22 52	<u>Turkey</u> 37.66 N 30.08 E
	eS	B	26(25)	H = 04 18 30.2 h = normal MB = 4.4
	LmH	C	30.9	D = 18.48 Az = 320.5 (NEIC)
	LmV	C	30.9	P2V A 1.6s 16.5nm M = 3.9
14.	LmH	B	09 07.4	<u>Java</u> 6.5 S 107.1 E
	LmV	B	08.0	H = 07 56 36.7 h = 10 km MB = 5.3 (NEIC)
				D = 98.5
14.	eP	A	10 10 42	<u>Bonin Islands Region</u> 27.72 N 141.90 E
				H = 09 57 38.1 h = 6 km MB = 5.6
				D = 90.46 Az = 330.9 (NEIC)
				PV A 1.2s 18.3nm M = 5.2

Day	Phase		h m s	Remarks
14.	eX	A	14 11 55.5	XV A 0.9s 19.5nm
	e	A	12 05	
14.	iPg	A	15 05 20	Hilders/Rhön, Fed. Rep. of Germany
	iSg	A	05 36	50°32.48' N 10°02.30' E
				H = 15 05 00.38 yield 8.4 t (Hannover)
				D = 1.0
14.	eP	A	16 53 31	Oaxaca, Mexico 16.25 N 93.98 W
				H = 16 40 54.0 h = 95 km MB = 4.8
				D = 87.13 Az = 37.9 (NEIC)
				PV A 0.8s 15.4nm M = 5.1
14.	eP	A	17 23 31.5	West Pakistan 25.13 N 68.11 E
				H = 17 14 40.9 h = 57 km MB = 4.5
				D = 49.89 Az = 316.0 (NEIC)
				PV A 1.2s 10.2nm M = 4.7
14.	eP	AB	22 22 38.5	Turkey 37.67 N 30.01 E
	eS	B	26 14	H = 22 18 23.0 h = normal MB = 4.5
	LmH	B	31.0	D = 18.43 Az = 320.6 (NEIC)
	LmV	B	31.6	PV A 1.6s 27.5nm M = 4.1
				LmH B 10 0.8 μm 4.3
				LmV B 10 0.6 μm 4.4
14.	eiP	AB	22 55 26	Turkey 37.41 N 29.53 E
	e	A	55 36	H = 22 51 10.7 h = normal MB = 4.7
	eS	B	59 00	D = 18.40 Az = 321.7 (NEIC)
	LmH	B	23 03.5	PV A 2.4s 138.0nm M = 4.7
	LmV	B	03.5	PV B 4.5 0.55 μm 5.1
				LmH B 11 2.1 μm 4.7
				LmV B 12 2.2 μm 4.8
15.	eP	A	05 28 18.5	Eastern Mediterranean Sea
				35.8 N 28.3 E
				H = 05 24 02 h = 113 km
				D = 19.10 Az = 326 (ISC)

Day	Phase		h m s	Remarks
15.	eP	A	07 55 24	Off East Coast of Honshu, Japan
	epP	A	55 34	40.59 N 143.77 E
	esP	A	55 40	H = 07 43 15.7 h = 23 km MB = 4.4
	LmH	B	08 29.5	D = 79.96 Az = 331.4 (NEIC)
	LmV	B	37.0	h = 46 km
				PV A 1.5s 10.0nm M = 4.6
				LmH B 18 0.8/um 5.1
				LmV B 16 0.4/um 4.9
15.	eP	A	08 45 42	Turkey 37.61 N 32.5 E
	e	A	45 49	H = 08 41 12 h = 51 km
				D = 19.79 Az = 318
15.	eP	A	12 24 13	Turkey 37.48 N 29.91 E
	eS	B	27 45	H = 12 19 56.9 h = normal MB = 4.3
	LmV	B	32.4	D = 18.53 Az = 321.0 (NEIC)
	LmH	B	32.5	PV A 2.7s 68.4nm M = 4.4
				LmH B 12 0.3/um 3.9
				LmV B 13.5 0.4/um 4.0
15.	eP	A	21 15 46	Kurile Islands 43.37 N 147.27 E
	epP	A	15 57	H = 21 03 45.5 h = 48 km MB = 4.7
	LmH	C	58.5	D = 78.72 Az = 333.0 (NEIC)
	LmV	C	59.3	h = 40 km
				LmV C 16s 0.4/um M = 4.9
16.	ePKP	A	05 30 12	Fiji Islands Region 18.37 S 178.03 W
				H = 05 11 34.3 h = 585 km MB = 5.2
				D = 146.89 Az = 348.7 (NEIC)
				PKPV A 1.6s 16.5nm
16.	eP1	A	05 32 06	Turkey 37.62 N 30.10 E
	eP2	A	32 09	H = 05 27 51.0 h = normal MB = 4.6
	LmH	C	40.0	D = 18.52 Az = 320.5 (NEIC)
	LmV	C	40.0	P1V A 1.4s 11.6nm M = 3.9
				P2V A 1.7 24.2nm 4.2

Day	Phase		h m s	Remarks
16.	eP1	AB	08 43 49	Turkey 37.54 N 29.91 E
	eP2	A	43 53	H = 08 39 34.1 h = normal MB = 4.5
	eS	B	47 23	D = 18.49 Az = 320.9 (NEIC)
	LmH	B	50.3	P1V A 2.2s 38.2nm M = 4.1
	LmV	B	52.8	P2V A 2.2 65.5nm 4.5
				LmH B 9.5 1.1/um 4.5
				LmV B 11 0.6/um 4.3
16.	eP1	A	09 29 18	Turkey 37.57 N 29.74 E
	eiP2	AB	29 20	H = 09 25 01.4 h = normal MB = 4.9
	eS	B	32 48	D = 18.38 Az = 321.1 (NEIC)
	LmH	B	35.6	P2V A 1.8s 105.0nm M = 4.8
	LmV	B	37.4	LmH B 14.5 2.3/um 4.6
				LmV B 11.5 1.9/um 4.8
16.	eP	A	12 09 32.5	Turkey 37.44 N 29.58 E
				H = 12 05 14 h = 4 km MB = 4.2
				D = 18.41 Az = 322 (ISC)
16.	eP	A	17 29 35	Kashmir-Sinkiang Border Region
				36.10 N 77.87 E
				H = 17 20 56.7 h = 84 km MB = 4.6
				D = 48.70 Az = 309.1 (NEIC)
16.	e(P)	A	20 22 43	Western Caucasus 42.51 N 43.37 E
				H = 20 17 36.2 h = 42 km MB = 4.2
				D = 23.15 Az = 302 (ISC)
				traces
16.	eP	A	23 32 22	Turkey 37.57 N 30.03 E
				H 23 28 01.9 h = 0 km MB = 4.4
				D = 18.53 Az = 321
17.	ePKP	A	06 06 09	Fiji Islands Region 16.18 S 179.98 W
				H = 05 46 29.9 h = normal MB = 4.3
				D = 144.38 Az = 347.3 (NEIC)
				traces

Day	Phase		h m s	Remarks
17.	+eiP1	AB	11 16 52	<u>Ecuador</u> 1.60 S 77.70 W
	eiP2	A	16 58	H = 11 04 07.1 h = 176 km MB = 5.7
	eiP3	A	17 02.5	D = 90.79 Az = 39.5 (NEIC)
	epP	B	17 34	P1V A 2.0s 278.0nm M = 5.9
	eipP	A	17 37	P2V A 2.0 274.0nm 5.9
	ePP	B	20 00	P3V A 1.8 2.16nm 5.9
	iS	B	27 11	pPV A 1.9 190.0nm
	ei	B	28 44	SH B 12 7.7/um 6.8
	eiSS	C	33 40	P'P'V A 2.5 92.2nm
	eP'P'	A	42 23	LmH B 20 3.3/um
	LmV	B	51.0	LmV B 20 5.1/um
	LmH	B	51.3	
17.	ePKIKP	A	12 08 29	<u>Solomon Islands</u> 6.84 S 156.21 E
				H = 11 49 40.7 h = 154 km MB = 4.6
				D = 127.34 Az = 332 (ISC)
				traces
17.	eP1	A	14 20 33	<u>Turkey</u> 37.67 N 29.88 E
	eP2	AB	20 38	H = 14 16 17.9 h = normal MB = 4.8
	eS	B	24 05	D = 18.38 Az = 320.7 (NEIC)
	LmV	B	28.6	P1V A 2.3s 61.0nm M = 4.4
	LmH	B	28.7	P2V A 1.9 75.8nm 4.6
				P2V B 4.5 0.72/um 5.2
				LmH B 11 1.7/um 4.6
				LmV B 12 1.7/um 4.7
18.	ePKIKP	A	01 05 08	<u>Kermadec Islands Region</u>
	ePKP2	A	05 43	29.66 S 178.74 E
	e	A	07 26	H = 00 46 20.1 h = 600 km MB = 5.4
	e	A	07 46	D = 156.91 Az = 338.8 (NEIC)
	ePP	A	09 21	PKP2V A 1.5s 90.5nm
	e	A	12 31	PPV A 1.6 27.5nm M = 5.0
	e	A	13 01.5	
18.	eP	A	02 08 12	<u>Turkey</u> 37.53 N 29.80 E
	e	A	08 18	H = 02 03 55.1 h = normal MB = 4.5

Day	Phase		h m s	Remarks
cont. 18.	eS	B	02 11 44	D = 18.44 Az = 321.1 (NEIC)
	LmH	B	14.5	PV A 1.6s 16.5nm M = 3.9
	LmV	B	17.6	LmH B 14 0.6/um 4.0
18.	eP	A	03 17 25	Turkey 37.8 N 29.8 E
	eS	C	20 59	H = 03 13 10 h = 0 km
	LmH	B	25.7	D = 18.23 Az = 321 (ISC)
	LmV	B	25.7	LmH B 14s 0.3/um M = 3.8
				LmV B 15 0.4/um 4.0
18.	e(PKP)	AB	06 30 58	La Rioja Province, Argentina
	ePP	BC	31 25	28.43 S 68.81 W
	e	B	31 37	H = 06 12 36.7 h = 95 km MB = 5.8
	e	A	31 41	D = 105.78 Az = 40.7 (NEIC)
	e(SKKS)	C	38 05	PPV B 5s 0.5/um M = 6.4
	e	C	40 10	LmH B 18 0.4/um
	ePS	C	40 50	LmV B 18 0.5/um
	e	C	45 30	
	e	C	45 50	
	LmH	B	07 15.0	
	LmV	B	15.0	
18.	ePKIKP	A	07 17 56	Tonga Islands 20.76 S 175.94 W
	ePKHKP	A	18 01	H = 06 58 30.8 h = 175 km MB = 5.1
	epPKP	A	18 54	D = 149.58 Az = 350.5 (NEIC)
				h = 227 km
				PKIKPV A 0.8s 11.5nm
				PKHKPV A 1.2 57.0nm
18.	eP1	AB	22 54 51	Eastern Siberia 63.95 N 146.11 E
	-iP2	A	54 54	H = 22 44 43.8 h = normal MB = 5.8
	iX1	AB	55 00.5	D = 60.34 Az = 328.5 (NEIC)
	iX2	A	55 21.8	P1V A 1.5s 55.2nm M = 5.4
	e	B	57 00	P1V B 5 4.56/um 6.8
	iPP	B	57 10	P2V A 2.3 1580.0nm 6.6
	eiPa	B	58 23	X1V A 1.9 545.0nm
	e	B	59 18	X2V A 1.8 406.0nm
	eiS	BC	23 03 00	PPV B 5 2.58/um 6.8

Day	Phase		h m s	Remarks
cont. 18.	iSS	B	23 07 18	PaV B 9s 3.38 µm
	eSa	B	08 24	SH B 10 5.2 µm M = 6.3
	LmH	B	19.5	SaH B 22 8.6 µm
	LmV	B	22.2	LmH B 21.5 138.8 µm 7.1
				LmV B 18 69.2 µm 6.9
19.	e	A	11 25 04	<u>Halmahera</u> 2.38 N 129.13 E
				H = 11 06 31.4 h = 45 km MB = 5.3
				D = 105.19 Az = 324.2 (NEIC)
				traces
19.	iPn	A	17 31 28.5	<u>Federal Rep. of Germany</u> 48.27 N 9.07 E
	i	A	31 31.3	H = 17 30 44.0 h = 17 km
	e	A	31 34.5	D = 2.90 Az = 34.0 (NEIC)
	iPg	AB	31 38.5	LmH B 5s 0.7 µm
	i	A	31 38.8	LmV B 5 0.7 µm
	i	A	31 44.2	
	eiSg	AB	32 14.5	
	LmH	B	32.6	
	LmV	B	32.6	
19.	ePKIKP	A	22 28 10	<u>New Hebrides Islands</u> 13.60 S 167.14 E
				H = 22 09 07.0 h = 199 km MB = 4.7
				D = 138.04 Az = 336.8 (NEIC)
				PKIKPV A 0.8s 3.8nm
				traces
20.	eP1	A	01 18 53	<u>Turkey</u> 37.60 N 29.88 E
	eP2	A	18 55	H = 01 14 37.4 h = normal MB = 4.4
	eS	C	22(28)	D = 18.43 Az = 320.9 (NEIC)
	LmV	B	26.9	P2V A 2.2s 38.2nm M = 4.2
	LmH	B	27.0	
20.	eP	A	01 39 32.5	<u>Greenland Sea</u> 75.01 N 9.0 E
	e	A	39 39	H = 01 34 14.8 h = 33 km M = 4.1
				D = 24.48 Az = 176 (ISC)
				PV A 1.2s 18.3nm M = 4.5

Day	Phase		h m s	Remarks
20.	eP	A	02 47 25	Andreanof Islands, Aleutian Is.
	e	A	47 41	52.22 N 173.33 W
				H = 02 35 36.7 h = 74 km MB = 5.5
				D = 77.43 Az = 356.8 (NEIC)
				PV A 1.5s 30.2nm M = 5.0
20.	eP1	A	03 11 02	Turkey 37.58 N 29.98 E
	eP2	AB	11 04	H = 03 06 45.3 h = normal MB = 4.8
	eS	B	14 34	D = 18.50 Az = 320.8 (NEIC)
	LmV	B	19.1	P1V A 1.0s 15.7nm M = 4.2
	LmH	B	19.2	P2V A 2.0 137.0nm 4.8
				P2V B 6 0.5 /um 4.9
				SH B 10.5 0.5/um 4.7
				LmH B 11 1.4/um 4.5
				LmV B 12 1.3/um 4.6
20.	e	A	08 13 26	Off Coast of Oregon 42.31 N 126.34 W
	e	A	13 36	H = 08 01 08.6 h = normal MB = 5.0
	LmH	B	50.0	D = 80.42 Az = 25.6 (NEIC)
	LmV	B	53.3	LmH B 19s 0.6/um M = 5.0
				LmV B 15 0.7/um 5.2
20.	eX	A	20 53 50	XV A 2.0s 38.4nm
	e	A	54 01	
20.	e	A	22 25 20	Southern Pacific Ocean 37.06 S 95.03 W
	LmH	C	23 10.0	H = 22 05 45.5 h = normal
	LmV	C	10.0	MB = 5.1 MS = 4.9 (NEIC)
				D = 127.5
21.	ePKIKP	A	00 49 40.5	Fiji Islands Region 19.23 S 177.24 W
	ePKHKP	A	49 44.5	H = 00 30 43.8 h = 396 km MB = 4.9
	ePKP2	A	49 47.5	D = 147.87 Az = 349.4 (NEIC)
	epPKP	A	51 19	PKHKPV A 1.6s 33.0nm
				PKP2V A 1.5 20.1nm
				pPKPV A 1.8 16.9nm

Day	Phase		h m s	Remarks
21.	eP1	AB	09 45 30	Turkey 37.51 N 29.65 E
	eP2	A	45 33	H = 09 41 12.2 h = 7 km MB = 4.9
	eS	B	49 00	D = 18.38 Az = 321.3 (NEIC)
	LmH	B	53.6	P1V A 1.6s 44.0nm M = 4.4
	LmV	B	53.6	P1V B 3.2 0.66 /um 5.3
				P2V A 2.5 284.0nm 5.0
				SH B 8 0.9 /um 5.0
				LmH B 12 1.8 /um 4.6
				LmV B 12 1.7 /um 4.7
21.	-eP	AB	19 08 34.5	Andreanof Islands, Aleutian Is.
	eS	B	18 26	52.55 N 173.22 W
	LmH	B	51.6	H = 18 56 43.7 h = 36 km MB = 5.7
	LmV	B	53.8	D = 77.10 Az = 356.8 (NEIC)
				PV A 1.5s 50.2nm M = 5.3
				PV B 9 0.8 /um 5.7
				LmH B 18 1.0 /um 5.2
				LmV B 16 0.6 /um 5.1
22.	e(pP)	A	10 19 11	Andreanof Islands, Aleutian Is.
				51.87 N 173.07 W
				H = 10 07 04.1 h = 48 km MB = 4.8
				D = 77.79 Az = 357.0 (NEIC)
22.	eP	A	14 09 12	Iran 35.62 N 58.25 E
	e	A	09 19	H = 14 02 07.5 h = 36 km MB = 4.8
				D = 36.55 Az = 309.0 (NEIC)
22.	eP1	A	16 49 10.5	Turkey 38.83 N 40.52 E
	+iP2	B	49 12	H = 16 43 58.7 h = 3 km MB = 6.0
	-iP3	A	49 14	D = 23.53 Az = 309.6 (NEIC)
	+iP4	A	49 16.5	P1V A 1.4s 79.0nm M = 5.1
	-iS	BN	53 32	P2V B 6.5 19.9 /um 6.8
	+iS	BE	53 32	P2H B 7 13.0 /um 6.7
	LmH	B	59.5	P3V A 2.4 297.0nm 5.4
	LmV	B	17 00.7	P4V A 2.3 2800.0nm 6.4
				SH B 12 207.0 /um 7.4
				LmH B 18 372.0 /um 6.9
				LmV B 16.5 319.0 /um 7.0

Day	Phase		h m s	Remarks
22.	eX	A	17 37 43	Turkey 38.93 N 40.53 E H = 17 32 29.1 h = normal MB = 4.5 D = 23.47 Az = 309.4 (NEIC) XV A 1.3s 8.7nm
22.	eP1	A	17 39 27	Turkey 38.77 N 40.59 E
	eP2	A	39 31	H = 17 34 15.1 h = 13 km MB = 4.8 D = 23.61 Az = 309.6 (NEIC)
22.	eP	A	18 40 38	Turkey 39.00 N 40.65 E H = 18 35 30.6 h = normal MB = 4.7 D = 23.50 Az = 309.2 (NEIC) PV A 1.7s 24.2nm M = 4.4
22.	eX	A	18 48 52.5	Turkey 38.99 N 40.54 E H = 18 43 38.8 h = normal MB = 4.4 D = 23.44 Az = 309.3 (NEIC) XV A 1.7s 18.2nm
22.	eP1	A	20 13 37	Tibet 32.39 N 92.12 E
	eP2	AB	13 39	H = 20 03 32.4 h = normal MB = 5.6
	ePcP	B	14 25	D = 60.00 Az = 313.5 (NEIC)
	ePP	B	15(46)	P1V A 1.5s 22.6nm M = 5.0
	e	B	17 30	P2V A 1.7 103.0nm 5.6
	eS	B	21 52	LmH B 18 22.4nm 6.4
	eSS	B	25 51	LmV B 14 13.5nm 6.3
	LmH	B	37.5	
	LmV	B	44.5	
23.	eP	A	01 07 13	Turkey 37.59 N 30.10 E
	eS	B	10 46	H = 01 02 54.7 h = normal MB = 4.4
	LmH	B	13.6	D = 18.54 Az = 320.6 (NEIC)
	LmV	B	15.4	PV A 1.6s 16.5nm M = 4.0 LmH B 14 0.8 µm 4.2 LmV B 12 0.5 µm 4.2

Day	Phase		h m s	Remarks
23.	ePKHKP	A	03 43 24	South of Fiji Islands 23.45 S 179.12 E
				H = 03 24 34.2 h = 580 km MB = 4.6
				D = 151.16 Az = 343.4 (NEIC)
				PKHKPV A 1.0s 11.8nm
23.	eP1	A	05 23 28	Turkey 37.61 N 30.10 E
	eP2	A	23 34.5	H = 05 19 08.1 h = 11 km MB = 4.5
	eS	B	27 00	D = 18.53 Az = 320.6 (NEIC)
	LmH	B	29.9	P1V A 2.0s 21.4nm M = 3.9
	LmV	B	31.7	P2V A 2.0 42.7nm 4.3
				LmH B 12 1.0 µm 4.3
				LmV B 10 0.8 µm 4.5
23.	eP	A	09 11 55	Peru 14.06 S 72.90 W
				H = 08 58 31.29 h = 94 km MB = 5.5 (NEIC)
				D = 97.3
				PV A 1.8s 13.5nm M = 5.2
23.	LmH	B	11 03.4	Java 6.12 S 107.07 E
	LmV	B	04.0	H = 09 56 34.2 h = normal MB = 4.6 (NEIC)
				D = 98.2
				LmH B 16s 0.3 µm M = 4.7
24.	e(P)	A	02 25 27	Turkey 38.76 N 40.50 E
	e	A	25 39	H = 02 20 13.9 h = normal MB = 4.7
	e	A	25 55	D = 23.56 Az = 309.7 (NEIC)
	eS	B	29 38	LmH B 14s 0.5 µm M = 4.1
	LmH	B	35.1	LmV B 13 0.3 µm 4.0
	LmV	B	37.6	
24.	eP	A	11 22 06	Turkey 37.52 N 30.03 E
				H = 11 17 49.6 h = normal MB = 4.4
				D = 18.57 Az = 320.8 (NEIC)
				PV A 1.8s 16.9nm M = 3.9
24.	LmV	C	18 48.4	LmV C 18s 0.3 µm

Day	Phase		h m s	Remarks
25.	iP	A	04 10 47.0	**Eastern Kazakh SSR** 49.80 N 78.21 E
				H = 04 02 57.7 h = 0 km MB = 5.2
				D = 41.30 Az = 297.7 (NEIC)
				PV A 0.6s 23.0nm M = 5.1
25.	eP	A	04 40 15.5	**Southern Iran** 27.70 N 55.39 E
	e	A	40 18	H = 04 32 36.9 h = 23 km MB = 5.1
	e	A	43 25	D = 40.16 Az = 316.9 (NEIC)
25.	eiP1	AB	05 47 27.5	**Turkey** 39.03 N 29.74 E
	-iP2	AB	47 30.3	H = 05 43 27.0 h = 24 km MB = 5.8
	eS	B	50 38	D = 17.27 Az = 318.2 (NEIC)
	LmH	B	53.7	P1V A 1.6s 60.5nm M = 4.5
	LmV	B	55.3	P2V A 2.7 2740.0nm 5.9
				P2H A 2.5 1531.0nm 5.8
				P2V B 6.0 7.4/um 6.0
				P2H B 6.0 6.9/um 6.1
				SH B 13 6.6/um 6.5
				LmH B 11.5 40.6/um 5.9
				LmV B 12 31.4/um 5.9
25.	eP	A	06 22 49	**Turkey** 39.01 N 29.66 E
				H = 06 18 45.7 h = normal MB = 4.4
				D = 17.25 Az = 318.3 (NEIC)
				PV A 1.7s 15.1nm M = 3.9
25.	eP	A	07 00 19	**Southern Iran** 27.30 N 53.38 E
				H = 06 52 48.9 h = 11 km MB = 4.8
				D = 39.25 Az = 317.9 (NEIC)
25.	eP	A	11 52 02	**Andreanof Islands, Aleutian Is.**
				51.64 N 174.28 W
				H = 11 40 04.7 h = 24 km MB = 4.9
				D = 77.96 Az = 356.2 (NEIC)
				PV A 0.8s 13.5nm M = 5.0
25.	eP1	AB	13 12 38	**Kyushu, Japan** 31.31 N 131.49 E
	-iP2	AB	12 48	H = 13 00 19.9 h = 52 km MB = 5.6

Day	Phase		h m s	Remarks
cont. 25.	ePP	B	13 15 47	D = 82.69 Az = 326.2 (NEIC)
	eS	C	23 07	P1V A 2.5s 184.0nm M = 5.6
	eSS	C	28 35	P1V B 5 1.0/um 6.1
	LmV	B	53.4	P2V A 2.0 214.0nm 5.8
	LmH	B	53.5	LmH B 18 13.7/um
				LmV B 18.5 11.0/um
26.	-iPKIKP	AB	00 30 57	South of Fiji Islands 24.91 S 179.15 W
	-iPKHKP	AB	31 05	H = 00 11 54.5 h = 413 km MB = 5.4
	iPKP2	AB	31 18	D = 152.99 Az = 344.8 (NEIC)
	epPKP	A	32 57	PKIKPV A 1.4s 53.5nm
	esPKP	B	33 35	PKHKPV A 1.2 73.3nm
	ePP	AB	34 50	PKP2V A 1.6 143.0nm
	epPP	A	36 13	pPKPV A 2.0 85.3nm
	esPP	B	37 04	PPV A 1.7 39.3nm M = 5.4
	eSS	B	53 38	
	esSS	B	56 30	
26.	eP	A	01 58 07.5	Kurile Islands 43.51 N 146.66 E
				H = 01 46 11.2 h = 59 km MB = 4.7
				D = 78.40 Az = 332.7 (NEIC)
				PV A 0.8s 17.3nm M = 5.0
26.	+iP1	AB	02 48 52	Iran 35.51 N 58.22 E
	iP2	A	48 57	H = 02 41 46.0 h = 26 km MB = 5.4
	iP3	A	49 00	D = 36.60 Az = 309.1 (NEIC)
	ePP	B	50 12	P1V A 1.5s 105.5nm M = 5.5
	eS	B	54 38	P2V A 1.1 72.6nm 5.5
	eSS	B	57 04	P3V A 1.3 137.5nm 5.7
	LmH	B	03 08.1	SH B 12 1.2/um 5.7
	LmV	B	11.7	LmH B 13.5 4.0/um 5.4
				LmV B 11 2.9/um 5.4
26.	+eP	AB	06 25 33	Kyushu, Japan 31.34 N 131.51 E
	e	A	25 43	H = 06 13 14.7 h = 50 km MB = 5.5 MS=5.7
	ePP	B	28 42	D = 82.66 Az = 326.2 (NEIC)
	ePPPP	C	32 12	PV A 1.9s 76.0nm M = 5.3
	eS	B	35 49	PV B 7 103/um 5.9

Day	Phase		h m s	Remarks
cont. 26.	ePS	C	06 36 32	LmH B 19s 18.6 µm M = 6.5
	eSS	C	41 34	LmV B 18 17.8 µm 6.5
	LmH	B	07 06.2	
	LmV	B	06.5	
26.	ePKHKP	A	07 38 40	<u>South of Fiji Islands</u> 25.34 S 177.13 W
	ePKP2	A	38 49	H = 07 18 47.1 h = 115 km MB = 4.8
				D = 153.85 Az = 347.3 (NEIC)
26.	ePKHKP	A	09 11 35	<u>South of Fiji Islands</u> 25.47 S 176.83 W
				H = 08 51 48.1 h = 113 km MB = 4.4
				D = 154.03 Az = 347.7 (NEIC)
				traces
26.	eP	A	16 51 20	<u>Eastern Caucasus</u> 43.6 N 45.5 E
	LmH	B	17 01.9	H = 16 46 01 MB = 5.1 MS = 4.3 (ANUSSR)
	LmV	B	03.6	D = 23.95
				PV A 2.0s 25.6nm M = 4.4
				LmH B 12.5 0.6 µm 4.3
				LmV B 10 0.4 µm 4.4
27.	e	A	00 24 31	
27.	eP	A	00 38 15	<u>Tadzhik SSR</u> 38.26 N 69.00 E
	LmH	C	56.5	H = 00 30 27.7 h = 36 km MB = 4.8
	LmV	C	59.3	D = 41.78 Az = 306.4 (NEIC)
				PV A 1.7s 15.1nm M = 4.5
				LmH C 16 0.2 µm 4.1
				LmV C 16 0.4 µm 4.5
27.				In the intervall $04^h 45^m - 07^h 14^m$ the short-period seismographs (type A) had been out of operation.
27.	iPn	A	15 05 30	<u>Lohne near Fritzlar, Fed.Rep. of Germany</u>
	e(Pg)	A	05 35.5	explosion
				51° 11.33' N 09° 14.73' E

Day	Phase		h m s	Remarks
cont. 27.	iSg	A	15 05 51.5	H = 15 05 00.24 yield 6.6 t (Hannover) D = 1.6
27.	ePKIKP	A	17 01 11.5	Banda Sea 7.39 S 128.40 E
	e	A	02 06.5	H = 16 42 52.6 h = 162 km MB = 5.4
	e	A	02 28	D = 112.52 Az = 322.0 (NEIC)
				PKIKPV A 1.4s 13.9nm
27.	ePKHKP	A	21 09 56.5	West of Macquarie Islands
	ePKP2	A	10 12.5	53.81 S 150.54 E
				H = 20 49 59.3 h = normal MB = 5.6
				D = 154.89 Az = 279.6 (NEIC)
				PKHKPV A 1.3s 19.6nm
				PKP2V A 1.5 32.7nm
28.	e	A	10 33 19.5	Komandorsky Islands Region
				55.76 N 166.61 E
				H = 10 21 43.4 h = normal MB = 4.5
				D = 71.95 Az = 343.6 (NEIC)
				PV A 1.4s 14.0nm M = 4.8
28.	eP	A	12 22 30	Queen Charlotte Islands Region
	LmH	C	52.0	52.82 N 132.89 W
	LmV	C	52.0	H = 12 11 02.9 h = normal MB = 4.7
				D = 72.65 Az = 22.8 (NEIC)
28.	eP	A	13 04 04	Mariana Islands 19.26 N 145.30 E
	ePP	A	08 11	H = 12 50 41.1 h = 165 km MB = 5.3 (NEIC)
				D = 99.3
				PPV A 1.6s 38.5nm M = 5.5
28.	eP	A	14 24 38	Guatemala 14.83 N 91.41 W
	epP	AB	25 09	H = 14 12 07.9 h = 124 km MB = 5.4 (NEIC)
	eS	C	35 00	D = 86.7 h = 120 km
	e	C	36 25	PV A 1.6s 55.0nm M = 5.4
	LmH	C	15 02	pPV A 1.4 46.5nm

Day	Phase		h m s	Remarks
28.	ePKHKP	A	14 25 02	Fiji Islands Region 21.18 S 178.95 W
				H = 14 06 19.9 h = 620 km MB = 4.6 (NEIC)
				D = 149.5
28.	iPg	A	15 05 30	Dorheim/Hessen, Federal Rep. of Germany,
	iSg	A	05 51.5	explosion
				50°57.88' N 09°13.16' E
				H = 15 05 00.84 yield 6.7 t (Hannover)
				D = 1.5
28.	LmV	C	21 14.7	El Salvador 13.7 N 89.8 W
				H = 20 19 25.7 h = 84 km MB = 4.4 (NEIC)
				D = 86.7
				LmV C 16s 0.3 μm
28.	e	A	22 49 41.5	Poland 50.32 N 18.87 E
				H = 22 47 11.8 M = 3.1 (WAR)
				D = 4.15
29.	+iP	A	00 27 31.5	Kurile Islands 44.62 N 149.12 E
	LmH	C	01 00.0	H = 00 15 42.4 h = 112 km MB = 5.3
				D = 78.20 Az = 333.9 (NEIC)
				PV A 0.8s 19.2nm M = 5.0
29.	+eP	AB	09 05 02.5	Kyushu, Japan 31.33 N 131.38 E
	eiX	A	05 10.5	H = 08 52 43.6 h = 48 km MB = 5.5
	ePP	AB	08 15	D = 82.61 Az = 326.1 (NEIC)
	eS	C	15 28	PV A 2.2s 130.8nm M = 5.5
	ePS	C	16 10	XV A 2.0 162.2nm
	eSS	C	20 55	PPV A 2.0 51.2nm 5.6
	LmV	V	45.8	LmH B 17 14.4nm 6.3
	LmH	B	45.9	LmV B 19 12.4nm 6.3
30.	ePKHKP	A	02 31 12	Tonga Islands 21.18 S 175.62 W
	ePKP2	A	31 19.5	H = 02 11 33.7 h = 115 km MB = 4.4
				D = 150.05 Az = 350.8 (NEIC)

Day	Phase		h m s	Remarks
30.	ePKP	A	03 37 00	New Hebrides Islands 20.54 S 169.90 E H = 03 17 25.2 h = 42 km MB = 4.7 D = 145.42 Az = 335.5 (NEIC) PKPV A 1.0s 17.7nm
30.	+iPKP	A	07 27 42.0	New Hebrides Islands 19.10 S 169.39 E
	e	A	27 52.5	H = 07 08 38.1 h = 255 km MB = 5.3
	epPKP	A	28 43	D = 143.92 Az = 335.9 (NEIC) h = 243 km PKPV A 1.1s 56.5nm
30.	eP	A	09 32 08	North of Franz Josef Land 87.03 N 52.53 E H = 09 24 56.7 h = normal MB = 4.6 D = 37.32 Az = 223.4 (NEIC) PV A 1.3s 8.7nm M = 4.4
30.	eP	A	12 06 55.5	Burma 25.32 N 96.38 E
	LmH	B	35.2	H = 11 55 59.9 h = normal MB = 4.9
	LmV	B	39.6	D = 67.63 Az = 316.7 (NEIC) PV A 1.8s 23.5nm M = 5.1 LmH B 19 0.7 µm 4.9 LmV B 14 0.6 µm 5.0
30.	eP1	AB	15 55 14	Burma 25.23 N 96.44 E
	eP2	AC	55 17	H = 15 44 15.7 h = 15 km MB = 5.8
	eiS	B	16 04 08	D = 67.74 Az = 316.8 (NEIC)
	eSS	B	08 22	P1V A 1.6s 33.0nm M = 5.2
	eSSS	B	11 48	P2V A 1.4 139.4nm 5.9
	LmH	B	23.6	SH B 12 2.1 µm 6.2
	LmV	B	28.1	LmH B 20 17.7 µm 6.3 LmV B 14 7.9 µm 6.1
30.	eP	A	21 49 56	Burma 25.35 N 96.39 E
	LmH	B	22 19.0	H = 21 39 00.5 h = normal MB = 4.9
	LmV	B	20.4	D = 67.62 Az = 316.7 (NEIC) PV A 2.0s 30.0nm M = 5.0

Day	Phase		h m s	Remarks
31.	ePKHKP	A	02 23 18	**Tonga Islands Region** 22.07 S 175.13 W
	e	A	23 23.5	H = 02 03 26.9 h = normal MB = 4.6
				D = 151.00 Az = 351.1 (NEIC)
				traces
31.	eP	A	03 44 54	**Norwegian Sea** 72.12 N 0.45 E
				H = 03 40 00.4 h = normal MB = 4.4
				D = 22.12 Az = 160.9 (NEIC)
				PV A 1.6s 33.0nm M = 4.5
31.	+iP1	AB	03 51 45.5	**Norwegian Sea** 72.21 N 1.17 E
	+iP2	AB	51 48.3	H = 03 46 52.2 h = normal MB = 5.5
	eS	B	55 46	D = 22.14 Az = 162.2 (NEIC)
	LmH	B	04 00.0	P1V A 1.2s 55.0nm M = 4.9
	LmV	B	01.7	P2V A 1.5 407.0nm 5.6
				P2V B 11 4.4/um 5.6
				SH B 12 14.4/um 6.3
				LmH B 20.5 18.2/um 5.5
				LmV B 15 11.9/um 5.6
31.	eP	A	04 19 04.5	**Norwegian Sea** 72.04 N 0.40 E
	e	A	19 20	H = 04 14 08.8 h = normal MB = 4.4
				D = 22.06 Az = 160.7 (NEIC)
				PV A 1.5s 22.6nm M = 4.4
31.	eP	A	04 28 49	**Norwegian Sea** 72.06 N 0.36 E
				H = 04 23 55.8 h = normal MB = 4.5
				D = 22.07 Az = 160.7 (NEIC)
				PV A 1.5s 15.1nm M = 4.2
31.	+eiP	AB	05 24 57	**Burma** 25.19 N 96.51 E
	ePP	A	27 25	H = 05 13 59.7 h = normal MB = 5.3
	eS	B	33 48	D = 67.81 Az = 316.8 (NEIC)
	eSS	B	38 08	PV A 2.0s 77.0nm M = 5.4
	eSSS	B	41 30	LmH B 18 9.0/um 6.0
	LmH	B	53.2	LmV B 13 3.7/um 5.9
	LmV	B	57.6	

Day	Phase		h m s	Remarks
31.	eP	AB	08 26 15	<u>North Atlantic Ridge</u> 23.84 N 45.00 W
	eS	B	33 35	H = 08 17 15.1 h = normal MB = 5.0
	LmH	B	45.0	D = 50.90 Az = 43.2 (NEIC)
	LmV	B	45.0	PV A 1.4s 11.6nm M = 4.7
				LmH B 17 0.8 μm 4.8
				LmV B 18 0.6 μm 4.8
31.	eP	A	09 23 43.5	<u>Northern Sumatra</u> 2.62 N 96.56 E
	e	A	23 51	H = 09 11 09.6 h = 21 km MB = 5.0
	LmH	C	10 04.5	D = 84.78 Az = 320.4 (NEIC)

Day	Phase		h m s	Remarks
1.	eP	A	00 38 56.5	<u>Luzon, Philippine Islands</u>
	LmH	C	01 21.0	18.02 N 120.30 E
	LmV	C	21.3	H = 00 26 13.6 h = 59 km MB = 5.0
				D = 87.53 Az = 322.9 (NEIC)
				LmV C 18s 0.4 μm
1.	ePKP	A	07 33 56.5	<u>Fiji Region</u> 19.6 S 178.3 W
				H = 07 15 10.2 h = 519 km MB = 4.1
				D = 148.01 Az = 348 (ISC)
1.	e	A	13 41 22	
				In the intervall 1d 16h 48m - 2d 05h 47m the short-period seismographs (type A) have been out of operation.
1.	LmV	C	20 20.4	<u>Philippine Islands Region</u>
	LmH	B	20.6	20.2 N 120.8 E
				H = 19 23 44.8 h = normal (NEIC)
				D = 86.1
				LmV C 16s 0.3 μm M = 4.8
				LmH B 15 0.3 μm 4.8
2.	LmH	C	04 09.0	<u>Western Iran</u> 34.7 N 47.0 E
	LmV	C	09.0	H = 03 47 53.6 h = 25 km MB = 4.3 (NEIC)
				D = 30.2
2.	eP	A	10 12 16	<u>Southern Iran</u> 29.39 N 51.58 E
				H = 10 05 09.3 h = 35 km MB = 4.8
				D = 36.64 Az = 316.7 (NEIC)
2.	eP	A	19 17 29.5	<u>Southern Alaska</u> 61.03 N 151.26 W
	e	A	17 42.5	H = 19 06 32.9 h = 29 km MB = 5.0
				D = 67.83 Az = 11.7 (NEIC)
				PV A 1.0s 11.8nm M = 5.0
2.	LmH	C	20 26.5	<u>Venezuela</u> 9.7 N 69.9 W
	LmV	C	30.0	H = 19 41 11.7 h = 39 km MB = 4.8 (NEIC)
				D = 77.2
				LmH C 22s 0.2 μm M = 4.5

Day	Phase		h m s	Remarks
3.	LmH	C	03 07.0	
	LmV	C	07.3	LmV C 20s 0.3/um
3.	e	A	10 43 11	PV A 1.3s 26.2nm
3.	LmH	B	14 21.7	East China Sea 29.9 N 127.9 E
	LmV	B	21.7	H = 13 28 22 MB = 4.8 (ANUSSR)
				D = 82.1
				LmH B 16s 0.4/um M = 4.9
				LmV B 15 0.3/um 4.9
4.	eP	A	08 04 34	North of Severnaya Zemlya 84.49 N 107.48 E
	e	A	04 40	H = 07 56 54.8 h = normal MB = 5.0
	e	A	04 47	D = 40.43 Az = 282.4 (NEIC)
				PV A 1.3s 13.1nm M = 4.5
4.	eP	A	09 17 40	North of Severnaya Zemlya 84.57 N 107.99 E
	i	A	17 47.5	H = 09 10 02.7 h = normal MB = 5.1
	e	A	17 52	D = 40.46 Az = 282.8 (NEIC)
	e	A	18 05	PV A 1.4s 27.9nm M = 4.8
	e	A	18 21	LmH B 18 0.5/um 4.4
	eS	B	24 50	LmV B 16 0.4/um 4.4
	LmH	B	40.0	
	LmV	B	40.6	
4.	-iP	AB	15 26 25	Ryukyu Islands 28.80 N 128.62 E
	eS	B	36 28	H = 15 14 10.6 h = 134 km MB = 5.2
	e(SP)	B	37 28	D = 83.34 Az = 325.2 (NEIC)
	LmH	B	16 12.5	PV A 1.8s 81.0nm M = 5.3
	LmV	B	12.5	LmH B 14 0.2/um
				LmV B 14 0.2/um
4.	eP	A	21 00 05	Tibet 32.23 N 92.13 E
				H = 20 49 58.3 h = normal MB = 5.0
				D = 60.12 Az = 313.6 (NEIC)
				traces

Day	Phase		h m s	Remarks
4.	ePn	A	21 53 09	Federal Rep. Germany 48.3 N 9.1 E
	ePg	A	53 16.5	H = 21 52 23 (BCIS)
	iSg	A	53 55	D = 2.83
4.	eP	A	23 34 47.5	Fox Islands, Aleutian Is.
	eS	C	44 50	51.62 N 171.26 W
	eSS	C	49(50)	H = 23 22 50.3 h = normal MB = 4.9
	LmH	C	24 14.4	D = 78.08 Az = 358.1 (NEIC)
	LmV	C	16.4	PV A 1.5s 20.1nm M = 4.9
				LmH C 16 0.3/um 4.7
				LmV C 16 0.3/um 4.8
5.	-eiP	A	01 50 02	Nicobar Islands Region 9.40 N 92.46 E
	ipP	A	50 14.3	H = 01 38 10.9 h = normal MB = 5.3 (NEIC)
	eS	C	02 00 00	D = 77.0 h = 46 km
	eSS	C	05 05	PV A 1.5s 37.7nm M = 5.2
	eSSS	C	08.4	pPV A 1.3 35.0nm
	LmH	B	29.3	LmV C 19 0.6/um 5.0
	LmV	C	29.1	LmH B 19 0.4/um 4.7
5.	e	A	08 23 38.5	
5.	eP	A	09 34 36.5	Mindanao, Philippine Islands
	eSKS	B	45 05	6.05 N 126.16 E
	eS	B	46 10	H = 09 20 54.0 h = 67 km MB = 5.5
	ePS	B	47 45	D = 100.50 Az = 323.9 (NEIC)
	LmV	B	10 24.3	PV A 1.5s 25.1nm M = 5.6
	LmH	B	24.4	
5.	LmH	B	11 04.1	Northeastern China 37.3 N 113.7 E
	LmV	B	04.3	H = 10 21 28.0 h = normal MB = 4.7 (NEIC)
				D = 68.8
				LmH B 14s 0.6/um M = 5.0
				LmV B 19 0.5/um 4.8
5.	e	A	13 33 18	
	e	A	34 36.5	

Day	Phase		h m s	Remarks
5.	eP	AB	14 33 25	**Costa Rica** 9.34 N 84.23 W
	e	AB	33 34	H = 14 20 42.3 h = 26 km MB = 5.4
	eS	B	43 51	D = 86.54 Az = 39.4 (NEIC)
	e	B	44 12	PV A 1.8s 67.6nm M = 5.6
	LmH	B	15 08.0	XV A 1.8 166.0nm
	LmV	B	13.4	XV B 3.2 0.44 μm
				LmH B 20 1.0 μm 5.2
				LmV B 19 0.7 μm 5.1
5.	e(pP)	A	19 59 29	**Greece** 38.50 N 21.94 E
	LmH	B	20 05.4	H = 19 55 55.1 h = 43 km MB = 4.4
	LmV	B	05.5	D = 14.18 Az = 332.2 (NEIC)
				LmH B 11s 0.5 μm M = 3.9
				LmV B 11 0.3 μm
5.	ePKIKP	A	21 07 19	**Sea of Okhotsk** 56.1 S 150.4 E
	e	A	07 39.5	H = 20 47 26 M = 5.5
	LmH	B	22 17.7	D = 155.1 (ANUSSR)
	LmV	B	(35)	LmH B 20s 0.3 μm M = 5.0
6.	eP	A	01 43 31	**Sea of Japan** 40.82 N 133.26 E
				H = 01 32 32.3 h = 451 km MB = 4.5
				D = 75.60 Az = 326.0 (NEIC)
6.	LmH	B	02 57.0	**Probably Near Coast of Oaxaca, Mexico**
	LmV	B	59.3	15.1 N 96.3 W
				H = 02 10 26.9 h = normal MB = 4.7 (NEIC)
				D = 68.9
				LmH B 16s 0.2 μm M = 4.5
				LmV B 12 0.2 μm 4.5
6.	+iP	A	04 10 43.5	**Eastern Kazakh SSR** 49.98 N 77.77 E
	ePn	A	12 15	H = 04 02 57.1 h = 0 km MB = 5.5
				D = 40.97 Az = 297.4 (NEIC)
				PV A 0.8s 73.0nm M = 5.5
6.	LmH	B	16 37.0	
	LmV	B	37.0	

Day	Phase		h m s	Remarks
6.	eP1	A	16 59 12	North Atlantic Ridge 50.46 N 29.62 W
	eP2	A	59 18	H = 16 53 41.8 h = normal MB = 4.7
	LmH	B	17 08.8	D = 25.97 Az = 73.4 (NEIC)
	LmV	B	14.0	P2V A 1.2s 14.2nm M = 4.4
				LmH B 16 0.4 µm 4.2
6.	eP	A	18 32 47	Fox Islands, Aleutian Is.
				52.97 N 170.86 W
				H = 18 21 04.2 h = 97 km MB = 4.7
				D = 76.74 Az = 358.4 (NEIC)
6.	ePn	A	22 00 56.5	Northern Italy 44.93 N 6.96 E
	ePg	A	01 20.5	H = 21 59 14.0 h = 17 km MB = 4.1
	eSg	A	02 50	D = 6.52 Az = 27.1 (NEIC)
	LmH	B	03.1	LmH B 7.5s 1.5 µm M = 4.0
	LmV	B	03.5	LmV B 6.5 0.9 µm
6.	LmV	B	25 00.0	South Sandwich Islands Region
	LmH	B	00.2	55.4 S 27.9 W
				H = 23 50 01.3 h = 126 km MB = 5.5 (NEIC)
				D = 110.5
				LmH B 16s 0.6 µm
				LmV B 16 0.4 µm
7.	eP	A	13 38 31	Mediterranean Sea 34.33 N 22.67 E
	LmH	B	47.0	H = 13 34 18.8 h = 36 km MB = 4.5
	LmV	B	47.0	D = 18.20 Az = 337.0 (NEIC)
				LmH B 14s 0.6 µm M = 4.0
				LmV B 14 0.7 µm 4.2
7.	eP	A	23 08 57.5	Kurile Islands 46.20 N 153.10 E
				H = 22 57 00.2 h = normal MB = 4.7
				D = 77.95 Az = 336.1 (NEIC)
8.	ePn	A	02 22 46.5	Federal Rep. Germany 48.61 N 9.19 E
	eiPg	A	22 53.3	H = 02 22 06.1 h = 31 km
	i	AB	23 56	D = 2.58 Az = 36.9 (NEIC)
	iSg	AB	23 32.3	

Day	Phase		h m s	Remarks
8.	e(Pn)	A	09 25 54	
	e(Sg)	A	27 00	
8.	e	A	15 53 37	Poland 50.33 N 18.89 E
				H = 15 51 03.7 (WAR)
				D = 4.7
8.	eP	B	17 03 42	Turkey 37.52 N 29.77 E
	eP	A	03 45	H = 16 59 24.8 h = 10 km MB = 4.7
	eS	B	07 13	D = 18.43 Az = 321.1 (NEIC)
	LmH	B	10.0	PV A 1.5s 30.2nm M = 4.2
	LmV	B	11.9	LmH B 13 1.4 µm 4.5
				LmV B 12 1.0 µm 4.5
8.	eP1	AC	23 47 11	Turkey 37.56 N 29.72 E
	eP2	AB	47 14	H = 23 42 55.2 h = 28 km MB = 4.8
	eiS	B	50 40	D = 18.38 Az = 321.2 (NEIC)
	LmH	B	55.3	P1V A 1.8s 40.6nm M = 4.3
	LmV	B	55.3	P2V A 1.9 121.0nm 4.7
				P2V B 6.5 1.63 µm 5.3
				SH B 10 2.1 µm 5.3
				SV B 10 3.2 µm
				LmH B 10 6.1 µm 5.2
				LmV B 9 4.7 µm 5.3
9.	e	A	01 44 14	
9.	e	A	02 36 19	
	e	A	36 29.5	
9.	eP	A	03 01 49	Turkey 37.57 N 29.74 E
	eS	C	05 15	H = 02 57 25.5 h = 12 km MB = 4.4
	LmH	B	09.8	D = 18.38 Az = 321.1 (NEIC)
	LmV	B	09.8	PV A 1.3s 10.9nm M = 3.8
				LmH B 11 0.4 µm 4.0
				LmV B 10 0.4 µm 4.2

Day	Phase		h m s	Remarks
9.	eP	A	07 59 53	<u>Nicobar Islands Region</u> 6.76 N 93.97 E
				H = 07 47 45.2 h = normal
				D = 79.96 Az = 320.1 (NEIC)
9.	ePKP	A	13 26 31	<u>Easter Island Cordillera</u>
	eX	A	26 34	56.43 S 123.95 W
	e	B	26 55	H = 13 06 33.0 h = normal MB = 5.1
	e(PP)	B	30 35	D = 153.35 Az = 83.5 (NEIC)
	eSS	B	49 55	PKPV A 1.5s 12.6nm
	LmV	B	46.7	XV A 1.2 16.3nm
	LmH	B	47.0	LmH B 19 0.7/um M = 5.4
				LmV B 18 0.4/um 5.3
9.	ePKP	A	17 56 27	<u>Tonga Islands</u> 16.11 S 174.15 W
				H = 17 36 57.4 h = 78 km MB = 4.8
				D = 145.24 Az = 353.6 (NEIC)
				PKPV A 1.4s 55.8nm
9.	+iPKP	AB	20 45 57.5	<u>Tonga Islands Region</u> 17.53 S 172.66 W
	ei	AB	45 28.5	H = 20 26 16.4 h = normal MB = 5.1
				D = 146.78 Az = 355.0 (NEIC)
				PKPV A 2.0s 94.0nm
9.	+iP	A	21 21 12	<u>Alaska Peninsula</u> 56.68 N 156.53 W
	e	A	21 32.5	H = 21 09 48.3 h = 56 km MB = 5.2
	LmH	B	55.0	D = 72.60 Az = 7.9 (NEIC)
	LmV	B	59.0	PV A 1.0s 29.5nm M = 5.2
				LmH B 16 0.2/um
				LmV B 16 0.2/um
9.	e	A	22 17 49	
	e	A	17 57	
10.	e	A	02 41 33	
10.	eP	A	07 00 58.5	<u>Peru</u> 10.90 S 76.23 W
	esP	A	01 30	H = 06 47 34.6 h = 99 km MB = 5.7
	ePP	B	04 51	D = 96.97 Az = 39.9 (NEIC) h = 85 km

Day	Phase		h m s	Remarks
cont.				
10.	esPP	B	07 05 30	PV A 1.6s 24.7nm M = 5.5
	eSKS	B	11 24	sPV A 2.5 154.0nm
	eS	BC	12 10	LmH B 19 0.9 µm
	e(SP)	B	13 25	LmV B 18 1.2 µm
	e	B	14 24	e A 07 02 21
	LmH	B	44.3	
	LmV	B	44.3	
10.	ePn	A	08 00 36	<u>Czechoslovakia</u> $49^{\circ}35.7'$N $13^{\circ}40.2'$E
	iPg	A	00 36.5	explosion, yield 20 to (PRU)
	eSg	A	00 59	D = 1.69
10.	ePKHKP	A	08 17 13.5	<u>South of Fiji Islands</u> 22.23 S 178.75 W
	ePKP2	A	17 22.5	H = 07 58 09.5 h = 418 km MB = 4.7
				D = 150.50 Az = 346.5 (NEIC)
				PKHKPV A 1.0s 33.4nm
10.	eP1	A	09 35 53	<u>Turkey</u> 39.12 N 29.60 E
	eiP2	A	35 55.5	H = 09 31 54.5 h = normal MB = 4.9
	LmH	B	43.4	D = 17.14 Az = 318.1 (NEIC)
	LmV	B	43.9	P2V A 1.4s 55.7nm M = 4.5
				LmH B 11 0.9 µm 4.2
				LmV B 10 1.0 µm 4.5
10.	eP	A	09 50 29	<u>Afghanistan - USSR Border Region</u>
				36.52 N 71.67 E
				H = 09 42 51.4 h = 383 km MB = 4.5
				D = 44.53 Az = 308.1 (NEIC)
				PV A 0.9s 7.8nm M = 4.0
10.	eP	A	10 33 14	<u>Near East Coast of Honshu, Japan</u>
				35.44 N 140.54 E
				H = 10 20 46.5 h = 30 km MB = 4.7
				D = 83.20 Az = 330.1 (NEIC)
				In the intervall 10d 12h 52m - 11d 04h 44m the short-period seismographs (type A) have been out of operation.

Day	Phase		h m s	Remarks
10.	LmH	B	13 09.5	
10.	eP	B	18 59 28	North Atlantic Ocean 59.54 N 30.26 W
	LmH	B	19 09.9	H = 18 54 04.6 h = normal MB = 4.4
	LmV	B	09.9	D = 25.17 Az = 92.0 (NEIC)
				PV B 6s 0.27/um M = 5.0
				LmH B 17 1.0/um 4.4
				LmV B 17 0.9/um 4.5
10.	eP	B	19 08 06	North Atlantic Ridge 59.43 N 30.40 W
	LmH	B	19.0	H = 19 02 41.6 h = normal MB = 4.5
	LmV	B	19.0	D = 25.23 Az = 91.7 (NEIC)
				PV B 5.5s 0.54/um M = 5.3
				LmH B 14 2.2/um 4.8
				LmV B 14 2.2/um 5.0
10.	-iP	B	20 11 24	Eastern Sea of Japan 41.13 N 138.42 E
	e(PcP)	B	11 32	H = 19 59 52.5 h = 226 km MB = 5.7
	e	B	12 17	D = 77.46 Az = 328.5 (NEIC)
	-ipP	B	12 24	PV B 4s 1.54/um M = 6.1
	e(pPcP)	B	12 36	PPV B 8 0.66/um 5.9
	ei(sP)	B	12 40	LmH B 14 0.9/um
	ePP	B	14 24	LmV B 13 1.0/um
	eiS	B	20 53	
	eSKS	B	21 15	
	esS	B	22 35	
	LmH	B	49.0	
	LmV	B	50.3	
10.	eP	B	21 38 03	North Atlantic Ocean 59.28 N 30.39 W
	eS	B	42 34	H = 21 32 38.4 h = normal MB = 4.9
	LmH	B	48.8	D = 25.23 Az = 91.4 (NEIC)
	LmV	B	48.9	PV B 6s 0.38/um M = 4.9
				LmH B 16 1.6/um 4.6
				LmV B 15 1.6/um 4.8
10.	e	B	23 50 50	North Atlantic Ocean 60.0 N 30.5 W
	LmH	B	24 01.1	H = 23 44 38.6 h = normal MB = 4.3 (NEIC)
	LmV	B	01.1	D = 25.3
				LmH B 17.5s 1.7/um M = 4.6
				LmV B 17 1.7/um 4.8

Day	Phase		h m s	Remarks
11.	eP	B	01 46 17	Peru-Ecuador Border Region
	eSKS	B	56 55	4.17 S 80.66 W
	eS	C	57 30	H = 01 32 58.98 h = 43 km MB = 5.4 (NEIC)
	LmH	B	02 25.0	D = 94.7
	LmV	B	26.0	LmH B 20s 0.4 µm M = 4.8
				LmV B 20 0.3 4.8
11.	LmH	B	05 13.7	Off East Coast of Kamchatka
	LmV	B	13.8	51.4 N 159.3 E
				H = 04 24 02.7 h = normal MB = 4.8 (NEIC)
				D = 74.7
11.	eP	A	06 46 54.5	North Atlantic Ocean 59.19 N 30.26 W
	e	A	47 09	H = 06 41 31.1 h = normal MB = 4.5
	LmH	B	57.9	D = 25.15 Az = 91.3 (NEIC)
	LmV	B	58.2	PV A 1.4s 14.0nm M = 4.3
				LmH B 13.5 0.4 µm 4.0
				LmV B 16 0.4 µm 4.1
11.	iPg	A	10 15 34.0	Federal Rep. Germany
	i	A	15 42.5	50°07.8'N 12°14.2'E
	i(Sg)	A	15 48	explosion yield 25 t (PRU)
				D = 0.68
11.	eP	A	10 34 05.5	North Atlantic Ocean 59.40 N 30.38 W
	e	A	34 18	H = 10 28 40.1 h = normal MB = 4.4
	LmH	B	45.0	D = 25.22 Az = 91.6 (NEIC)
	LmV	B	45.0	LmH B 14s 0.4 µm M = 4.0
				LmV B 14 0.3 µm 4.1
11.	eP	A	11 00 27.5	Kurile Islands 49.95 N 156.47 E
	LmV	B	38.0	H = 10 48 47.9 h = 55 km MB = 5.2
	LmH	B	38.4	D = 75.39 Az = 337.7 (NEIC)
				PV A 1.5s 25.1nm M = 4.9
				LmH B 16 0.2 µm
11.	eiP1	AB	13 07 17.5	Dominican Republic Region
	eiP2	A	07 23	17.97 N 69.78 W

Day	Phase		h m s	Remarks
cont. 11.	iP3	B	13 07 40	H = 12 56 04.3 h = 57 km MB = 6.1
	ePP	B	10 02	D = 70.91 Az = 41.8 (NEIC)
	e	B	10 22	P1V A 1.8s 115.0nm M = 5.5
	iS	B	16 36	P1V B 7 1.85 /um 5.9
	eiPS	B	17 10	P2V A 1.6 448.0nm 6.1
	eSS	B	20 58	P3V B 8 3.92 /um 6.7
	LmH	B	32.6	SH B 14 6.35 /um 6.4
	LmV	B	32.8	P'P'V A 3.0 184.0nm
	eP'P'	A	35 09	LmH B 22.5 42.7 /um
				LmV B 22 34.2 /um
11.	eiP1	AB	14 10 31	Rat Islands, Aleutian Is.
	eiP2	A	10 47.5	51.49 N 176.08 E
	ePP	B	13 36	H = 13 58 37.7 h = 32 km MB = 5.9 MS = 6.5
	eS	B	20 24	D = 77.39 Az = 349.9 (NEIC)
	eiPS	BC	20 52	P1V A 0.9s 85.6nm M = 5.8
	eSS	C	25 10	P1V B 11 5.3 /um 6.5
	LmH	B	46.5	P2V A 1.7 291.0nm 6.0
	LmV	B	54.5	SH B 14 8.28 /um 6.7
				LmH B 19.5 35.0 /um 6.7
				LmV B 15.5 23.2 /um 6.6
11.	ePn	A	14 45 29	explosion Adelebsen, Federal Rep. Germany
	eSg	A	45 50	51°36.5' N 9°44.8' E
				H = 14 45 00 yield 6.2 t (Hannover)
				D = 1.45
11.	eP	A	15 33 23.5	Andreanof Islands, Aleutian Is.
				51.26 N 179.26 W
				H = 15 21 27.9 h = 51 km MB = 5.1
				D = 78.05 Az = 352.9 (NEIC)
11.	LmH	B	22 23.4	LmH B 16s 0.2 /um
11.	eP	A	22 54 59	Taiwan Region 23.97 N 122.48 E
	e	A	55 22	H = 22 42 31.35 MB = 5.1 (NEIC)
	LmH	B	23 36.8	D = 84.1
	LmV	B	36.9	LmH B 15s 0.3 /um M = 4.9
				LmV B 14 0.4 /um 5.0

Day	Phase		h m s	Remarks
12.	LmH	B	02 00.3	Ryukyu Islands 28.5 N 128.3 E
	LmV	B	07.6	H = 01 13 21.2 h = 69 km MB = 4.7 (NEIC)
				D = 83.4
				LmH B 17s 1.2/um
				LmV B 14 0.4/um
12.	eP	A	03 06 14	Near East Coast of Honshu, Japan 35.34 N 140.48 E
				H = 02 53 47.5 h = 35 km MB = 4.8
				D = 83.26 Az = 330.1 (NEIC)
12.	ePKP	A	22 59 42.5	Fiji Islands Region 18.76 S 176.51 W
				H = 22 40 35.3 h = 329 km MB = 5.2
				D = 147.53 Az = 350.3 (NEIC)
				PKPV A 1.1s 34.3nm
12.	ePKIKP	A	23 55 46.5	Fiji Islands Region 20.79 S 178.90 W
	-iPKHKP	A	55 52	H = 23 37 10.1 h = 605 km MB = 4.9
	ePKP2	A	55 58.5	D = 149.07 Az = 346.9 (NEIC)
	e	A	58 19	PKIKPV A 0.9s 11.7nm
				PKHKPV A 1.3 61.1nm
				PKP2V A 1.2 20.3nm
13.	+iP	AB	04 18 10.5	Near East Coast of Honshu, Japan
	epP	A	18 28	36.23 N 140.89 E
	esP	A	18 37.5	H = 04 05 51.7 h = 55 km MB = 5.5
	LmH	B	05 00.0	D = 82.65 Az = 330.2 (NEIC) h = 70 km
	LmV	B	00.0	PV A 1.0s 46.5nm M = 5.4
				pPV A 1.5 47.7nm
				sPV A 1.8 54.0nm
				LmH B 16 0.4/um
				LmV B 14 0.5/um
13.	ePKP	A	18 56 19	Tonga Islands 15.61 S 173.11 W
				H = 18 36 43.2 h = 28 km MB = 5.0 (NEIC)
				D = 144.85 Az = 354.8
14.	eP	A	11 37 37	Hokkaido, Japan Region 41.83 N 141.32 E
				H = 11 25 49.6 h = 115 km MB = 5.2
				D = 77.98 Az = 329.9 (NEIC)

Day	Phase		h m s	Remarks
14.				In the intervall 13h 30m - 14h 01m the short-period seismographs (type A) have been out of operation.
14.	eP	B	13 58 58	<u>Eastern Russia</u> 56.17 N 123.58 E
	ePP	B	14 01(16)	H = 13 48 55.7 h = normal MB = 5.6
	eS	B	07 08	MS = 5.7 (NEIC)
	eSKS	B	08 48	D = 59.6
	eSS	C	11 05	LmH B 12.5s 5.2/um M = 6.0
	eSSS	B	13 21	LmV B 14 3.2/um 5.8
	LmH	B	27.1	
	LmV	B	29.8	
15.	ePKP	A	03 38 48	<u>New Hebrides Islands</u> 19.45 S 169.82 E
				H = 03 19 14.9 h = 25 km MB = 4.8
				D = 144.39 Az = 336.0 (NEIC)
				PKPV A 2.0s 42.7nm
15.	eP	A	05 40 17	<u>Philippine Islands Region</u>
	LmH	B	06 22.6	18.04 N 119.54 E
				H = 05 27 34.9 h = 52 km MB = 5.1
				D = 87.08 Az = 322.7 (NEIC)
				PV A 1.9s 30.3nm M = 5.2
				LmH B 16 0.4/um
15.	+iP1	AB	07 48 02	<u>Kirgiz-Sinkiang Border Region</u>
	eiP2	A	48 05.5	41.46 N 79.40 E
	ePP	AB	49 48	H = 07 39 37.1 h = normal MB = 5.6
	eS	B	54 50	D = 46.41 Az = 305.5 (NEIC)
	eSS	B	58 20	P1V A 1.4s 102.0nm M = 5.6
	LmH	B	08 04.9	P2V A 1.2 81.2nm 5.5
	LmV	B	08.7	PPV A 1.9 83.3nm 5.3
				LmH B 7.0 2.1/um 5.5
				LmV B 12 1.5/um 5.3
15.	eP	A	14 15 41	<u>Off East Coast of Kamchatka</u>
	e	A	15 50	52.84 N 160.79 E
	e	A	16 05	H = 14 04 08.5 h = 55 km MB = 5.1
	e	A	16 13.5	D = 73.65 Az = 340.1 (NEIC)

Day	Phase		h m s	Remarks
cont. 15.	Lm1H	B	14 46.0	Lm1H B 20s 0.4 µm
	Lm2H	B	53.8	Lm2H B 15.5 0.5 µm
	LmV	B	55.2	LmV B 15 0.5 µm
15.	e	A	16 31 46	Turkey 37.6 N 29.8 E
	LmH	B	41.2	H = 16 27 21 (ATH)
	LmV	B	41.2	D = 18.40
				LmH B 10s 0.2 µm M = 3.7
				LmV B 10 0.3 4.0
15.	eP	A	18 12 56	Andreanof Islands, Aleutian Is. 51.14 N 177.97 W
				H = 18 00 59.5 h = 46 km MB = 4.8
				D = 78.27 Az = 353.8 (NEIC)
				PV A 1.4s 18.6nm M = 4.9
15.	+iP1	AB	22 12 37.5	Kirgiz-Sinkiang Border Region
	eiP2	AB	12 44	41.52 N 79.35 E
	e	B	14 29	H = 22 04 13.4 h = normal MB = 5.6
	iPP1	AB	14 32.5	D = 46.34 Az = 305.5 (NEIC)
	iPP2	AB	14 36.5	P1V A 1.2s 44.7nm M = 5.3
	e	B	17 12	P2V A 1.6 428.0nm 6.1
	eS	B	19 28	PP1V A 1.5 191.0nm 5.7
	eSS	B	22 54	PP2V A 1.4 205.0nm 5.7
	LmH	B	29.8	LmH B 12 8.1 µm 5.9
	LmV	B	33.4	LmV B 11 9.0 µm 6.1
15.	eP1	A	22 24 15.5	Kirgiz-Sinkiang Border Region
	eP2	A	24 19	41.40 N 79.35 E
				H = 22 15 49.7 h = normal MB = 5.1
				D = 46.41 Az = 305.6 (NEIC)
				P1V A 1.2s 32.5nm M = 5.1
				P2V A 1.5 40.2 5.1
15.	eP1	A	23 00 00	Turkey 37.05 N 28.96 E
	eP2	A	00 03	H = 22 55 45.5 h = normal MB = 4.6
				D = 18.41 Az = 323.1 (NEIC)
				P1V A 1.5s 15.1nm M = 3.9
				P2V A 1.4 32.6nm 4.3

Day	Phase		h m s	Remarks
15.	eP	A	23 15 37	Kirgiz-Sinkiang Border Region
				41.5 N 79.3 E
				H = 23 07 08 MS = 4.1
				D = 46.3 (ANUSSR)
				PV A 1.2s 12.2nm M = 4.7
15.	eP1	A	23 25 57.5	Kirgiz-Sinkiang Border Region
	eP2	A	26 03	41.62 N 79.21 E
				H = 23 17 33.9 h = normal MB = 4.9
				D = 46.19 Az = 305.4 (NEIC)
				P1V A 0.8s 13.4nm M = 4.9
				P2V A 1.3 19.7nm 4.9
16.	+iP1	AB	01 07 02.3	Kirgiz-Sinkiang Border Region
	eiP2	A	07 07	41.53 N 79.39 E
	ePP1	AB	08 50	H = 00 58 37.4 h = normal MB = 5.4
	ePP2	A	08 57	D = 46.36 Az = 305.5 (NEIC)
	eS	B	13 52	P1V A 1.6s 115.2nm M = 5.6
	eSS	B	17 18	P2V A 1.8 142.0nm 5.6
	LmV	B	27.7	PP1V A 2.0 59.9nm 5.0
	LmH	B	27.8	PP2V A 1.4 37.2nm 5.0
				LmH B 12.5 3.9/um 5.6
				LmV B 11 3.7/um 5.7
16.	eP	A	01 27 56	Kirgiz-Sinkiang Border Region
				41.0 N 79.3 E
				H = 01 19 28 MB = 4.8
				D = 46.6 (ANUSSR)
16.	e(Pn)	A	05 20 48	
	e	A	21 14.5	
	e(Sg)	A	21 23	
16.	eP	A	05 50 26	Kirgiz-Sinkiang Border Region
				41.2N 79.4 E
				H = 05 42 01 MB = 4.6
				D = 46.6 (ANUSSR)

Day	Phase		h m s	Remarks
16.	+iPn	A	06 14 40.5	Central Italy 43.94 N 12.48 E
	iX	A	14 42	H = 06 13 02.7 h = 39 km MB = 4.4
	ePg	A	15 15.5	D = 6.73 Az = 355.3 (NEIC)
	iSn	A	15 51	PnV A 0.5s 21.2nm
	eSg	A	16 52	XV A 0.6 46.0nm
	LmH	B	17.3	
16.	eP1	A	11 05 04	Kirgiz-Sinkiang Border Region
	eiP2	A	05 06.5	41.42 N 79.34 E
	LmH	B	25.0	H = 10 56 38.7 h = normal MB = 5.0
	LmV	B	28.5	D = 46.39 Az = 305.5 (NEIC)
				P2V A 0.7s 21.1nm M = 5.2
16.	+eP	A	13 55 16	Kirgiz-Sinkiang Border Region
	LmV	B	14 15.6	41.32 N 79.32 E
	LmH	B	17.3	H = 13 46 50.5 h = normal MB = 5.1
				D = 46.44 Az = 305.6 (NEIC)
				PV A 1.2s 24.4nm M = 5.0
				LmH B 8 0.3 µm 4.6
				LmV B 10 0.3 µm 4.7
17.	ePn	A	07 41(40)	Switzerland 47.9 N 8.1 E
	ePg	A	41 48	H = 07 40 49 h = 20 km
	e	A	41 54	D = 3.55 (ISC)
	iSn	A	42 15	
	iSg	A	42 34	
17.	iPg	A	08 01 20	Czechoslovakia 50°17.8' N 12°35.4' E
	iSg	A	01 30.5	yield 10 t (PRU)
				D = 0.7
17.	e	A	09 44 14	Kurile Islands 44.41 N 148.91 E
				H = 09 32 05.4 h = normal MB = 4.9
				D = 78.32 Az = 333.8 (NEIC)
17.	eP1	A	15 28 36	Kirgiz-Sinkiang Border Region
	eP2	C	28 43	41.32 N 79.40 E
				H = 15 20 12.0 h = normal MB = 4.9
				D = 46.48 Az = 305.6 (NEIC)
				P2C A 1.1s 14.1nm M = 4.8

Day	Phase		h m s	Remarks
17.	eP	A	17 39 29	<u>Kirgiz-Sinkiang Border Region</u>
				41.2 N 79.3 E
				H = 17 31 03 MB = 4.8
				D = 46.5 (ANUSSR)
17.	LmH	C	19 59.0	
	LmV	C	20 00.9	
17.	eP	AB	21 14 34.5	<u>Northern Chile</u> 25.48 S 69.15 W
	epP	AB	15 05	H = 21 00 40.9 h = 93 km MB = 6.3
	ePP	AB	18 54	D = 103.74 Az = 40.3 (NEIC) h = 117 km
	eipP	A	19 16	LmH B 20s 47.6/um M = 7.0
	eisP	A	19 28	LmV B 21 50.9/um 7.0
	iSKS	B	25 10	e A 21 18 06.5 e AB 18 33 e B 20 00
	eiS	B	25 55	
	esS	B	27 10	
	eiSP	B	27 54	
	ePPS	B	28 50	
	eSS	B	33 30	
	LmV	B	58.7	
	LmH	B	58.8	
17.	ePKP2	A	22 43 46	<u>Kermadec Islands Region</u> 30.97 S 179.85 E
				H = 22 23 50.1 h = 323 km MB = 4.5 (NEIC)
				D = 158.6
				PKP2 V A 1.4s 18.6nm
18.	e(Sg)	A	11 11 40.5	<u>Czechoslovakia</u> 49°05' N 15°27' E
				explosion yield 8.8 t (PRU)
18.	+ePKP	AB	15 28 44	<u>Loyalty Islands Region</u> 21.93 S 170.11 E
	e	B	29 44	H = 15 09 03.5 h = 28 km MB = 5.1
				D = 146.75 Az = 334.8 (NEIC)
				PKPV A 1.1s 54.4nm
19.	eP	A	00 31 35	<u>Turkey</u> 37.12 N 29.37 E
	e	A	31 37.5	H = 00 27 15.7 h = normal MB = 4.7
	LmH	B	38.8	D = 18.56 Az = 322.4 (NEIC)
	LmV	B	39.6	LmH B 12s 0.2/um M = 3.7
				LmV B 11 0.3/um 4.0

Day	Phase		h m s	Remarks
19.	e	A	03 59 15	Poland 50.27 N 18.92 E H = 03 56 46.2 (WAR) D = 4.7
19.	+iP	A	04 11 44	Eastern Kazakh SSR 50.01 N 77.74 E
	ePn	A	13 16	H = 04 03 57.6 h = 0 km MB = 5.5 D = 40.94 Az = 297.3 (NEIC) PV A 0.7s 55.5nm M = 5.4
19.	eiPKP	AB	11 15 06.5	Loyalty Islands 20.79 S 168.49 E
	+ipPKP	AB	15 19	H = 10 55 30.3 h = 24 km MB = 5.1 D = 145.07 Az = 334.1 (NEIC) h = 45 km PKPV A 1.2s 69.1nm pPKPV A 1.0 67.0nm
19.	eP	AB	17 31 27	Kirgiz-Sinkiang Border Region
	ePP	B	33 09	41.50 N 79.31 E
	eS	B	38 30	H = 17 23 02.8 h = normal
	eSS	B	41 45	MB = 5.2 MS = 5.3
	LmV	B	52.2	D = 46.32 Az = 305.5 (NEIC)
	LmH	B	53.5	PV A 1.6s 49.4nm M = 5.2 LmH B 10 3.5/um 5.6 LmV B 12 3.6/um 5.6
19.	LmH	C	24 11.4	Revilla Gigedo Islands Region
	LmV	C	13.9	19.8 N 109.1 W H = 23 19 15.2 h = normal MB = 5.8 MS = 5.5 (NEIC) D = 92.7
20.	e(P)	A	02 07 26.5	Greece 39.06 N 21.85 E
	LmH	B	13.0	H = 02 04 07.4 h = 35 km MB = 4.3
	LmV	B	13.0	D = 13.65 Az = 331 (ISC)
20.	eiPKP	A	03 10 58	New Hebrides Islands 18.98 S 169.65 E
	e	A	11 05.5	H = 02 51 25.4 h = 19 km MB = 4.9 D= 143.90 Az = 336.2 (NEIC) PKPV A 1.3s 43.6nm

Day	Phase		h m s	Remarks
20.	e(P)	A	06 00 27.5	<u>Near West Coast of Albania</u> 41.47 N 19.84 E
	e	A	03 35	H = 05 57 43.4 h = 0 km
	LmH	B	04.0	D = 10.80 Az = 331 (ISC)
	LmV	B	05.2	LmH B 9s 0.3 μm
20.	eP	A	19 02 13	<u>Mid-Indian Rise</u> 8.97 S 67.53 E
	e(PcP)	A	02 28	H = 18 50 19.4 h = normal MB = 5.2
	eS	B	11 56	D = 76.52 Az = 327.2 (NEIC)
	eSS	B	16 46	PV A 1.5s 25.1nm M = 5.0
	LmV	C	24.3	LmH C 60 1.4 μm 4.8
	LmH	C	24.5	LmV C 40 0.6 μm 4.6
20.	e	A	20 15 15	
20.	eP	A	23 09 26	<u>Honshu, Japan</u> 36.16 N 139.77 E
				H = 22 57 11.0 h = 70 km MB = 4.9
				D = 82.26 Az = 329.7 (NEIC)
				PV A 1.3s 13.1nm M = 4.7
21.	ePn	A	07 26 56	<u>France</u> 46.36 N 5.73 E
	iPg	A	27 20	H = 07 25 28.5 h = 6 km MB = 4.5
	e	A	28 17.5	D = 5.80 Az = 40.2 (NEIC)
	eiSg	A	28 36	LmH B 7s 0.7 μm M = 3.6
	LmH	B	29.2	LmV B 8 0.8 μm
	LmV	B	29.2	
22.	ePn	A	01 00(50)	<u>Switzerland</u> 46.16 N 8.04 E
	ePg	A	01 12	H = 00 59 38.8 h = normal MB = 4.0
	e	A	01 16.5	D = 5.08 Az = 26.7 (NEIC)
	eSg	A	02 16.5	
22.	eP	A	06 37 50	<u>Hindu Kush Region</u> 36.20 N 69.85 E
				H = 06 30 00.3 h = 166 km MB = 4.9
				D = 43.57 Az = 308.2 (NEIC)
22.	ePKIKP	A	07 08 48.5	<u>Fiji Islands Region</u> 21.09 S 178.78 W
	ePKHKP	A	08 54.5	H = 06 50 08.9 h = 584km MB = 4.7
	ePKP2	A	09 01	D = 149.39 Az = 347.0 (NEIC)

Day	Phase		h m s	Remarks
22.				In the intervall 08h 55m - 11h 08m the short-period seismographs (type A) have been out of operation.
22.	eX	A	20 36 29	XV A 1.3s 13.1nm
23.	eP	A	18 07 56.5	Andaman Islands Region 10.60 N 93.66 E H = 17 56 13.4 h = 90 km MB = 5.0 D = 76.85 Az = 319.6 (NEIC)
23.	LmH	C	20 28.0	Probably off Coast of Southern Chile
	LmV	C	28.0	45.3 S 76.3 W H = 19 19 53.6 h = normal MB = 5 (NEIC) D = 122.0 LmH C 22s 0.15 µm M = 4.6 LmV C 22 0.26 µm 4.8
23.	ePKIKP	A	24 11 45	West of Macquarie Island 54.37 S 147.02 E
	+ePKHKP	AB	11 53	H = 23 51 59.2 h = normal MB = 5.4
	e	A	11 59	D = 152.95 Az = 280.6 (NEIC)
	ePKP2	AB	12 05	PKHKPV A 1.7s 57.6nm
	ePP	A	15 42	
	LmV	C	25 32.0	
24.	eP	A	06 02 59.5	Dodecanese Islands 35.22 N 27.82 E H = 05 58 34.1 h = 50 km MB = 4.3 D = 19.39 Az = 327.6 (NEIC) PV A(1.5)s 17.6nm M = 4.1
24.	iPg	A	11 29 36.5	Czechoslovakia 50°35.2' N 14°03.2' E
	eSg	A	29 57	explosion yield 12 t (PRU) D = 1.54
24.	+iP	AB	14 04 52.5	Southwestern Ryukyu Islands
	epP	AB	05 02.5	23.53 N 123.71 E
	esP	AB	05 08.5	H = 13 52 15.2 h = 9 km MB = 5.6
	eS	B	15 14	D = 85.05 Az = 323.7 (NEIC) h = 40 km
	e	B	15 30	PV A 1.9s 159.0nm M = 5.9

Day	Phase		h m s	Remarks
cont. 24.	LmH	B	14 47.5	pPV A 1.9s 83.4nm
	LmV	B	47.6	sPV A 1.9 83.4nm
				LmH B 18.5 2.8/um M = 5.7
				LmV B 19 3.2/um 5.8
24.	eP	A	14 12 19	<u>Southern Nevada</u> 37.15 N 116.97 W
				H = 14 00 00.2 h = 0 km MB = 5.2
				D = 81.22 Az = 30.7 (NEIC)
				37°08'48.1"N 116°04'00.6"W
				Nevada test site. "Harebell" (AEC)
				PV A 1.4s 11.6nm M = 4.8
24.	LmH	B	18 22.5	
24.	e(Pg)	A	21 10 21	<u>Austria</u> 47.5 N 15.4 E
	eSg	A	11 03	H = 21 08 54 (VIE)
				D = 3.95
24.	LmH	C	25 28.0	<u>Banda Sea</u> 4.2 S 127.8 E
	LmV	C	28.0	H = 23 19 33.9 h = normal MB = 5.1 (NEIC)
				D = 109.7
				LmH C 22s 0.24/um M = 4.7
				LmV C 23 0.46/um 5.0
25.	ePKP2	A	00 25 23	<u>New Hebrides Islands Region</u>
				21.10 S 174.18 E
				H = 00 05 34.8 h = 30 km MB = 4.5
				D = 147.47 Az = 339.2 (NEIC)
25.	eP	A	11 10 22	<u>Tadzhik SSR</u> 37.57 N 72.15 E
				H = 11 02 24.9 h = 140 km MB = 4.6
				D = 44.19 Az = 307.3 (NEIC)
				PV A (1.5)s 12.6nm M = 4.4
26.	eP	A	02 27 25	<u>Burma-India Border Region</u>
	epP	A	27 44.5	24.58 N 94.74 E
				H = 02 16 37.0 h = 76 km MB = 5.2
				D = 67.15 Az = 316.7 (NEIC)
				pPV A 1.1s 22.2nm

Day	Phase		h m s	Remarks
26.	eP	A	04 52 45	Albania 40.5 N 19.4 E
	e	A	55 14	H = 04 50 02 h = 0 km
	LmH	B	56.3	D = 11.49 Az = 334 (ISC)
	LmV	B	57.0	PV A 1.0s 5.9nm
26.	eP	A	09 08 28.5	Near East Coast of Honshu, Japan
	epP	A	08 42	40.12 N 142.41 E
				H = 08 56 23.7 h = 49 km MB = 4.9
				D = 79.86 Az = 330.7 (NEIC) h = 50 km
				PV A 1.0s 7.9nm M = 4.6
				pPV A 1.2 12.2nm
26.	ePKP	A	09 44 32	New Hebrides Islands
				15.13 S 167.37 E
				H = 09 25 21.9 h = 130 km MB = 4.9
				D = 139.52 Az = 336.3 (NEIC)
26.	eP	A	15 58 37	North Atlantic Ocean
	LmH	B	16 24.0	19.02 N 68.02 W
	LmV	B	24.0	H = 15 47 32.8 h = normal MB = 5.3
				D = 69.01 Az = 42.1 (NEIC)
				PV A 1.8s 40.6nm M = 5.2
				LmH B 20 0.3 µm 4.5
				LmV B 18 0.3 µm 4.6
26.	eiP	AB	19 40 18.5	Southwest of Sumatra 5.25 S 96.93 E
	ePP	B	43 58	H = 19 27 13.9 h = 25 km MB = 5.8
	ePPP	B	45 36	D = 91.05 Az = 320.6 (NEIC)
	ePPPP	B	47 38	PV A 1.7s 121.0nm M = 6.0
	eSKS	B	50 48	SH B 14 7.0 µm 6.7
	eiS	B	51 16	LmH B 18 13.4 µm 6.4
	ePS	B	52 20	LmV B 16 11.7 µm 6.4
	iSS	B	57 26	
	eSSS	B	20 01 04	
	eiSSSS	C	03 48	
	LmH	B	24.5	
	LmV	B	29.7	

Day	Phase		h m s	Remarks
26.	e	A	22 23 26.5	Auckland Islands Region 63.51 S 167.6 E
				H = 22 02 18.2 h = 12 km
				D = 164.88 Az = 277 (ISC)
				traces
26.	eP	A	22 31 30.5	Afghanistan-USSR Border Region
				36.26 N 71.43 E
				H = 22 23 29.00 h = 127 km MB = 5.0
				D = 44.53 Az = 308.3 (NEIC)
				PV A 1.5s 40.2nm M = 4.9
27.	eP	A	08 42 21	Mona Passage 19.07 N 67.92 W
				H = 08 31 18.9 h = 48 km MB = 4.9
				D = 68.91 Az = 42.1 (NEIC)
				PV A 2.0s 25.6nm M = 4.9
27.	ePKIKP	AB	14 41 35	South of Fiji Islands 25.19 S 178.23 E
	ePKHKP	A	41 43	H = 14 22 53.1 h = 607 km MB = 5.3
	+iPKP2	A	41 57	D = 152.56 Az = 341.3 (NEIC)
	epPKIKP	AB	43 48	PKIKPV A 1.2s 12.2nm
	e	A	43 56	PKHKPV A 1.2 34.6nm
	e	A	43 02	PKP2V A 1.3 48.0nm
27.	eP	A	17 04 30.5	Afghanistan-USSR Border Region
				36.76 N 71.13 E
				H = 16 56 33.7 h = 137 km MB = 4.7
				D = 44.03 Az = 307.9 (NEIC)
27.	eP	A	17 27 36.5	Fox Islands, Aleutian Is.
	e	A	27 42	52.01 N 170.45 W
	e(pP)	A	27 47	H = 17 15 39.19 h = 18 km MB = 5.2
	LmH	B	18 06.4	D = 77.71 Az = 358.7 (NEIC)
	LmV	B	11.0	PV A 1.3s 32.8nm M = 5.2
				LmH B 18 0.4/um 4.8
				LmV B 14 0.2/um 4.7
27.	ePKIKP	A	18 09(40)	Near North Coast of New Guinea
	e	A	09 43	3.49 S 145.47 E

Day	Phase		h m s	Remarks
cont. 27.	LmV	B	19 03.5	H = 17 50 50.7 h = normal MB = 5.4
	LmH	B	03.6	D = 119.13 Az = 328.3 (NEIC)
				LmV B 19s 1.0/um M = 5.5
				LmH B 19 1.0/um 5.5
27.	e	A	18 13 58	
27.	LmH	B	22 03.0	LmH B 16s 0.36/um
	LmV	B	03.0	LmV B 20 0.29/um
28.	eP1	A	05 12 23.5	Northern China 37.92 N 106.16 E
	eP2	A	12 26.5	H = 05 01 48.6 h = normal MB = 5.2
	LmH	B	41.3	D = 64.41 Az = 315.3 (NEIC)
	LmV	B	43.0	P1V A 1.0s 19.7nm M = 5.2
				P2V A 1.3 26.2nm 5.2
				LmH B 14 0.4/um 4.8
				LmV B 12 0.5/um 5.0
28.	eP	A	19 58 50	Western Caucasus 42.44 N 43.28 E
	LmH	B	20 10.5	H = 19 53 45.7 h = 34 km MB = 4.6
				D = 23.13 Az = 301.7 (NEIC)
				PV A 1.6s 13.7nm M = 4.1
28.	ePKP	A	20 56 55.5	Loyalty Islands Region 22.20 S 170.28 E
	e	A	57 07	H = 20 37 11.5 h = 14 km MB = 4.7
				D = 147.06 Az = 334.8 (NEIC)
				PKPV A 1.3s 21.8nm
28.	eP	AB	23 41 58.5	Turkey 37.55 N 29.82 E
	eS	B	45 29	H = 23 37 39.9 h = 7 km MB = 4.9
	LmH	B	48.4	D = 18.44 Az = 321.0 (NEIC)
	LmV	B	50.9	PV A 2.0s 115.4nm M = 4.7
				PV B 8 0.87/um 5.0
				SH B 10 1.22/um 5.1
				LmH B 14 4.1/um 4.9
				LmV B 10 2.3/um 4.9

Day	Phase		h m s	Remarks
29.	eP	AB	04 30 47.5	Turkey 37.46 N 29.89 E
	eS	B	34 16	H = 04 26 31.2 h = 28 km MB = 4.9
	LmH	B	37.2	D = 18.54 Az = 321.1 (NEIC)
	LmV	B	39.8	PV A 2.0s 162.0nm M = 4.9
				PV B 6 1.2/um 5.3
				SH B 10 1.53/um 5.1
				LmH B 14.5 4.3/um 4.9
				LmV B 11 2.2/um 4.9
29.	eP	AB	09 13 10	Turkey 37.15 N 36.83 E
	iS	B	17 18	H = 09 08 11.7 h = 35 km MB = 5.0
	LmH	B	23.0	D = 22.50 Az = 314.9 (NEIC)
	LmV	B	24.7	PV A 2.5s 169.0nm M = 5.1
				SH B 11.5 4.3/um 5.7
				LmH B 14 2.3/um 4.8
				LmV B 15.5 2.7/um 4.9
29.	ePKHKP	A	09 46 19	Tonga Islands 21.05 S 173.77 W
	ePKP2	A	46 25	H = 09 26 34.9 h = 69 km MB = 5.2
				D = 150.16 Az = 353.1 (NEIC)
				PKHKPV A 1.3s 32.8nm
29.	eP	A	11 18 40	Turkey 37.26 N 36.88 E
	eS	B	22 48	H = 11 13 42.3 h = 31 km MB = 4.6
	LmH	B	30.0	D = 22.45 Az = 314.6 (NEIC)
				PV A 2.0s 21.4nm M = 4.3
				LmH B 16 0.14/um 3.5
29.	eP	A	14 15 00	Alaska Peninsula 54.65 N 161.59 W
	LmH	B	55.0	H = 14 03 19.0 h = 24 km MB = 5.2
				D = 74.93 Az = 4.5 (NEIC)
29.	eP	A	15 23 29	Kurile Islands 45.33 N 151.69 E
	LmH	B	57.6	H = 15 11 32.7 h = 50 km MB = 4.8
				D = 78.34 Az = 335.3 (NEIC)
				PV A 1.3s 26.2nm M = 5.0
				LmH B 16 0.24/um 4.6

Day	Phase		h m s	Remarks
29.	eP	A	15 25 12	Mid-Indian Rise 21.02 S 68.62 E
	epP	A	25 18	H = 15 12 27.1 h = normal MB = 5.0
	esP	A	25 22	D = 87.19 Az = 327.7 (NEIC) h = 25 km
				PV A 1.1s 24.2nm M = 5.4
29.	eP	A	18 42 19	Southern Nevada 37.10 N 116.21 W
				H = 18 30 02.6 h = 25 km MB = 4.9
				D = 81.32 Az = 31 (ISC)
29.	eFKIKP	A	20 24 32	Santa Cruz Islands 11.05 S 165.92 E
	LmV	B	21 26.7	H = 20 05 13.8 h = 46 km MB = 5.0
	LmH	B	27.7	D = 135.24 Az = 336.9 (NEIC)
				PKIKPV A 1.1s 10.1nm
				LmH B 20 0.4 μm M = 5.1
				LmV B 20 0.4 μm 5.2
30.	+eiP	A	04 04 49.5	Eastern Kazakh SSR 49.98 N 79.05 E
	ePn	A	06 21	H = 03 56 57.2 h = 0 km MB = 5.4
	ePP	A	06 28.5	D = 41.70 Az = 297.9 (NEIC)
				PV A 1.0s 31.5nm M = 5.0
30.	eP	A	13 44 30	Central Kazakhstan 41.7 N 79.5 E
				H = 13 36 00 h = 10 km (ANUSSR)
				D = 46.4

Day	Phase		h m s	Remarks
1.	ePKIKP	A	01 34 39	Banda Sea 6.38 S 130.26 E
	ePP	A	35 26	H = 01 16 16.8 h = 133 km MB = 5.8
	ePP	B	35 30	D = 112.86 Az = 322.7 (NEIC)
	eX	A	35 34	PKIKPV A 1.0s 7.9nm
	e(SKKS)	B	42 12	PPV A 1.4 11.6nm M = 5.2
	eSS	B	55 00	PPV B 5 0.33 µm 6.1
	LmH	B	02 18.7	XV A 1.9 37.9nm
	LmV	B	24.3	LmH B 17 0.36 µm
1.	LmH	B	05 37.0	LmH B 14s 0.3 µm
	LmV	B	37.0	LmV B 14 0.3 µm
1.	ePKHKP	A	05 59 48	Loyalty Islands Region 22.14 S 170.14 E
				H = 05 40 06.2 h = 24 km
				D = 146.96 Az = 334.7 (NEIC)
				PKHKPV A 1.2s 12.2nm
1.	eP1	A	12 19(09)	Iraq 36.39 N 43.38 E
	eP2	A	19 11	H = 12 13 27.6 h = 16 km MB = 5.1
	e	C	23 50	D = 26.84 Az = 312.1 (NEIC)
	e	C	25 00	P2V A 2.0s 34.2nm M = 4.7
	LmH	B	32.3	LmH B 14 0.5 µm 4.2
	LmV	B	34.2	LmV B 12 0.3 µm 4.2
1.	eP	A	14 45 18	Hindu Kush Region 36.74 N 68.28 E
				H = 14 37 25.7 h = normal MB = 4.6
				D = 42.24 Az = 307.7 (NEIC)
1.	ePn	A	15 43 49	Hungary 47.05 N 17.86 E
	ePg	A	44 11.5	H = 15 42 27.0 h = normal MB = 4.0
	e(Sn)	A	44 51	D = 5.47 Az = 313.4 (NEIC)
	e	A	45 05	LmH B 7.5s 0.8 µm
	e	A	45 17	LmV B 6 0.3 µm
	eSg	AB	45 22	
	LmH	B	45.7	
	LmV	B	46.4	
1.	e	A	18 07 49	

Day	Phase		h m s	Remarks
1.	LmV	C	19 09	LmV C 22s 0.5/um
	LmH	B	15	LmH B 20 0.18/um
1.	eP	A	23 20 08.5	Turkey 37.18 N 36.88 E
	eS	BC	25 14	H = 23 15 07.1 h = normal MB = 4.5
	LmH	C	31	D = 22.50 Az = 314.8 (NEIC)
	LmV	C	31.5	SH B 7s 0.23/um M = 4.6
2.	eP	A	01 34 29	Luzon, Philippine Islands
				15.88 N 120.22 E
				H = 01 21 38.0 h = 58 km MB = 4.9
				D = 89.19 Az = 322.9 (NEIC)
				traces
2.	eP	A	03 43 03.5	North Atlantic Ridge 35.25 N 36.34 W
	eS	B	48 55	H = 03 35 48.6 h = normal MB = 4.7
	LmH	B	56.8	D = 37.63 Az = 50.8 (NEIC)
	LmV	B	57.0	SH B 16s 0.54/um M = 5.1
				LmH B 16.5 1.0/um 4.7
				LmV B 18 1.1/um 4.8
2.	eiPKP	AB	05 29 19	Tonga Islands 16.14 S 174.39 W
	epPKP	A	30 05.5	H = 05 09 59.3 h = 160 km MB = 4.4
				D = 145.24 Az = 353.3 (NEIC) h = 181 km
				PKPV A 1.0s 28.5nm
2.	eP	A	05 47 44	Luzon, Philippine Islands
	e	A	47 51.5	12.27 N 123.89 E
	ePP	A	51 40	H = 05 34 27.3 h = 35 km MB = 5.3
	eS	BC	59 52	D = 94.19 Az = 323.8 (NEIC)
	LmH	B	25.5	PPV A 1.4s 14.0nm M = 5.2
	LmV	B	31.2	SH B 14.5 0.96/um 5.9
				LmH B 25 12.5/um 6.3
				LmV B 17 4.7/um 5.9
2.	eP	A	06 46 25	Near Islands, Aleutian Is. 52.98 N 171.38 E
				H = 06 34 44.4 h = 39 km MB = 5.1
				D = 75.34 Az = 346.8 (NEIC)
				PV A 1.2s 36.6nm M = 5.3

Day	Phase		h m s	Remarks
2.	ePn	A	14 05 47.5	<u>Eschenlohe, Federal Rep. Germany</u>
	ePg	A	06(00)	47°37.9' N 11°08.8' E
	eSg	A	06 40	H = 14 05 00 explosion yield 14.7 t
				(München)
				D = 3.0
2.	eP	A	21 15 50.5	<u>Morocco</u> 34.14 N 5.24 W
	eS	B	20 00	H = 21 11 07.7 h = 7 km MB = 4.6
	LmH	B	24.3	D = 20.58 Az = 31.7 (NEIC)
	LmV	B	25.1	PV A 1.0s 15.8nm M = 4.4
				LmH B 12 1.0/um 4.4
				LmV B 11.5 0.8/um 4.5
3.	+iPKP	AB	01 10 14	<u>Tonga Islands</u> 17.71 S 173.23 W
	LmV	B	02 17.8	H = 00 50 32.1 h = normal MB = 5.2
	LmH	B	18.2	D = 146.91 Az = 354.3 (NEIC)
				PKPV A 1.8s 94.5nm
				PKPV B 8 0.49/um
				LmH B 20 0.2/um M = 4.9
				LmV B 20 0.2/um 4.9
3.	eP	AB	04 10 20	<u>Dodecanese Islands</u> 35.11 N 27.81 E
	eX	A	10 29	H = 04 05 53.8 h = 34 km MB = 4.7
	eS	C	13 52	D = 19.48 Az = 327.8 (NEIC)
	LmH	B	19.0	PV A 1.6s 63.0nm M = 4.6
	LmV	B	19.0	XV A 1.4 95.4nm
				LmH B 11.5 1.7/um 4.6
				LmV B 12 1.7/um 4.7
3.	ePKIKP	A	04 26 22.5	<u>Fiji Islands Region</u> 21.49 S 179.06 W
	ePKHKP	AB	26 28.5	H = 04 07 44.3 h = 600 km MB = 5.3
	ePKP2	A	26 36.5	D = 149.71 Az = 346.5 (NEIC)
	epPKP	A	28 48	PKIKPV A 1.2s 14.2nm
	e	A	28 54	PKHKPV A 1.4 111.5nm
				PKP2V A 1.4 37.2nm

Day	Phase		h m s	Remarks
3.	eP	A	04 34 49.5	Kirgiz-Sinkiang Border Region
				41.35 N 79.26 E
				H = 04 26 22.1 h = 17 km MB = 4.9
				D = 46.38 Az = 305.6 (NEIC)
				PV A 1.1s 24.2nm M = 5.0
3.	ePKIKP	A	12 31 19	South of Fiji Islands 23.48 S 179.96 W
	ePKHKP	A	31 26	H = 12 12 32.3 h = 543 km MB = 5.4
	ePKP2	A	31 36.5	D = 151.43 Az = 344.5 (NEIC)
				PKHKPV A 1.4s 32.6nm
3.	eP	A	17 31 04	Mid-Indian Rise 21.05 S 68.78 E
				H = 17 18 19.2 h = normal MB = 4.6
				D = 87.30 Az = 327.6 (NEIC)
				traces
3.	ePKIKP	A	23 03 46	Fiji Islands Region 19.54 S 177.51 W
	ePKHKP	A	03 50	H = 22 45 08.2 h = 580 km MB = 4.8
	ePKP2	A	03 55	D = 148.12 Az = 349.0 (NEIC)
				PKHKPV A 1.3s 39.3nm
3.	eP	A	23 58 37	Chile-Argentina Border Region
	epPP	A	24 03 13	24.13 S 68.93 W
	eSKS	C	09 10	H = 23 44 50.6 h = 103km MB = 5.4
	eSP	C	11 42	D = 102.59 Az = 40.1 (NEIC)
	eSS	C	17 20	pPPV A 1.5s 20.1nm
	LmH	C	37.7	LmH C 38 0.36 μm
	LmV	C	39.7	LmV C 30 0.49 μm
4.	e(P)	A	09 16 09.5	Mariana Islands Region 21.23 N 146.18 E
	ePP	A	20(00)	H = 09 02 30.8 h = 67 km MB = 5.2
	eSS	C	34 12	D = 98.00 Az = 332.7 (NEIC)
	LmH	B	59.2	(P)V A 1.1s 10.1nm M = 5.3
	LmV	B	10 07.3	LmH B 16 0.8 μm
				LmV B 15 0.5 μm
4.	eP	A	09 58 52	South of Honshu, Japan 33.78 N 138.64 E
				H = 09 46 19.1 h = 3km MB = 5.0
				D = 83.83 Az = 329.3 (NEIC)

Day	Phase		h m s	Remarks
4.	eP	AB	11 43 51	<u>Luzon, Philippine Islands</u>
	ePP	A	47 23	15.60 N 121.88 E
	e	A	47 34	H = 11 30 51.1 h = 30 km MB = 5.5
	eSKS	B	54 18	D = 90.36 Az = 323.3 (NEIC)
	eS	B	54 37	PV A 2.0s 42.7nm M = 5.4
	eSP	B	55 42	LmH B 18 2.7/um 5.7
	eSS	C	12 00.4	LmV B 20 2.7/um 5.7
	eSSSS	C	07 38	e A 11 46 19
	LmV	B	28	
	LmH	B	29.5	
4.	eP	A	14 54 28	<u>Andreanof Islands, Aleutian Is.</u>
	e	A	54 42	51.68 N 174.12 W
				H = 14 42 30.2 h = 26 km MB = 4.9
				D = 77.94 Az = 356.3 (NEIC)
				PV A 1.0s 17.7nm M = 5.1
4.	eP	A	18 45 19	<u>Kurile Islands</u> 43.61 N 147.90 E
	epP	A	45 31	H = 18 33 18.0 h = normal MB = 4.6
	esP	A	45 37	D = 78.71 Az = 333.3 (NEIC) h = 44 km
	LmH	C	19 18.0	PV A 1.2s 14.2nm M = 4.9
	LmV	C	24.4	LmH C 20 0.28/um 4.6
				LmV C 18 0.38/um 4.8
4.	eP	A	18 53 35	<u>Andreanof Islands, Aleutian Is.</u>
	epP	A	53 47	51.60 N 174.14 W
	LmH	B	19 25.3	H = 18 41 38.3 h = 31 km MB = 4.8
	LmV	B	25.3	D = 78.01 Az = 356.3 (NEIC) h = 44 km
				PV A 1.4s 16.3nm M = 4.9
				LmH B 15 0.3/um 4.7
				LmV B 16 0.2/um 4.7
4.	LmH	C	18 59.4	
5.	LmV	C	01 30	<u>Near East Coast of Honshu, Japan</u>
	LmH	C	35	35.43 N 140.58 E
				H = 00 38 03.0 h = 40 km MB = 4.8 (ISC)
				D = 83.2
				LmV C 20s 0.26/um M = 4.6

Day	Phase		h m s	Remarks
5.	e	A	01 41 45	<u>South of Kermadec Islands</u>
	LmH	B	03 13.5	32.99 S 178.64 W
	LmV	B	15.9	H = 01 21 05.1 h = normal MB = 5.1 (NEIC)
				D = 158.0
				LmH B 18s 0.3 µm M = 5.1
				LmV B 18 0.4 µm 5.3
5.	eP	A	16 56 51	<u>Turkey</u> 41.76 N 32.49 E
	e	A	56 54	H = 16 52 47.7 h = 6 km MB = 4.3
	LmH	B	17 08.2	D = 16.92 Az = 308.8 (NEIC)
	LmV	B	08.2	LmH B 11s 0.2 µm M = 3.6
				LmV B 10 0.2 µm 3.8
5.	eP	A	22 30 05	<u>Off East Coast of Honshu, Japan</u>
				34.1 N 143.3 E
				H = 22 17 29 MB = 4.8 (ANUSSR)
				D = 85.4
5.	eX1	A	23 50 04	X1V A 1.8s 20.3nm
	e	A	50 10	X2V A 1.5 40.2nm
	e	A	50 14	LmH B 18 0.3 µm
	eX2	A	51 41	LmV B 18 0.2 µm
	e	A	51 46	
	LmH	B	24 55.5	
	LmV	B	56.0	
6.				In the intervall 08h 05m - 11h 34m the short-period seismographs (type A) have been out of operation.
7.	eP	A	04 00 57.5	<u>Tadzhik-Sinkiang Border Region</u>
	epP	A	01 27	38.59 N 73.11 E
				H = 03 52 53.1 h = 62 km MB = 4.5
				D = 44.18 Az = 306.6 (NEIC) h = 141 km
				epPV A 1.2s 10.2nm
7.	LmV	C	07 00.0	<u>Aegean Sea</u> 38.47 N 26.07 E
				H = 06 52 32.3 h = 15 km MB = 3.8 (NEIC)

Day	Phase		h m s	Remarks
cont. 7.	LmV	B	07 00.1	D = 15.91
				LmV C 15s 0.5 /um M = 4.0
				LmH B 14.5 0.2 /um 3.4
7.	eP	A	11 36 36	North Atlantic Ridge 23.54 N 44.85 W
	eS	B	44 00	H = 11 27 34.7 h = normal MB = 4.6
	LmV	B	55.4	D = 51.02 Az = 43.1 (NEIC)
	LmH	B	55.5	PV A 1.0s 11.8nm M = 4.8
				SH B 14 0.29 /um 5.0
				LmH B 18 0.3 /um 4.3
				LmV B 18 0.4 /um 4.5
7.	ePg	A	22 37 27	Rockburst, mine region of Kladno / ČSSR
	eSg	A	37 42	(PRU)
				D = 1.1
8.	eP	A	06 05 14	North Atlantic Ocean 19.08 N 68.03 W
	LmH	B	30	H = 05 54 12.4 h = 48 km MB = 5.0
	LmV	B	30	D = 68.98 Az = 42.1 (NEIC)
				LmH B 20s 0.2 /um M = 4.4
8.	eP	A	06 39 39.5	Turkey 37.03 N 29.27 E
				H = 06 35 22.1 h = 26 km MB = 4.3
				D = 18.57 Az = 322.7 (NEIC)
8.	eP	A	07 43 12	Kurile Islands 45.47 N 147.69 E
				H = 07 31 30.6 h = 120 km MB = 4.8
				D = 76.99 Az = 333.0 (NEIC)
				PV A 1.4s 14.0nm M = 4.5
8.	ePg	A	09 00 34.5	Czechoslovakia 49°20' N 14°03' E
	iSg	A	00 56	explosion yield 6 t (PRU)
				D = 2.05
8.	eP	A	12 06 11	Hokkaido, Japan Region 45.64 N 142.29 E
				H = 11 45 38.8 h = 251 km MB = 4.5 (NEIC)
				D = 75.0
				traces

Day	Phase		h m s	Remarks
8.	e	C	12 08 44	LmH B 16s 0.2/um
	LmH	B	14.9	LmV B 12 0.2/um
	LmV	B	19.8	
8.	eP	A	14 09 30	Sea of Okhotsk 53.02 N 152.98 E
				H = 13 58 59.6 h = 519 km MB = 4.4
				D = 71.69 Az = 335.2 (NEIC)
				traces
8.	+eP	A	14 12 18	Southern Nevada 37.11 N 116.05 W
				H = 14 00 00.1 h = 0 km MB = 5.5
				D = 81.24 Az = 30.7 (NEIC)
				Nuclear explosion "Miniata" (AEC)
				PV A 1.3s 30.6nm M = 5.3
8.	eP diff	AB	19 21 41.5	Banda Sea 6.94 S 129.56 E
	ePKIKP	A	25 35.5	H = 19 07 00.5 h = 33 km MB = 6.3
	ePP	B	26 25	D = 112.87 Az = 322.4 (NEIC)
	ePP	A	26 32	P diff V A 1.6s 38.5nm
	ePPPP	B	30 44	PKIKPV A 1.4 69.7nm
	eS diff	B	33 56	PPV A 1.8 230.0nm M = 6.6
	eSKP	B	35 48	PPV B 10 2.8/um 7.0
	ePS	B	35 56	LmH B 21 4.7/um 6.1
	e	B	36 32	LmV B 17 2.4/um 5.9
	eSPP	B	37 00	
	e(PKKS)	B	40 12	
	LmH	B	20 13.2	
	LmV	B	18.1	
8.	eP	A	23 46 51.5	United Arab Republic 27.63 N 33.81 E
				H = 23 40 56.5 h = 32 km MB = 4.8
				D = 28.51 Az = 329.7 (NEIC)
				PV A 2.0s 38.5nm M = 4.8
9.	+iP diff	AB	03 17 44	Near Coast of Central Chile
	e	A	21 04	32.54 S 71.15 W
	ePKIKP	A	21 45	H = 03 03 18.7 h = 58 km MB = 6.6
	ePP	A	22 04	D = 110.19 Az = 42.3 (NEIC)

Day	Phase		h m s	Remarks
cont. 9.	ePP	B	03 22 08	P diff B 8s 4.35 /um
	e	A	22 17	P diff B 20 8.4 /um
	eSKS	B	28 44	PPV B 8 15.1 /um M = 7.7
	e(SKKS)	B	29 42	LmH B 21 347.0 /um 7.9
	iPS	B	32 00	LmV B 19 453.0 /um 8.1
	ePKKP	A	32 54	(LmV off scale)
	iPPS	B	33 24	
	e	B	35 52	
	iSS	B	38 04	
	eSSS	B	42 04	
	LmH	B	04 07.1	
	LmV	B	07.2	
9.	eP	A	06 51 46	<u>Kurile Islands</u> 43.49 N 147.72 E
	eX	A	52 04.5	H = 06 39 47.0 h = 49 km MB = 5.1
				D = 78.76 Az = 333.3 (NEIC)
				PV A 1.2s 22.4nm M = 5.0
				XV A 1.2 20.3nm
9.	eP	A	08 18 33	<u>Kurile Islands</u> 43.44 N 147.74 E
	e	A	18 52	H = 08 06 35.2 h = 54 km MB = 4.9
				D = 78.81 Az = 333.3 (NEIC)
9.	ePKIKP	A	08 58 00.5	<u>Fiji Islands Region</u> 20.71 S 177.69 W
	ePKHKP	A	58 06	H = 08 39 13.4 h = 500 km MB = 4.6
	e	A	58 19	D = 149.23 Az = 348.4 (NEIC)
				PKHKPV A 0.9s 27.2nm
9.	e	A	10 09 50	<u>Solomon Islands</u> 5.9 S 154.4 E
	ePP	AB	10 30	H = 09 49 43.3 h = 62 km MB = 5.3 (NEIC)
	e	A	10 47	D = 125.7
	LmV	B	11 03.7	LmH B 19s 0.6 /um M = 5.3
	LmH	B	08.0	LmV B 23 0.8 /um 5.3
9.	ePKHKP	A	12 46 11	<u>Fiji Islands Region</u> 20.50 S 178.11 W
	ePKP2	A	46 16.5	H = 12 27 24.5 h = 550 km MB = 5.1
	e	A	46 24.5	D = 148.95 Az = 348.0 (NEIC)
	epPKP	A	48 22	PKHKPV A 1.3s 21.8nm

Day	Phase		h m s	Remarks
9.	eP	A	16 56 16	Kurile Islands 43.48 N 147.73 E
	eX	A	56 22	H = 16 44 15.8 h = 46 km MB = 4.9
	LmH	B	17 28.5	D = 78.78 Az = 333.3 (NEIC)
	LmV	B	35.9	PV A 1.4s 21.0nm M = 4.9
				XV A 1.4 23.3nm
				LmH B 22 1.3 μm 5.2
				LmV B 16 0.65 μm 5.1
9.	eP	A	18 06 51.5	Kurile Islands 43.67 N 147.68 E
				H = 17 54 52.5 h = 42 km MB = 4.6
				D = 78.58 Az = 333.2 (NEIC)
				PV A 1.3s 10.9nm M = 4.6
9.	eSKS	C	19 40 40	Near Coast of Central Chile
	eS diff	C	42 25	32.2 S 71.7 W
	ePS	BC	44 10	H = 19 15 33.9 h = 53 km MB = 5.3 (NEIC)
	eSS	C	50 00	D = 110.2
	eSSS	C	55 10	
	LmV	B	20 21.5	
	LmH	B	21.7	
9.	ePg	A	21 09 02	Northern Italy 45.9 N 8.8 E
	iSn	A	09 39	H = 21 07 16 (BCIS)
	eSg	A	10 12	D = 5.11
9.	eP	AB	22 32 03	North of Ascension Islands
	eS	B	39 49	0.27 N 16.98 W
	LmV	B	56.1	H = 22 22 27.6 h = normal MB = 5.3
	LmH	B	23 00.9	D = 55.77 Az = 21.6 (NEIC)
				PV A 2.0s 85.5nm M = 5.4
				PV B 6 0.38 μm 5.6
				LmH B 14.5 1.0 μm 5.0
				LmV B 14.5 1.1 μm 5.2
10.	eP	A	03 17 01	Kurile Islands 43.63 N 147.68 E
	LmV	B	56.5	H = 03 05 00.5 h = 36 km MB = 4.8
	LmH	B	56.6	D = 78.63 Az = 333.2 (NEIC)
				LmH B 12s 0.35 μm M = 4.9
				LmV B 16 0.24 μm 4.7

Day	Phase		h m s	Remarks
10.	e	A	05 43 16	<u>Solomon Islands</u> 6.02 S 154.64 E
	e	A	45 21	H = 05 23 51.6 h = 59 km MB = 5.8
	LmH	B	06 40.5	Γ = 125.89 Az = 331.8 (NEIC)
	LmV	B	40.5	
10.	eP	A	09 13 33	<u>Kurile Islands</u> 44.99 N 150.48 E
				H = 09 01 34.7 h = normal MB = 4.6
				D = 78.29 Az = 334.7 (NEIC)
				PV A 1.2s 8.1nm M = 4.6
10.	ePKIKP	A	12 15 12	<u>New Guinea</u> 5.87 S 144.44 E
	LmH	C	13 02	H = 11 56 26.1 h = 68 km MB = 5.4
				D = 120.58 Az = 327.2 (NEIC)
				PKIKPV A 1.0s 7.9nm
10.	e(Pn)	A	13 19 46	
	i	A	19 49.5	
	e	A	20 58	
	e	A	21 25	
	e(Sg)	A	22 01	
10.	LmH	B	14 10	<u>Near Coast of Central Chile</u>
	LmV	B	10	32.6 S 71.5 W
				H = 13 04 43.3 h = 55 km MB = 5.1 (NEIC)
				D = 110.4
				LmH B 18s 0.22 μm
				LmV B 19 0.69 μm
10.	eP	A	14 40 57	<u>Kurile Islands</u> 43.54 N 147.77 E
				H = 14 28 56.4 h = 55 km MB = 4.8
				D = 78.73 Az = 333.3 (NEIC)
10.	ePKP	A	16 00 02.5	<u>Tonga Islands</u> 15.95 S 173.96 W
				H = 15 40 34.9 h = 92 km MB = 4.6
				D = 145.10 Az = 353.8 (NEIC)
				PKPV A 1.0s 15.8nm

Day	Phase		h m s	Remarks
10.	eP	A	17 05 38.5	Ural Mountains Region 64.17 N 55.18 E
	e(Pn)	A	05 50	H = 16 59 59.3 h = 0 km MB = 5.3
				D = 26.45 Az = 260.1 (NEIC)
				PV A 1.1s 38.3nm M = 5.0
				underground explosion
10.	LmH	B	22 33	Near Coast of Central Chile 32.8 S 71.6 W
	LmV	B	33	H = 21 23 03.9 h = 50 km MB = 4.5 (NEIC)
				D = 110.5
				LmH B 16s 0.2 µm M = 4.8
				LmV B 18 0.19 µm 4.7
11.				In the intervall 04h 54m - 08h 40m the short-period seismographs (type A) have been out of operation.
11.	eP diff	B	10 05 06	Near Coast of Central Chile
	ePKIKP	A	10 10	32.27 S 71.81 W
	ePP	AB	10 36	H = 09 51 37.8 h = 36 km MB = 5.9
	e	AB	10 52	D = 110.36 Az = 42.4 (NEIC)
	eiPS	B	20 12	PPV A 4s 622.0nm M = 6.7
	ePKKP	A	21 19.5	LmH B 19.5 13.7 µm 6.5
	eSS	B	26 12	LmV B 19.5 16.3 µm 6.6
	LmH	B	55.9	
	LmV	B	56.4	
11.	eP	A	13 01 32	Central Mid-Atlantic Ridge
	e	A	01 45	8.21 N 37.97 W
				H = 12 51 33.7 h = normal MB = 4.5
				D = 58.81 Az = 34.5 (NEIC)
11.	eP	A	17 14 58.5	Off East Coast of Honshu, Japan
	e	A	15 13	39.37 N 143.32 E
	LmH	B	50	H = 17 02 48.4 h = 54 km MB = 4.7
	LmV	B	56.3	D = 80.86 Az = 331.2 (NEIC)
				PV A 1.4s 11.6nm M = 4.6
				LmH B 16 0.2 µm
				LmV B 14 0.3 µm

Day	Phase		h m s	Remarks
11.	e	A	20 16 08	X1V A 1.0s 11.8nm
	eX1	A	16 09.5	X2V A 1.2 12.2nm
	eX2	A	16 18.5	
11.	+eP1	AB	20 17 55.5	<u>Turkey</u> 37.17 N 36.80 E
	-iP2	AB	18 05.5	H = 20 12 54.8 h = 9 km MB = 5.2
	iS	B	22 02	D = 22.46 Az = 314.8 (NEIC)
	LmH	B	26.4	P1V A 1.9s 144.0nm M = 5.1
	LmV	B	29.5	P1V B 4 0.49/um 5.3
				P2V A 1.8 290.0nm 5.4
				SH B 11.5 5.7/um 5.8
				LmH B 16.5 6.3/um 5.1
				LmV B 12 4.1/um 5.2
11.	LmH	B	23 06.0	<u>Near Coast of Central Chile</u>
	LmV	B	06.0	33.0 S 71.9 W
				H = 21 59 57.2 h = 49 km MB = 5.1 (NEIC)
				D = 110.0
				LmH B 18s 0.5/um M = 5.2
				LmV B 18 0.6/um 5.2
11.	LmH	B	23 43.5	<u>Near Coast of Central Chile</u>
	LmV	B	43.5	32.1 S 71.7 W
				H = 22 38 09.3 h = 45 km MB = 5.1 (NEIC)
				D = 110.0
				LmH B 18s 0.6/um M = 5.2
				LmV B 18 0.7/um 5.3
12.	LmV	C	01 22.0	<u>Near Coast of Central Chile</u>
				33.1 S 72.1 W
				H = 00 18 16.3 h = 51 km MB = 4.5 (NEIC)
				D = 110.2
				LmV C 20s 0.37/um
12.	eP	A	02 24 00	<u>Near East Coast of Kamchatka</u>
				53.07 N 160.02 E
				H = 02 12 29.5 h = normal MB = 4.9
				D = 73.28 Az = 339.6 (NEIC)

Day	Phase		h m s	Remarks
12.	ePn	A	07 11 26	Federal Rep. Germany 49.51 N 8.57 E
	iPg	A	11 30.8	H = 07 10 50.6 h = normal
	iSn	A	11 51.8	D = 2.27 Az = 58.7 (NEIC)
	iSg	AB	11 57.3	
12.	eP	A	08 43 45	Volcano Islands Region 21.98 N 143.03 E
				H = 08 30 43.1 h = 220 km MB = 4.7
				D = 95.97 Az = 331.3 (NEIC)
12.	LmH	B	09 56.0	Baja California 24.0 N 111.2 W
	LmV	B	56.0	H = 09 02 09.2 h = normal MB = 4.8 (NEIC)
				D = 90.2
				LmH B 17s 1.2/um M = 5.4
				LmV B 16 1.2/um 5.4
12.	ePKIKP	A	13 20 58	New Britain Region 6.36 S 148.72 E
	epPKIKP	A	21 11.5	H = 13 02 02.0 h = 33 km MB = 5.1
	LmH	B	14 19.0	D = 123.26 Az = 328.8 (NEIC) h = 50 km
	LmV	B	19.3	PKIKPV A 1.1s 12.1nm
				LmH B 18 1.0/um M = 5.5
				LmV B 18 1.3/um 5.6
12.	-iP	A	16 01 01.8	Near Islands, Aleutian Islands
	e	A	01 10.5	52.05 N 173.18 E
	e	A	01 18.5	H = 15 49 11.4 h = 25 km MB = 5.3
	eS	B	10 45	D = 76.50 Az = 348.1 (NEIC)
	LmH	B	37.4	PV A 0.6s 26.8nm M = 5.4
	LmV	B	43.2	LmH B 17 0.5/um 4.9
				LmV B 14 0.5/um 5.0
12.	+eP	A	23 09 06	Andreanof Islands, Aleutian Is.
	e	A	09 19	51.64 N 174.13 W
				H = 22 57 06.9 h = 14 km MB = 5.0
				D = 77.98 Az = 356.3 (NEIC)
				PV A 1.0s 27.6nm M = 5.2

Day	Phase		h m s	Remarks
13.	ePKHKP	A	10 44 57	<u>Fiji Islands Region</u> 20.50 S 178.27 W H = 10 26 08.1 h = 520 km MB = 4.7 D = 148.79 Az = 347.1 (NEIC) PKHKPV A 1.1s 16.1nm
13.	LmH	B	11 12.5	
	LmV	B	18.0	
14.	eP	A	00 15 18	<u>Mindanao, Philippine Islands</u>
	LmH	C	55.0	9.33 N 126.17 E
	LmV	C	57.7	H = 00 01 49.3 h = 73 km MB = 5.4 D = 97.87 Az = 324.2 (NEIC) PV A 1.4s 9.3nm M = 5.2 LmH C 26 0.33/um LmV C 32 0.58/um
14.	+eP diff	B	06 27 00	<u>New Ireland Region</u> 5.47 S 153.89 E
	+iPKIKP	B	30 25	H = 06 11 29.1 h = 47 km MS = 7.9
	e	A	30 27	D = 125.05 Az = 331.6 (NEIC)
	ePP	B	32 16	P diff V B 16s 2.11/um
	iPS	B	42 20	PKIKPV B 14 4.75/um
	iPPS	B	43 48	PPV B 16 29.3/um M = 7.3
	LmH	B	07 27.8	LmH B 19 294.0/um 8.0
	LmV	B	27.8	LmV B 19 351.0/um 8.1
14.	ePKP	A	07 35 54.5	<u>New Ireland Region</u> 5.80 S 153.23 E H = 07 16 53.6 h = normal MB = 4.9 D = 125.02 Az = 331.1 (NEIC)
14.	ePKIKP	A	07 56 13	<u>New Ireland Region</u> 4.83 S 153.57 E
	+iPP	A	58 03.8	H = 07 37 15.9 h = normal MB = 5.6 D = 124.34 Az = 331.6 (NEIC) PPV A 1.7s 91.0nm M = 5.9
14.	ePKIKP	A	08 00 09.5	<u>New Ireland Region</u> 5.46 S 153.84 E H = 07 41 09.3 h = normal MB = 5.7 D = 125.02 Az = 331.6 (NEIC)

Day	Phase		h m s	Remarks
14.	ePKIKP	A	08 01 30	<u>New Ireland Region</u> 5.6 S 153.8 E H = 08 42 30.1 h = normal MB = 5.0 (NEIC) D = 127.4
14.	ePKIKP	A	08 29 34	<u>New Ireland Region</u> 5.46 S 153.30 E H = 08 10 35 h = 39 km MB = 4.7 D = 124.76 Az = 331 (ISC)
14.	ePKIKP e	A A	09 23 08 23 20	<u>New Ireland Region</u> 5.11 S 153.75 E H = 09 04 10.2 h = normal MB = 5.1 D = 124.67 Az = 331.6 (NEIC) traces
14.	ePKIKP	A	10 07 42.5	<u>New Ireland Region</u> 5.44 S 153.59 E H = 09 48 42.4 h = normal MB = 4.6 D = 124.88 Az = 331.4 (NEIC)
14.	ePKIKP	A	11 05 50	<u>New Ireland Region</u> 4.68 S 153.46 E H = 10 46 48.0 h = 0 km MB = 4.7 D = 124.16 Az = 332 (ISC)
14.	ePKIKP ePP	A A	11 48 45.5 50 42	<u>New Ireland Region</u> 5.15 S 153.40 E H = 11 29 47.3 h = normal MB = 4.9 D = 124.54 Az = 331.4 (NEIC) PKIKPV A 1.2s 10.2nm
14.	ePKIKP	A	12 02 28	<u>New Ireland Region</u> 4.86 S 153.19 E H = 11 43 29.9 h = normal MB = 5.1 D = 124.19 Az = 331.4 (NEIC)
14.	ePKIKP ePP	A A	12 09 18 11 13	<u>New Ireland Region</u> 5.12 S 153.26 E H = 11 50 16 h = 16 km MB = 4.8 D = 124.45 Az = 331 (ISC) PKIKPV A 1.0s 33.5nm
14.	ePKIKP	A	12 46 33	<u>New Ireland Region</u> 4.48 S 152.98 E H = 12 27 31.9 h = normal MB = 5.1 D = 123.75 Az = 331.4 (NEIC) PKIKPV A 1.2s 20.3nm

Day	Phase		h m s	Remarks
14.	e	A	12 59 28	New Ireland Region 5.49 S 153.17 E H = 12 40 13 h = 44 km MB = 4.6 D = 124.72 Az = 331 (ISC)
14.	ePKIKP	A	13 04 38	New Ireland Region 5.13 S 153.02 E
	e	A	04 50	H = 12 45 36 h = 30 km MB = 4.7
	e	A	04 59	D = 124.34 Az = 331 (ISC)
14.	ePKIKP	A	13 27 08.5	New Ireland Region 5.48 S 153.53 E H = 13 08 08.4 h = normal MB = 4.8 D = 124.89 Az = 331.4 (NEIC) PKIKPV A 0.8s 13.5nm
14.	ePKIKP	A	14 06 57	New Ireland Region 4.67 S 153.49 E H = 13 47 59.5 h = normal MB = 4.5 D = 124.16 Az = 331.6 (NEIC)
14.				In the intervall 14h 54m - 15h 11m the short-period seismographs (type A) have been out of operation.
14.	ePKIKP	A	15 17 52	New Ireland Region 5.41 S 153.10 E H = 14 58 52 h = 11 km MB = 5.2 D = 124.62 Az = 331 (ISC) PKIKPV A 1.3s 10.9nm
14.	e	A	15 52 04	Southern Alaska 59.98 N 152.70 W H = 15 40 48.6 h = 82 km MB = 4.0 D = 69.00 Az = 10.6 (NEIC)
14.	ePKIKP	A	16 06 04.5	New Ireland Region 5.33 S 153.16 E
	ePP	A	07 53	H = 15 47 06.4 h = normal MB = 4.9 D = 124.58 Az = 331.3 (NEIC) PKIKPV A 2.0s 34.2nm PPV A 1.7 24.2nm M = 5.3
14.	ePKIKP	A	16 56 56	New Ireland Region 5.18 S 153.07 E H = 16 37 58.1 h = normal MB = 5.3 D = 124.41 Az = 331.3 (NEIC)

Day	Phase	h m s	Remarks
14.			In the intervall 16h 58m - 18h 19m the short-period seismographs (type A) have been out of operation.
14.	+ePKIKP AB	18 46 42.5	New Ireland Region 5.20 S 153.35 E
	ePP B	48 30	H = 18 27 43.7 h = normal MB = 5.6
	e(PcPPKP)A	19 00 06	D = 124.56 Az = 331.4 (NEIC)
	LmH B	44.7	PKIKPV A 1.8s 54.0nm
	LmV B	46.4	LmH B 19 12.3/um M = 6.6
			LmV B 18.5 11.5/um 6.6
14.	e A	20 44 09	New Ireland Region 5.73 S 153.31 E
	ePP A	45 53	H = 20 24 54.0 h = normal MB = 4.7
			D = 125.00 Az = 331.2 (NEIC)
14.	e A	22 07 29	Northern Italy 44.8 N 10.3 E
	ePg A	07 39.5	H = 22 05 40 (BCIS)
	e A	08 44	D = 5.94
	eSg A	08 56	
14.	e A	22 37 55	Northern Italy 44.8 N 10.3 E
	ePg A	38 10	H = 22 36 01 (BCIS)
	eSg A	39 13	D = 5.94
14.	e A	23 18 36	Halmahera 2.63 N 129.32 E
			H = 23 00 05.9 h = 41 km MB = 5.5
			D = 105.10 Az = 324.3 (NEIC)
15.	eiP A	00 35 28	Queen Charlotte Islands Region
			54.22 N 133.73 W
			H = 00 24 02.3 h = normal MB = 5.2
			D = 71.55 Az = 22.4 (NEIC)
			PV A 1.6s 27.5nm M = 5.0
15.	ePn1 AB	01 34 50.5	Northern Italy 44.78 N 10.34 E
	ePn2 A	34 51.5	H = 01 33 22.3 h = 8 km MB = 5.2
	ePn3 A	34 53.5	D = 5.94 Az = 7.9 (NEIC)
	ePn4 A	34 54.5	Pn1V A 0.8s 15.4nm M = 4.8
	iPg AB	35 15	Pn2V A 0.8 30.8nm 5.1

Day	Phase		h m s	Remarks
ont. 15.	iSn	B	01 35 56	Pn3V A 0.6s 95.8nm M = 5.7
	i	B	36 28	Pn4V A 1.0 122.0nm 5.6
	eSg	B	36 33	LmH B 11 89.8 μm 5.5
	LmH	B	37.6	LmV B 12 101.0 μm
	LmV	B	37.7	
15.	eP	A	06 20 30	<u>Turkey</u> 37.23 N 36.75 E
	eS	B	24 25	H = 06 15 31.4 h = normal MB = 4.6
	LmV	B	33.6	D = 22.39 Az = 314.8 (NEIC)
	LmH	B	33.5	PV A 1.6s 22.0nm M = 4.4
				SH B 8 0.37 μm 4.7
				LmH B 11 0.28 μm 3.9
				LmV B 12 0.22 μm 3.9
15.	ePg	A	07 35 18.5	<u>Northern Italy</u> 44.8 N 10.4 E
	i	A	35 23	H = 07 33 26 (BCIS)
	e	A	36 22	D = 5.94
	eSg	A	36 43.5	PgV A 1.5s 40.2nm
15.				In the intervall 10h 28m - 15h 46m the short-period seismographs (type A) have been out of operation.
15.	LmH	B	15 15.5	<u>New Ireland Region</u> 5.1 S 153.4 E
	LmV	B	15.7	H = 13 56 52.9 h = normal MB = 5.0
				MS = 5.3 (NEIC)
				D = 124.5
				LmH B 18s 0.6 μm M = 5.3
				LmV B 17 0.9 μm 5.5
15.	e	A	15 45 51	<u>Northern Italy</u> 44.8 N 10.4 E
	eSg	A	46 54	H = 15 43 36 (BCIS)
				D = 5.94
15.	ePKIKP	A	18 34 29	<u>South of Fiji Islands</u> 25.22 S 178.42 E
	eiPKHKP	A	34 36	H = 18 15 44.86 h = 590 km MB = 5.3
	+iPKP2	A	34 50.3	D = 152.65 Az = 341.6 (NEIC)
				PKIKPV A 1.0s 13.8nm
				PKHKPV A 1.2 32.6nm
				PKP2V A 1.2 36.6nm

Day	Phase		h m s	Remarks
15.	ePn	A	18 50 37	Northern Italy 44.78 N 10.25 E
	ePg	A	51 02	H = 18 49 07.3 h = 19 km MB = 4.0
	eSn	A	51 40	D = 5.94 Az = 8.4 (NEIC)
	eSg	A	52 20	LmH B 12s 0.7/um M = 4.0
	LmH	B	53.4	LmV B 12 0.65/um 4.2
	LmV	B	53.4	
16.	ePKIKP	A	02 17 35	New Ireland Region 4.88 S 153.42 E
				H = 01 58 37.4 h = normal MB = 5.0
				D = 124.31 Az = 331.5 (NEIC)
16.	eP	A	05 54 27	Crete 35.01 N 23.10 E
	LmH	B	06 01.6	H = 05 50 23.7 h = 42 km MB = 4.5
	LmV	B	03.0	D = 17.72 Az = 335.4 (NEIC)
				PV A 1.5s 25.1nm M = 4.1
				LmH B 13 0.9/um 4.2
				LmV B 11 0.8/um 4.4
16.	ePKIKP	A	08 06 41	New Ireland Region 5.77 S 153.34 E
	LmH	B	09 07.0	H = 07 47 38.6 h = 40 km MB = 5.1
	LmV	B	07.2	D = 125.05 Az = 331.2 (NEIC)
				PKIKPV A 2.2s 43.6nm
				LmH B 18 0.5/um
				LmV B 17 0.7/um
16.	ePKIKP	A	13 38 30.5	New Ireland Region 5.66 S 153.32 E
	LmH	C	14 37.0	H = 13 19 32.1 h = 41 km MB = 5.1
				D = 124.95 Az = 331.2 (NEIC)
16.	iPg	A	13 58 51.2	Czechoslovakia quarry blast
	iSg	A	59 06.2	50°23.6' N 13°55.6' E
				H = 13 58 47.5 yield 31.0 t (PRU)
				D = 1.2
16.	eP	A	15 11 49	Off East Coast of Honshu, Japan
				39.02 N 143.44 E
				H = 14 59 35.2 h = 37 km MB = 4.9
				D = 81.22 Az = 331.3 (NEIC)

Day	Phase		h m s	Remarks
16.	eP	A	21 53 16	Oaxaca, Mexico 16.72 N 96.08 W
				H = 21 40 23.1 h = 14 km MB = 5.2
				D = 87.99 Az = 37.4 (NEIC)
				traces
16.	eP	AB	22 27 54.5	Off East Coast of Honshu, Japan
	e	A	28 08	39.03 N 143.49 E
	LmH	B	23 06.9	H = 22 15 39.1 h = 22 km MB = 5.2
	LmV	B	07.1	D = 81.22 Az = 331.3 (NEIC)
				PV A 1.5s 42.7nm M = 5.3
				LmH B 15.5 1.5 μm 5.5
				LmV B 16 1.1 μm 5.3
17.	LmV	B	04 02.7	North Atlantic Ridge 15.3 N 45.3 W
	LmH	B	04.0	H = 03 23 53.6 h = normal MB = 4.6 (NEIC)
				D = 57.5
				LmV B 16s 0.41 μm M = 4.7
17.	eP	A	04 52 56	Off East Coast of Honshu, Japan
				39.04 N 143.38 E
				H = 04 40 42.0 h = 35 km MB = 4.6
				D = 81.17 Az = 331.3 (NEIC)
17.	+iP	AB	05 44 39.5	Nicobar Islands Region 7.01 N 94.69 E
	epP	B	45 14	H = 05 32 42.9 h = 138 km MB = 5.8
	epPP	B	48 31	D = 80.23 Az = 320.1 (NEIC) h = 142 km
	ePPP	B	49 33	PV A 1.7s 218.0nm M = 5.6
	ePPPP	B	50 50	PV B 2.5 0.66 μm 5.9
	iS	B	54 28	SH B 6.5 3.1 μm 6.6
	e	B	55 00	
	isS	B	55 40	
	LmH	B	06 05.2	
	eSKPPKP	A	14 41	
	LmV	B	15.5	
17.	ePKIKP	A	06 57 20	New Ireland Region 5.14 S 153.21 E
	LmH	B	07 55.3	H = 06 38 20.0 h = normal MB = 4.8

Day	Phase		h m s	Remarks
cont. 17.	LmV	B	07 55.4	D = 124.44 Az = 331.3 (NEIC)
				PKIKPV A 1.8s 23.6nm
				LmH B 19 0.8/um M = 5.4
				LmV B 20 0.8/um 5.4
17.	ePKIKP	A	09 05 09.5	<u>New Ireland Region</u> 4.77 S 153.37 E
				H = 08 46 17.4 h = 63 km MB = 4.9
				D = 124.19 Az = 331.5 (NEIC)
17.				In the intervall 11h 09m - 12h 00m the short-period seismographs (type A) have been out of operation.
17.	+iP	AB	15 11 31	<u>Eastern India</u> 26.47 N 93.23 E
	epP	A	11 42.5	H = 15 00 55.4 h = 49 km MB = 5.3
	esP	A	11 48.5	D = 64.85 Az = 315.9 (NEIC) h = 49 km
	X	A	12 36	PV A 1.0s 33.5nm M = 5.3
	eS	B	20 08	pPV A 1.4 25.6nm
	ePS	BC	20 26	sPV A 1.0 29.6nm
	eSSS	C	27 25	XV A 1.3 21.8nm
	LmH	B	38.3	LmH B 20.5 1.5/um 5.2
	LmV	B	43.0	LmV B 19 1.2/um 5.2
17.	e	A	17 59 17	
17.	LmH	C	18 09.5	LmH C 14s 0.17/um
	LmV	C	09.5	LmV C 14 0.24/um
17.	eP	A	19 42 07	<u>Off East Coast of Honshu, Japan</u>
	LmV	C	20 21.0	39.15 N 143.32 E
	LmH	B	21.2	H = 19 29 52.6 h = 25 km MB = 4.7
				D = 81.05 Az = 331.2 (NEIC)
				LmV C 16s 0.4/um M = 4.9
				LmH B 14 0.31/um 4.8
17.	eP	A	21 50 30	<u>Turkey</u> 38.30 N 39.85 E
	LmH	B	22 04.5	H = 21 45 23.6 h = normal MB = 4.5
	LmV	C	04.5	D = 23.47 Az = 310.9 (NEIC)
				LmV C 18s 0.33/um M = 4.0

Day	Phase		h m s	Remarks
18.	ePg	A	05 31 23	Northern Italy 44.8 N 10.3 E
	eSn	A	32 03.5	H = 05 29 33 (BCIS)
	eSg	A	32 49	D = 5.94
	LmH	B	33.7	
	LmV	B	33.8	
18.	ePKIKP	A	09 46 02	New Ireland Region 5.10 S 153.34 E
	LmV	C	10 44.0	H = 09 27 03.8 h = normal MB = 5.4
	LmH	B	44.3	D = 124.46 Az = 331.4 (NEIC)
				LmV C 20s 0.37/um M = 5.0
18.	ePKIKP	A	11 23 13	Fiji Islands Region 20.83 S 178.76 W
	eiPKHKP	A	23 17.5	H = 11 04 37.6 h = 628 km MB = 4.9
	ePKP2	A	23 24.5	D = 149.14 Az = 347.1 (NEIC)
				PKHKPV A 1.4s 34.8nm
18.	eP diff	AB	14 46 49.5	New Ireland Region 4.83 S 153.30 E
	+iPKIKP	AB	50 13.5	H = 14 31 17.5 h = 46 km MB = 5.7
	e	A	50 23.5	D = 124.21 Az = 331.5 (NEIC)
	ePP	B	51 52	PKIKPV A 1.0s 141.5nm
	eSKS	B	57 15	PKIKPV B 5 0.93/um
	e(PKKP)	B	15 00 04	LmH B 19 9.8/um M = 6.5
	eSKSP	B	01 52	LmV B 20 10.8/um 6.5
	ePPS	B	03 29	
	eiSS	B	08 48	
	LmH	B	45.7	
	LmV	B	46.7	
18.	eP	A	16 20 59	Romania 45.71 N 26.31 E
				H = 16 18 22.8 h = 137 km MB = 4.6
				D = 10.98 Az = 302.0 (NEIC)
19.	eP diff	B	00 30 20	New Ireland Region 5.69 S 153.80 E
	ePKIKP1	AB	33 45	H = 00 14 45.3 h = 42 km MB = 5.8
	ePKIKP2	A	33 54	D = 125.20 Az = 331.5 (NEIC)
	ePP	B	35 36	PKIKP1V A 2.0s 38.5nm
	eSKKS	B	42 28	PKIKP2V A 1.7 94.0nm
	ePPS	B	47 10	LmH B 19.5 21.0/um M = 6.8

Day	Phase		h m s	Remarks
cont. 19.	eSS	B	00 52 37	LmV B 19s 18.3 µm M = 6.8
	eSSS	B	56 52	
	LmH	B	01 22.3	
	LmV	B	32.6	
19.	ePKIKP	A	06 14 29	<u>New Ireland Region</u> 5.11 S 153.32 E
	e	A	14 39	H = 05 55 32.9 h = 50 km MB = 5.4
	LmH	B	07 14.5	D = 124.47 Az = 331.4 (NEIC)
	LmV	B	14.5	LmH B 18s 0.4 µm M = 5.1
				LmV B 18 0.5 µm 5.2
19.	eP	A	12 35 21.5	<u>Off East Coast of Honshu, Japan</u>
	LmV	C	13 14.3	39.11 N 143.28 E
	LmH	C	15.4	H = 12 23 08.9 h = 34 km MB = 4.8
				D = 81.07 Az = 331.2 (NEIC)
19.	eiPKIKP	A	15 07 26	<u>Near North Coast of New Guinea</u>
	epPKIKP	A	07 44	4.89 S 144.50 E
	ePKKP	A	17 37	H = 14 48 42.3 h = 74 km MB = 5.5
	LmH	B	59.7	D = 119.80 Az = 327.5 (NEIC) h = 65 km
	LmV	B	16 00.0	PKIKPV A 1.0s 33.5nm
				pPKIKPV A 1.0 11.8nm
				PKKPV A 1.5 15.1nm
				LmH B 22 1.7 µm
				LmV B 20 1.6 µm
19.	ePKIKP	A	15 56 45	<u>New Britain Region</u> 5.52 S 150.60 E
	ePP	B	58 25	H = 15 37 46.1 h = normal MB = 5.3
	ePP	A	58 28	D = 123.49 Az = 329.9 (NEIC)
	ePPS	C	16 09 40	PKIKPV A 1.7s 18.2nm
	eSS	C	15 00	PPV A 1.7 142.0nm M = 5.2
	LmH	B	50.0	LmH B 20.5 8.2 µm 6.4
	LmV	B	50.0	LmV B 20 7.1 µm 6.3
19.	eP	A	17 10 33	<u>Taiwan Region</u> 24.25 N 122.38 E
				H = 16 58 09.3 h = 62 km MB = 5.4
				D = 83.74 Az = 323.2 (NEIC)
				PV A 1.5s 22.6nm M = 5.0

Day	Phase		h m s	Remarks
19.	e	A	17 29 04.5	PV A 1.6s 22.0nm
	e	A	29 21	
19.	ePg	A	22 00 04	Northern Italy 44.8 N 10.5 E
	eSg	A	01 27	H = 21 58 08 h = 0 km (BCIS)
				D = 5.92
19.	LmH	C	22 35	LmV C 16s 0.23/um
	LmV	C	35.3	
20.	eP	A	08 35 56.5	North Sea 56.54 N 0.68 E
				H = 08 33 48.2 h = 0 km MB = 4.8
				D = 8.78 Az = 127.7 (NEIC)
				explosion yield 10.3 t (ISC)
20.	ePKP	AB	10 11 23.5	Tonga Islands Region 17.51 S 172.68 W
	e	A	11 36	H = 09 51 42.0 h = normal MB = 4.8
	LmH	C	11 24.3	D = 146.76 Az = 355.0 (NEIC)
				PKPV A 1.5s 25.1nm
20.	eiP	A	10 47 08	Luzon, Philippine Islands
	LmH	B	11 35.8	15.27 N 120.26 E
	LmV	B	36.9	H = 10 34 11.5 h = normal MB = 5.4
				D = 89.70 Az = 322.9 (NEIC)
				PV A 1.1s 32.3nm M = 5.6
				LmH B 18 0.3/um 4.8
				LmV B 14 0.3/um 5.0
				In the intervall 20d 10h 58m - 21d 08h 58m the short-period seismographs (type A) have been out of operation.
20.	ePKP	B	17 54 00	South of Fiji Islands 19.57 S 177.30 E
	eX	B	54 22	H = 17 34 20.1 h = normal MB = 5.5
	eSS	B	18 16 30	D = 146.98 Az = 343.2 (NEIC)
	e	B	17 51	PKPV B 5s 0.5/um
	LmH	B	19 08.9	XV B 8 1.58/um
	LmV	B	09.4	LmH B 18 0.8/um M = 5.5
				LmV B 18 0.9/um 5.6

Day	Phase		h m s	Remarks
20.	LmH	B	22 45.5	**New Ireland Region** 5.9 S 153.9 E
	LmV	B	45.5	H = 21 25 55.9 h = 53 km MB = 5.4 (NEIC)
				D = 125.3
				LmH B 19s 0.5 µm M = 5.2
				LmV B 18 0.5 µm 5.2
20.	ePKP	B	23 42 43	**Tonga Islands Region** 17.4 S 172.9 W
	eX	B	43 08	H = 23 23 03.3 h = normal MB = 4.9 (NEIC)
				D = 146.6
				XV B 5s 0.39 µm
21.	eP	A	11 05 19.5	**Ryukyu Islands** 26.37 N 128.12 E
	LmH	B	42.0	H = 10 52 44.5 h = 26 km MB = 4.5
	LmV	B	47.8	D = 85.08 Az = 325.1 (NEIC)
				LmH B 16s 0.42 µm M = 4.9
				LmV B 16 0.18 µm 4.6
21.	eP	A	12 15 36.5	**Fox Islands, Aleutian Is.**
				52.00 N 170.51 W
				H = 12 03 41.4 h = 39 km MB = 4.9
				D = 77.72 Az = 358.6 (NEIC)
				PV A 1.4s 16.3nm M = 4.9
21.	ePKIKP	A	12 22 29	**Solomon Islands** 5.88 S 153.99 E
	ePP	A	24 30	H = 12 03 27.3 h = 27 km MB = 5.5
	PPm	A	24 40	D = 125.46 Az = 331.5 (NEIC)
	eSS	C	41 20	PPV A 2.5s 69.1nm M = 5.2
	LmH	B	13 19.8	LmH B 20.5 0.9 µm 5.4
	LmV	B	20.0	LmV B 20 0.9 µm 5.4
22.	+eP	A	22 19 41	**Honshu, Japan** 35.49 N 138.78 E
	epP	A	19 48	H = 22 07 19.3 h = 24 km MB = 5.4
	LmH	B	56.8	D = 82.43 Az = 329.2 (NEIC) h = 26 km
	LmV	B	23 02.0	PV A 1.7s 42.4nm M = 5.3
				pPV A 1.4 41.9nm
				LmH B 16 0.39 µm 4.9
				LmV B 13 0.55 µm 5.1

Day	Phase		h m s	Remarks
22.	eP	A	23 08 21	Off Coast of Hokkaido, Japan
	eX	A	08 33.5	42.63 N 148.55 E
				H = 22 56 30.8 h = 166 km MB = 4.0
				D = 79.80 Az = 333.8 (NEIC)
				XV A 1.2s 16.3nm
23.	ePKIKP	A	24 08(13)	New Hebrides Islands 14.62 S 167.34 E
				H = 23 49 05.0 h = 118 km
				D = 139.04 Az = 336.5 (NEIC)
				traces
23.	ePKIKP	A	24 14 20	New Ireland Region 4.66 S 153.10 E
	LmH	B	25 09.8	H = 23 55 25.1 h = 61 km MB = 5.2
	LmV	B	10.6	D = 123.97 Az = 331.4 (NEIC)
23.	ePKIKP	AB	07 04 23	Fiji Islands Region 17.63 S 178.89 W
	ePKHKP	AB	04 25	H = 06 45 45.1 h = 547 km MB = 5.2
	ePKP2	A	04(28)	D = 146.00 Az = 348.0 (NEIC)
	e	A	06 35	PKIKPV A 0.8s 13.5nm
				PKHKPV A 2.0 154.0nm
				PKHKPV B 5 0.99 μm
23.	eP	A	08 12 56	Off East Coast of Kamchatka
	LmH	B	47.7	52.68 N 160.68 E
	LmV	B	51.7	H = 08 01 22.8 h = normal MB = 4.8
				D = 73.79 Az = 340.1 (NEIC)
				PV A 1.0s 19.7nm M = 5.1
				LmH B 16 0.8 μm 5.1
				LmV B 14 0.4 μm 4.9
23.	e	A	13 44 23	
23.	e(P)	A	15 09 35	Turkey 40.0 N 26.1 E
	LmH	B	15.7	H = 15 06 00 (BCIS)
	LmV	B	15.7	D = 14.75
				LmH B 10s 0.28 μm M = 3.6
				LmV B 10 0.27 μm

Day	Phase		h m s	Remarks
23.	ePn	A	20 15 23.5	D c. 5.9
	ePg	A	15 52	
	eSn	A	16 28	
	eSg	A	16 57	
	LmH	B	17.5	
	LmV	B	18.3	
24.	eP	A	00 57 01	Iran 30.42 N 59.87 E
	LmH	B	01 19.0	H = 00 49 21.5 h = 56 km MB = 5.0
	LmV	B	21.8	D = 41.01 Az = 313.6 (NEIC)
				LmH B 15s 0.2 μm
				LmV B 14 0.1 μm
24.	eP	A	02 13 44	Hokkaido, Japan Region 41.54 N 141.88 E
				H = 02 01 50.7 h = 73 km MB = 4.7
				D = 78.43 Az = 330.3 (NEIC)
				traces
24.	e	A	05 28 04	
	e	A	28 44	
24.	eP	AB	11 51 42	Tadzhik-Sinkiang Border Region
	ePP	AB	53 27.5	39.46 N 73.19 E
	eS	B	58 18	H = 11 43 38.8 h = normal MB = 5.6
	eSS	B	12 01 36	D = 43.72 Az = 305.9 (NEIC)
	LmH	B	12.0	PV A 1.0s 55.1nm M = 5.3
	LmV	B	14.0	PV B 3.5 0.28 μm 5.4
				PPV A 1.8 94.5nm 5.3
				LmH B 16.5 2.0 μm 5.1
				LmV B 12 1.6 μm 5.2
24.	ePKP	AB	13 08 53	Tonga Islands Region 17.94 S 172.55 W
	eX	A	09 03	H = 12 49 11.2 h = normal MB = 5.4
	LmH	B	14 18.0	D = 147.20 Az = 355.1 (NEIC)
	LmV	B	18.0	PKPV A 1.8s 182.0nm
				XV A 1.5 65.3nm
				LmH B 20 0.3 μm M = 5.0
				LmV B 18 0.19 μm 4.9

Day	Phase		h m s	Remarks
25.	eP	A	01 23 22	<u>Hindu Kush Region</u> 36.38 N 70.75 E
				H = 01 15 34.1 h = 213 km MB = 4.9
				D = 44.03 Az = 308.2 (NEIC)
25.	eP	AB	03 56 38	<u>Off East Coast of Kamchatka</u>
	e	A	56 49	52.56 N 160.69 E
	LmH	B	04 31.3	H = 03 45 04.6 h = normal MB = 4.9
	LmV	C	35.4	D = 73.90 Az = 340.1 (NEIC)
				PV A 1.1s 16.1nm M = 4.9
				LmH B 15.5 0.8 μm 5.1
				LmV C 13 0.8 μm 5.2
25.	eP	AB	13 04 58	<u>Luzon, Philippine Islands</u>
	eS	BC	16 08	12.36 N 123.69 E
	eSS	C	22 20	H = 12 51 42.0 h = normal MB = 5.6
	LmH	B	48.2	D = 94.00 Az = 323.7 (NEIC)
	LmV	B	53.4	PV A 2.0s 68.4nm M = 5.7
				SH B 13 2.42 μm 6.3
				LmH B 21 23.6 μm 6.6
				LmV B 17 20.5 μm 6.7
25.	e	A	14 32 52	<u>Banda Sea</u> 5.94 S 129.79 E
	e	A	33 50	H = 14 14 23.4 h = 151 km MB = 5.7
				D = 112.23 Az = 322.7 (NEIC)
				traces
25.	eP	A	15 45 50	<u>Near Islands, Aleutian Is.</u>
				51.99 N 172.91 E
				H = 15 33 58.9 h = 16 km MB = 4.8 (NEIC)
				D = 76.6
				PV A 1.2s 10.2nm M = 4.7
25.	+iP	AB	15 53 10	<u>Near Islands, Aleutian Is.</u>
	epP	AB	53 20	52.15 N 173.10 E
	eS	AB	16 02 54	H = 15 41 21.3 h = 28 km MB = 5.8
	ePS	C	03 28	D = 76.39 Az = 348.0 (NEIC) h = 37 km
	eSS	C	07 55	PV A 1.9s 280.0nm M = 6.0

Day	Phase		h m s	Remarks
cont. 25.	LmH	B	16 30.9	PV B 9s 3.11 µm M = 6.3
	LmV	B	35.2	pPV A 2.5 614.0nm
				LmH B 17 11.8 µm 6.3
				LmV B 15 6.19 µm 6.2
25.	e	A	16 03 03	<u>Near Islands, Aleutian Is.</u>
				51.90 N 173.29 E
				H = 15 51 04.6 h = normal MB = 4.8
				D = 76.66 Az = 348.1 (NEIC)
25.	e	A	16 08 04	<u>Near Islands, Aleutian Is.</u>
				51.97 N 173.22 E
				H = 15 56 04.4 h = 22 km MB = 4.7
				D = 76.58 Az = 348.1 (NEIC)
25.	eiPKP	A	16 25 13	<u>Fiji Islands Region</u> 16.83 S 178.30 W
				H = 16 06 26.0 h = 450 km MB = 5.1
				D = 145.34 Az = 348.9 (NEIC)
				PKPV A 1.1s 40.3nm
25.	eP	A	16 51 09	<u>Kurile Islands</u> 45.12 N 150.76 E
				H = 16 39 15.0 h = 70 km MB = 4.7
				D = 78.25 Az = 335 (ISC)
				PV A 0.8s 7.7nm
25.	ePKIKP	A	17 03 35	<u>New Ireland Region</u> 5.57 S 152.99 E
	e	A	03 38.5	H = 16 44 36.5 h = 49 km MB = 5.0
				D = 124.71 Az = 331.1 (NEIC)
25.	e	A	20 04 05	
	e	A	04 18	
25.	eP	A	21 01 42	<u>Near Islands, Aleutian Is.</u>
				52.05 N 173.02 E
				H = 20 49 53.4 h = 27 km MB = 4.8
				D = 76.47 Az = 347.9 (NEIC)
				traces

Day	Phase		h m s	Remarks
25.	ePKIKP	A	22 43 25	New Ireland Region 5.22 S 153.52 E H = 22 24 26.0 h = 35 km MB = 4.9 D = 124.66 Az = 331.5 (NEIC)
25.	eP	A	23 32 50	Near Islands, Aleutian Is. 51.97 N 172.95 E H = 23 20 59.5 h = 25 km MB = 4.6 D = 76.54 Az = 347.9 (NEIC) traces
25.	e	A	23 59 12	Near Islands, Aleutian Is. 51.96 N 173.20 E H = 23 47 16.0 h = 36 km MB = 4.7 D = 76.58 Az = 348.1 (NEIC)
26.	e(P)	A	00 55 12.5	XV A 1.5s 20.1nm
	eX	A	55 15.5	
26.	eP diff	AB	01 38 52.5	New Ireland Region 4.94 S 153.17 E
	+ePKIKP	AB	42 17	H = 01 23 21.3 h = 48 km MB = 6.3
	PKIKPm	A	42 27	D = 124.25 Az = 331.4 (NEIC)
	ePP	B	44 00	P diff V A 1.0s 11.8nm
	e(SKKS)	B	51 17	P diff V B 13 1.97 µm
	e	B	53 49	PKIKPVm A 1.1 434.0nm
	ePS	B	54 12	PKIKPV B 7 5.0 µm
	eSS	B	02 01 25	PPV B 11 29.9 µm M = 7.5
	eiSSS	B	06 00	LmH B 20.5 342.0 µm 8.0
	LmH	B	41.2	LmV B 20 366.0 µm 8.0
	LmV	B	41.2	
26.	-iP	A	01 56 55.5	Southern Sinkiang Prov., China 39.91 N 77.17 E H = 01 48 33.9 h = normal MB = 6.0 D = 45.94 Az = 306.2 (NEIC) PV A 1.5s 146.0nm M = 5.7

Day	Phase		h m s	Remarks
26.	+iPKIKP	A	02 09 39	New Britain Region 4.72 S 152.8 E H = 01 50 36.7 h = 0 km D = 123.89 Az = 331 (ISC) PKIKPV A 2.3s 256.0nm
26.	+iPKIKP	A	02 12 41.5	New Ireland Region 4.66 S 153.26 E H = 01 53 45.1 h = normal MB = 5.6 D = 124.05 Az = 331.5 (NEIC)
26.	ePKIKP	A	02 18 07	New Britain Region 5.22 S 152.81 E H = 01 59 16 h = 87 km D = 124.32 Az = 331 (ISC)
26.	ePKIKP i	A A	02 21 39.5 21 45	New Britain Region 5.40 S 152.07 E H = 02 02 47 h = 11 km MB = 5.9 D = 124.12 Az = 331 (ISC)
26.	ePKIKP	A	02 38 20.5	New Britain Region 5.46 S 151.66 E H = 02 19 21.4 h = normal MB = 5.8 D = 123.97 Az = 330.5 (NEIC)
26.	+iPKIKP	A	02 43 54.5	New Britain Region 5.17 S 152.22 E H = 02 24 57.0 h = normal MB = 6.4 D = 123.99 Az = 330.8 (NEIC) PKIKPV A 1.5s 80.4nm
26.	eP e	A A	02 47 54 48 04	Southern Sinkiang Provinz 39.80 N 77.34 E H = 02 39 26.8 h = 0 km MB = 5.0 D = 46.11 Az = 306 (ISC)
26.	ePKIKP e	A A	02 48 35.5 48 42	New Britain Region 5.73 S 151.94 E H = 02 29 38.2 h = normal MB = 6.2 D = 124.34 Az = 330.5 (NEIC) PKIKPV A 2.0s 51.3nm
26.	ePKIKP	A	03 36 11	New Britain Region 5.64 S 151.70 E H = 03 17 14.7 h = normal MB = 5.1 D = 124.14 Az = 330.4 (NEIC)

Day	Phase		h m s	Remarks
26.	ePKIKP	A	03 45 24	<u>New Britain Region</u> 5.81 S 151.72 E H = 03 26 27.8 h = normal MB = 5.3 D = 124.30 Az = 330.4 (NEIC) PKIKPV A 1.7s 24.2nm
26.	ePKIKP	A	04 00 03	<u>New Britain Region</u> 5.05 S 152.80 E H = 03 41 03.7 h = normal MB = 4.8 D = 124.16 Az = 331.2 (NEIC)
26.	e	A	04 11 06	<u>New Ireland Region</u> 5.16 S 152.96 E H = 03 51 58.6 h = normal MB = 4.5 D = 124.34 Az = 331.2 (NEIC)
26.	ePKIKP	A	04 38 13	<u>New Britain Region</u> 5.4 S 152.6 E H = 04 19 15 MB = 6.1 D = 124.4 (ANUSSR)
26.	ePKIKP	A	05 04 20	<u>New Britain Region</u> 5.44 S 152.86 E H = 04 45 21.4 h = normal MB = 4.9 D = 124.53 Az = 331.1 (NEIC)
26.	ePKIKP iPKHKP iPKP2 e(pPKP)	A A A A	06 47 30 47 34.5 47 40 49 53	<u>West of Tonga</u> 20.0 S 176.5 W H = 06 27 51 MB = 5.5 D = 148.8 (ANUSSR) PKIKPV A 1.4s 9.7nm PKHKPV A 1.0 57.2nm PKP2V A 1.5 52.8nm
26.	iX	A	06 51 29.5	<u>Solomon Islands</u> 7.0 S 154.3 E H = 06 32 (RAB) D = 126.58 Az = 331 (ISC) XV A 0.6s 19.2nm
26.	ePKIKP	A	07 11 59	<u>New Britain Region</u> 5.57 S 151.65 E H = 06 53 02.4 h = normal MB = 5.3 D = 124.05 Az = 330.4 (NEIC)

Day	Phase		h m s	Remarks
26.	ePKIKP	A	07 16 41	New Britain Region 5.05 S 152.86 E
				H = 06 57 42.0 h = normal MB = 4.8
				D = 124.20 Az = 331.2 (NEIC)
				traces
26.	ePKIKP	A	07 18 31	New Britain Region 5.49 S 152.11 E
	e	A	18 38	H = 06 59 28.3 h = normal MB = 5.7
	eX	A	18 43	D = 124.22 Az = 330.7 (NEIC)
	ePP	B	20 15	XV A 1.7s 45.5nm
	ePS	B	30 14	PPV B 9 0.76 µm M = 6.1
	eSSS	B	42 00	LmH B 19 2.6 µm 5.9
	LmH	B	08 10.3	LmV B 18 2.0 µm 5.8
	LmV	B	17.0	
26.	ePKIKP	A	09 53 48	New Britain Region 5.73 S 151.60 E
				H = 09 34 51.5 h = normal MB = 5.3
				D = 124.17 Az = 330.3 (NEIC)
				PKIKPV A 1.5s 30.2nm
26.	ePKIKP	A	09 56 10	New Britain Region 5.91 S 151.57 E
				H = 09 37 12.3 h = normal MB = 5.4
				D = 124.31 Az = 330.3 (NEIC)
				PKIKPV A 1.5s 25.1nm
26.	ePKIKP	A	11 15 39	New Ireland Region 4.98 S 153.16 E
				H = 10 56 40.6 h = normal MB = 5.1
				D = 124.28 Az = 331.4 (NEIC)
				PKIKPV A 1.2s 12.2nm
26.	ePKIKP	A	12 23 55	New Britain Region 5.37 S 151.42 E
				H = 12 04 59.1 h = normal MB = 5.6
				D = 123.77 Az = 330.4 (NEIC)
26.	ePKIKP	A	13 15 40	New Britain Region 5.94 S 151.69 E
				H = 12 56 42.8 h = normal MB = 5.1
				D = 124.39 Az = 330.3 (NEIC)

Day	Phase		h m s	Remarks
26.	e	A	14 25 37.5	New Britain Region 5.8 S 152.8 E
				H = 14 05 49 h = 58 km MB = 5.0
				D = 124.86 Az = 331 (ISC)
26.	ePKIKP	A	15 08 41	New Ireland Region 4.80 S 153.25 E
				H = 14 49 44.0 h = 43 km MB = 4.8
				D = 124.16 Az = 331.5 (NEIC)
26.	ePKIKP	A	15 49 41	New Britain Region 5.64 S 151.50 E
				H = 15 30 43.7 h = normal MB = 5.0
				D = 123.99 Az = 330.3 (NEIC)
				traces
26.	ePKIKP	AB	15 54 34	New Britain Region 5.63 S 151.41 E
	eX	A	54 48	H = 15 35 37.8 h = normal MB = 5.4
	ePP	B	56 20	D = 123.99 Az = 330.3 (NEIC)
	ePS	B	16 06 15	PKIKPV A 1.2s 24.4nm
	eSS	B	13 05	PPV B 9 0.82/um M = 6.2
	LmH	B	43.4	LmH B 23 9.2/um 6.4
	LmV	B	51.9	LmV B 19 6.2/um 6.3
26.	ePKIKP	A	17 03 57	New Britain Region 5.99 S 151.50 E
	eX	A	04 07	H = 16 45 00.0 h = normal MB = 5.5
				D = 124.35 Az = 330.2 (NEIC)
				PKIKPV A 1.4s 37.2nm
				XV A 1.4 30.2nm
26.	e	A	17 25 42	New Ireland Region 4.88 S 153.12 E
				H = 17 06 37.0 h = normal MB = 4.5
				D = 124.17 Az = 331.4 (NEIC)
26.	ePKIKP	A	18 08 58	New Ireland Region 5.19 S 153.14 E
	eX	A	09 11	H = 17 49 59.7 h = normal MB = 5.3
				D = 124.45 Az = 331.3 (NEIC)
				XV A 1.5s 25.2nm
26.	ePKIKP1	A	19 34 27.5	New Britain Region 5.86 S 151.31 E
	+ePKIKP2	AB	34 31.5	H = 19 15 31.4 h = normal MB = 5.6
	+ePKIKP3	A	34 37.5	D = 124.14 Az = 330.2 (NEIC)

Day	Phase		h m s	Remarks
cont. 26.	+ePKIKP4	A	19 34 42.5	PKIKP1 A 1.3s 17.5nm
	ePP	B	36 15	PKIKP2 A 1.9 98.5nm
	ePKKP	A	44 17	PKIKP3 A 1.8 74.4nm
	eSS	C	53 00	PKIKP4 A 1.5 55.4nm
	LmH	B	20 28.5	PPV B 9 1.09/um M = 6.3
	LmV	B	28.5	PKKPV A 1.6 16.5nm
				LmH B 20 3.1/um 6.0
				LmV B 18 2.1/um 5.8
				e A 19 44 24
26.	e(Pg)	A	21 01 06.5	
26.	ePKIKP	A	21 19 33	<u>New Ireland Region</u> 4.40 S 153.04 E
				H = 21 00 41.0 h = 66 km MB = 5.1
				D = 123.71 Az = 331.5 (NEIC)
26.	eP	A	23 20 57	<u>Near East Coast of Honshu, Japan</u>
				35.95 N 140.02 E
				H = 23 08 41.1 h = 78 km MB = 5.0
				D = 82.54 Az = 329.8 (NEIC)
				PV A 1.5s 15.1nm M = 4.7
27.	ePKIKP	A	00 41 23	<u>New Britain Region</u> 5.45 S 152.38 E
	e	A	41 39	H = 00 22 25.3 h = normal MB = 5.4
	e	A	41 48	D = 124.31 Az = 330.8 (NEIC)
	LmH	B	01 42.5	LmH B 19s 0.95/um M = 5.5
	LmV	B	42.5	LmV B 18 0.83/um 5.4
27.	ePKIKP	A	01 23 12.5	<u>New Ireland Region</u> 4.63 S 153.24 E
				H = 01 04 19.2 h = 67 km MB = 4.6
				D = 124.01 Az = 331.5 (NEIC)
				PKIKPV A 1.0s 11.8nm
27.	-eiP1	AB	02 15 42	<u>Peru-Ecuador Border Region</u>
	eiP2	AB	15 54	2.75 S 77.43 W
	e	B	25 16	H = 02 02 49.6 h = 135 km MB = 6.3
	iSKS	AB	26 10	D = 91.50 Az = 39.6 (NEIC)
	eiS	B	26 28	P1V A 2.1s 1090.0nm M = 6.6
	e	A	26 37	P2V A 2.2 1420.0nm 6.7

Day	Phase		h m s	Remarks
cont. 27.	esS	B	02 27 32	P2V B 10s 10.15 µm M = 6.9
	eSS	B	33 00	pPKKPV A 1.8 283.0nm
	epPKKP	A	33 43	P'P'2V A 2.5 307.0nm
	eP'P'1	A	41 05	pP'P'1V A 2.2 109.0nm
	eP'P'2	A	41 19	pP'P'2V A 2.8 365.0nm
	epP'P'1	A	41 35	P'P'P'V A 2.5 107.5nm
	epP'P'2	A	41 47	LmH B 35
	LmH	B	42.5	LmV B 40
	LmV	B	47.0	
	eP'P'P'	A	03 02 10	
27.	eP	A	03 29 34	Near Islands, Aleutian Is. 52.11 N 173.05 E H = 03 17 46.1 h = 38 km MB = 4.8 D = 76.42 Az = 348.0 (NEIC)
27.	ePKIKP	A	04 23 02.5	New Britain Region 5.57 S 151.53 E H = 04 04 07.4 h = 54 km MB = 5.1 D = 124.00 Az = 330.4 (NEIC)
27.	eX	A	05 53 08	New Britain Region 6.02 S 151.58 E H = 05 33 58.7 h = normal MB = 5.0 D = 124.41 Az = 330.2 (NEIC) XV A 2.3s 30.4nm
27.	e	A	06 51 08	New Britain Region 5.89 S 151.78 E H = 06 31 45.4 h = normal MB = 5.4 D = 124.40 Az = 330.4 (NEIC)
27.	e	A	07 04 35	New Britain Region 5.40 S 152.89 E H = 06 45 20.4 h = 34 km MB = 5.5 D = 124.51 Az = 331.1 (NEIC)
27.	ePKIKP	A	07 54 39	New Britain Region 5.89 S 151.56 E
	LmH	B	08 53.5	H = 07 35 41.1 h = 40 km MB = 5.4
	LmV	B	54.5	D = 124.29 Az = 330.3 (NEIC)
				PKIKPV A 1.9s 30.3nm
				LmH B 18 1.33 µm M = 5.6
				LmV B 18 1.3 µm 5.6

Day	Phase		h m s	Remarks
27.	ePKIKP	A	09 37 33	New Britain Region 5.35 S 152.70 E H = 09 18 35.0 h = normal MB = 5.3 D = 124.38 Az = 331.0 (NEIC) PKIKPV A 1.8s 20.3nm
27.	ePKIKP	A	13 06 33	New Ireland Region 5.36 S 153.11 E H = 12 47 36.8 h = 51 km D = 124.59 Az = 331 (ISC)
27.	ePKIKP	A	13 37 40	New Britain Region 5.72 S 151.87 E H = 13 18 44.5 h = 47 km MB = 5.0 D = 124.30 Az = 330.4 (ISC)
27.	e	A	13 50 52	New Britain Region 5.03 S 152.80 E H = 13 31 46.8 h = normal MB = 4.7 D = 124.15 Az = 331.2 (NEIC)
27.	e	A	14 21 28	New Britain Region 5.09 S 152.86 E H = 14 02 25.3 h = 53 km MB = 4.6 D = 124.22 Az = 331.2 (NEIC)
27.	eP	AB	14 37 15	Andaman Islands Region 13.80 N 95.84 E
	eS	B	46 59	H = 14 25 31.2 h = 36 km MB = 5.4
	eSS	B	52 12	D = 75.82 Az = 319.2 (NEIC)
	eSSS	C	55 15	PV A 2.0s 85.5nm M = 5.4
	LmH	B	15 17.5	PV B 8 0.60 μm 5.7
	LmV	B	17.5	LmH B 18.5 3.1 μm 5.7
				LmV B 19 3.3 μm 5.7
27.	ePKIKP	A	18 27 36	New Britain Region 5.63 S 151.38 E
	LmH	B	19 24.5	H = 18 08 41.2 h = 46 km MB = 5.4
	LmV	B	24.5	D = 123.98 Az = 330.3 (NEIC)
				PKIKPV A 1.5s 20.0nm
				LmH B 20 0.72 μm M = 5.3
				LmV B 18 0.89 μm 5.5
27.	ePKIKP	A	21 06 55	New Ireland Region 5.15 S 152.97 E
	eX	A	06 03.5	H = 20 47 56.5 h = 43 km MB = 5.5
	e	A	06 13	D = 124.34 Az = 331.2 (NEIC)

Day	Phase		h m s	Remarks
cont. 27.	ePP	B	21 08 40	PKIKPV A 1.9s 60.6nm
	e	A	08 46	XV A 1.5 44.2nm
	eSP	B	18 44	PPV B 10 1.78/um M = 6.4
	eSPP	B	19 59	LmH B 20 9.1/um 6.4
	e	BZ	24 00	LmV B 18 7.0/um 6.4
	e	BZ	24 24	
	eSS	B	25 38	
	ePSPS	B	26 00	
	e	B	26 50	
	eSSS	B	30 30	
	e	C	31 28	
	LmH	B	54.0	
	LmV	B	22 06.5	
27.	ePKIKP	A	21 12 43	New Britain Region 5.11 S 152.95 E
				H = 20 53 39.9 h = 0 km MB = 5.2
				D = 124.29 Az = 331 (ISC)
				PKIKPV A 1.6s 22.0nm
27.	eX	A	21 20 35	XV A 2.0s 34.2nm
27.	ePKIKP	A	23 54 19	New Britain Region 6.34 S 151.15 E
	ePP	A	56 00	H = 23 35 21.5 h = normal MB = 5.2
	LmH	B	24 48.7	D = 124.48 Az = 329.9 (NEIC)
	LmV	B	54.9	PKIKPV A 1.6s 33.0nm
				LmH B 17 0.97/um M = 5.5
				LmV B 18 1.0/um 5.5
28.	ePKIKP	A	00 34 51	New Britain Region 6.05 S 151.55 E
				H = 00 15 55.2 h = 53 km MB = 4.9
				D = 124.42 Az = 330.2 (NEIC)
28.	ePKIKP	AB	01 29 21.5	New Britain Region 5.14 S 152.88 E
	ePP	AB	31 09	H = 01 10 24.1 h = normal MB = 6.0
	eSKS	B	36 24	D = 124.28 Az = 331.2 (NEIC)
	eiSKKS	B	38 04	PKIKPV A 2.0s 77.0nm
	ePS	B	41 00	PKIKPV B 4 3.3/um
	ePPS	B	42 25	PPV A 2.0 102.5nm M = 5.9
	eSS	C	48 00	PPV B 9 0.76/um 6.1

Day	Phase		h m s	Remarks
cont. 28.	eSSS	C	01 53 00	LmH B 17 7.9 /um M = 6.4
	LmH	B	02 24.9	LmV B 18 7.8 /um 6.4
	LmV	B	27.0	
28.	ePKHKP	A	02 48 46	Fiji Islands Region 20.75 S 177.34 W
				H = 02 29 41.9 h = 411 km MB = 4.6
				D = 149.34 Az = 348.8 (NEIC)
28.	ePKIKP	A	03 06 18	New Britain Region 5.11 S 152.74 E
				H = 02 47 23.7 h = 61 km MB = 5.1
				D = 124.19 Az = 331.1 (NEIC)
				PKIKPV A 1.2s 8.1nm
28.	ePKIKP	A	06 02 17	Fiji Islands Region 20.34 S 177.91 W
	iPKHKP	A	02 22.0	H = 05 43 35.0 h = 550 km MB = 4.7
	ePKP2	A	02 27.5	D = 148.83 Az = 348.3 (NEIC)
				PKHKPV A 1.2s 20.4nm
28.	ePKIKP	A	08 07 58	New Britain Region 5.81 S 151.71 E
	ePP	B	09 40	H = 07 48 58.2 h = normal MB = 5.2
	e	A	09 48	D = 124.29 Az = 330.4 (NEIC)
	LmH	B	09 00.3	LmH B 20s 0.66 /um M = 5.3
	LmV	B	11.5	LmV B 16 0.71 /um 5.4
28.	e(Pg)	A	10 07 53.5	
	e(Sg)	A	09 26.5	
28.	ePKIKP	A	10 36 27.5	New Britain Region 5.56 S 152.47 E
				H = 10 17 29.4 h = normal MB = 5.7
				D = 124.45 Az = 330.8 (NEIC)
				PKIKPV A 1.7s 18.2nm
28.	ePKIKP	A	14 21 49	New Britain Region 5.95 S 151.44 E
				H = 14 02 52.4 h = 37 km MB = 5.0
				D = 124.28 Az = 330.2 (NEIC)
				PKIKPV A 1.3s 10.9nm

Day	Phase		h m s	Remarks
29.	ePKIKP	A	03 59(57)	New Ireland Region 5.40 S 153.10 E
				H = 03 40 57.9 h = 43 km MB = 4.3
				D = 124.62 Az = 331.2 (NEIC)
29.	ePn	A	10 27 25	Northern Italy 44.70 N 10.19 E
	e	A	27 48	H = 10 25 55.4 h = normal MB = 4.3
	ePg	A	27 50	D = 6.03 Az = 8.7 (NEIC)
	ei	A	27 57.5	
	eSn	A	28 35	
	eSg	A	29 12	
	LmH	B	30.3	
	LmV	B	30.3	
29.	ePKIKP	A	10 37 35	New Ireland Region 5.32 S 152.99 E
	LmH	B	11 36.5	H = 10 18 38.2 h = 44 km MB = 4.9
	LmV	B	37.3	D = 124.50 Az = 331.2 (NEIC)
				PKIKPV A 2.0s 25.6nm
				LmV B 18 0.26/um M = 4.9
29.	+iP	AB	22 30 07.5	Near Islands, Aleutian Is.
	eipP	AB	30 17.5	52.08 N 173.38 E
	ePP	BC	33 00	H = 22 18 20.1 h = 42 km MB = 5.6
	eSS	B	40 00	D = 76.49 Az = 348.2 (NEIC)
	ePS	C	40 32	PV A 1.4s 60.5nm M = 5.4
	eSS	C	44 55	pPV A 1.6 121.0nm
	LmH	B	23 06.4	LmH B 17 1.25/um 5.3
	LmV	B	12.4	LmV B 15 0.69/um 5.1
30.	+ePKIKP	B	00 19 07	Near Islands, Aleutian Is.
	ePKIKP	A	19 10	52.10 N 173.46 E
	epPKIKP	B	19 16	H = 00 07 22.4 h = 45 km MB = 4.9
				D = 76.49 Az = 348.2 (NEIC) h = 32 km
				PKIKPV B 4s 0.45/um
				PKIKPV A 2.0 34.2nm
30.	e	A	05 37 05	
	e(Sg)	A	38 17	

Day	Phase		h m s	Remarks
30.	ePKIKP	AB	06 37 07	New Britain Region 4.54 S 152.87 E
	+iX	A	37 24	H = 06 18 12.0 h = 42 km MB = 5.7
	ePP	B	38 50	D = 123.75 Az = 331.4 (NEIC)
	e	A	38 52	PKIKPV A 1.1s 44.4nm
	e	A	44 32	XV A 1.4 37.2nm
	e	A	45 12	PPV B 8 0.38 /um M = 5.8
	ePS	B	38 40	PPV A 1.4 23.3nm 5.3
	ePPS	C	50 25	LmH B 20 1.53 /um 5.7
	eSS	C	51 20	LmV B 20 1.5 /um 5.6
	eSSS	C	07 00 40	
	LmH	B	32.2	
	LmV	B	36.2	
30.	iPn	A	08 30 11	Czechoslovakia $50^{o}36.3'$ N $14^{o}09.5'$ E
	eSg	A	30 33	explosion yield 6.5 t (PRU)
				D = 1.60
30.	ePKIKP	A	13 42 55.5	New Britain Region 6.04 S 151.49 E
	e	A	43 05	H = 13 23 58.3 h = normal MB = 5.3
	ePP	B	44 38	D = 124.38 Az = 330.2 (NEIC)
	ePS	B	54 30	PKIKPV A 1.5s 20.1nm
	ePPS	B	56 00	PPV B 10 0.32 /um M = 5.7
	eSS	C	14 01 22	LmH B 16 0.63 /um 5.4
	eSSS	C	06 30	LmV B 16 0.59 /um 5.3
	LmH	B	44	
	LmV	B	45	
30.	eSS	C	19 07.8	New Ireland Region 5.7 S 153.8 E
	LmH	B	48.0	H = 18 30 06.6 h = 36 km MB = 5.1
	LmV	B	48.0	MS = 5.4 (NEIC)
				D = 125.3
				LmH B 17s 0.26 /um M = 5.0
				LmV B 18 0.26 /um 4.9
30.	e(PKP)	A	20 21 40	Near Coast of Central Chile
	eX	A	21 43	32.16 S 71.53 W
				H = 20 03 39.2 h = 52 km MB = 4.8
				D = 110.12 Az = 42.3 (NEIC)
				(PKP)V A 1.2s 8.1nm
				XV A 1.2 18.3nm

Day	Phase		h m s	Remarks
30.	ePKIKP	A	21 05 57	<u>New Ireland Region</u> 5.25 S 153.27 E
	ePP	B	07 45	H = 20 47 01.56 h = normal
	ePS	C	17 30	MB = 5.5 MS = 5.3 (NEIC)
	eSS	C	24 40	D = 124.7
	eSSS	C	29 35	LmH B 18s 0.53 /um M = 5.2
	LmH	B	22 03.0	LmV B 17 0.62 /um 5.3
	LmV	B	07.5	
31.	e	A	02 04 56	<u>Northern Italy</u> 44.8 N 10.3 E
	ePg	A	05 05	H = 02 03 05 (BCIS)
	eSg	A	06 21	D = 5.94
	LmH	B	07.5	
	LmV	B	07.5	
31.	e	A	09 15 21	1.2s 10.2nm
31.	ePKIKP	A	21 41 22	<u>New Britain Region</u> 6.1 S 153.0 E
	e	A	41 35	H = 21 22 19 h = 0 km
				D = 125.17 Az = 331 (ISC)

Day	Phase		h m s	Remarks
1.	eiP	AB	02 17 48	Kurile Islands 50.39 N 156.82 E
	e	A	17 54	H = 02 06 06.6 h = 20 km MB = 5.6
	e	A	18 07.5	D = 75.06 Az = 337.9 (NEIC)
	e	A	18 14	PV A 1.2s 67.0nm M = 5.5
	e	A	19 08.5	LmH B 17 1.74/um 5.4
	e	A	20 17	LmV B 17 1.48/um 5.4
	eS	B	27 22	
	ePS	B	28 00	
	LmH	B	55.6	
	LmV	B	56.4	
1.	eP	A	10 54 28.5	Mid-Indian Rise 16.20 S 67.07 E
	e	A	54 36	H = 10 42 05.9 h = normal MB = 4.9
	e	A	54 39.5	D = 82.35 Az = 328.1 (NEIC)
1.	ePKIKP	A	16 22(44)	New Britain Region 6.20 S 151.45 E
	e	A	22 49	H = 16 03 39.2 h = normal MB = 5.0
	LmH	C	17 15.0	D = 124.50 Az = 330.1 (NEIC)
	LmV	C	19.2	
1.	eP	A	19 07 10	Kurile Islands 44.42 N 148.92 E
				H = 18 55 10.6 h = normal MB = 4.5
				D = 78.31 Az = 333.8 (NEIC)
				PV A 1.6s 19.2nm M = 4.9
1.	e	A	22 08 49	New Britain Region 5.34 S 152.66 E
	LmH	B	23 05.9	H = 21 49 46.3 h = 38 km MB = 5.0
	LmV	B	06.1	D = 124.35 Az = 331.0 (NEIC)
				LmH B 20s 0.53/um M = 5.2
				LmV B 20 0.52/um 5.2
2.	ePKIKP	A	00 38 52.5	New Britain Region 5.56 S 152.61 E
	e	B	39 00	H = 00 19 54.2 h = 36 km MB = 5.6
	ePP	B	40 40	D = 124.52 Az = 330.9 (NEIC)
	ePS	C	50 28	PKIKPV A 1.4s 18.6nm
	ePPS	C	51 55	PPV B 8 0.55/um M = 6.0
	eSS	C	57 32	LmH B 18.5 1.18/um 5.6
	eSSS	C	01 02 32	LmV B 17 1.35/um 5.7
	LmH	B	38.4	
	LmV	B	40.4	

Day	Phase		h m s	Remarks
2.	e	A	04 12 07	Upper Silesia, Poland 50.28 N 18.97 E
	e	A	12 12	H = 04 09 42.7 (WAR)
	eSg	A	12 16	D = 4.72
2.	ePn	A	04 18 21	Northern Italy 44.58 N 10.38 E
	e	A	18 37	H = 04 16 46.3 h = 56 km
	ePg	A	18 48	D = 6.13 Az = 7.4 (NEIC)
	eSg	A	20 00	LmH B 10s 0.35 µm M = 3.1
	LmH	B	21.2	LmV B 8 0.38 µm
	LmV	B	21.3	
2.	ePg	A	05 48 13	Northern Italy 44.64 N 10.33 E
	eSg	A	49 32	H = 05 46 06.6 h = 0 km (ISC)
				D = 6.05
2.	-iP	AB	07 36 57.3	Hokkaido, Japan Region 41.38 N 143.46 E
	i	B	37 17	H = 07 24 56.8 h = 51 km MB = 6.6
	ePP	B	40 00	D = 79.15 Az = 331.1 (NEIC)
	iS	BC	46 47	PV A 1.7s 1575.0nm M = 6.7
	e(SKS)	B	47 20	PV B 11 17.9 µm 7.0
	eiSS	B	52 00	PPV B 14 5.75 µm 6.5
	eSSS	B	55 15	SH B 11.5 38.8 µm 7.4
	ePKKP	A	55 38	PKKPV A 2.4 124.0nm
	e	A	55 49.5	P'P'V A 1.6 30.2nm
	eP'P'	A	08 03 37.5	XV A 1.8 108.0nm
	e	A	03 51	LmH B 19 190.0 µm (NS-component off scale)
	e	A	03 51	LmV B 20 222.0 µm
	e	A	03 58	P'P'P'V A 2.9 236.0nm
	eX	A	04 03	
	LmH	B	15.5	
	LmV	B	15.5	
	eP'P'P'	A	23 58.5	
2.	eP	A	09 15 11	Molucca Passage 1.71 N 126.53 E
				H = 09 01 09.2 h = normal MB = 5.2
				D = 104.20 Az = 323.4 (NEIC)
				traces

Day	Phase		h m s	Remarks
2.	ePKIKP	A	11 47 11	New Ireland Region 5.14 S 153.33 E H = 11 28 14.4 h = 54 km MB = 4.6 D = 124.50 Az = 331.4 (NEIC) PKIKPV A 1.2s 12.2nm
2.	eP	A	13 12 02.5	Hokkaido, Japan Region 41.30 N 143.47 E
	e	A	12 09	H = 13 00 01.5 h = 54 km MB = 5.6
	e	AB	12 17	D = 79.23 Az = 331.1 (NEIC)
	ePP	A	15 00	PV A 1.2s 28.4nm M = 5.1
	eSS	B	21 56	PPV A 1.5 32.6nm 5.3
	LmH	B	50.4	LmH B 20 1.2/um
	LmV	B	50.4	LmV B 20 1.03/um
2.	eX	A	14 29 22	XV A 0.8s 11.5nm
2.	e	A	17 08 28	Yellow Sea 35.7 N 124.0 E H = 16 56 41 h = 52 km D = 75.46 Az = 323 (ISC)
2.	LmH	B	17 39.4	Republic of South Africa 27.8 S 27.0 E
	LmV	B	45.2	H = 16 52 28.7 h = normal MB = 5.1 (NEIC) D = 79.2 LmH B 18s 0.30/um M = 4.7 LmV B 12 0.22/um 4.7
2.	e(PKIKP)	A	19 23 12	New Ireland Region 5.72 S 153.48 E H = 19 04 11.5 h = 31 km MB = 4.6 D = 125.07 Az = 331.3 (NEIC) PKIKPV A 0.8s 9.6nm
2.	eP	A	19 26 06	Southern Greece 36.9 N 21.98 E H = 19 22 26 h = 64 km MB = 4.3 D = 15.63 Az = 335 (ISC)
2.	ePKIKP	A	20 29 03	New Britain Region 6.26 S 151.45 E
	e	A	29 10	H = 20 10 04.3 h = 39 km MB = 4.9 D = 124.55 Az = 330.1 (NEIC)

Day	Phase		h m s	Remarks
2.	ePKP	A	22 11(31)	Fiji Islands Region 15.15 S 178.61 W
				H = 21 52 36.4 h = 320 km MB = 4.8
				D = 143.65 Az = 349.0 (NEIC)
				traces
3.	LmH	B	01 56.5	LmH B 17s 0.21/um
	LmV	B	56.5	LmV B 18 0.32/um
3.	eP	A	05 42 33	North Atlantic Ridge 28.39 N 39.18 W
	e	A	42 39	H = 05 34 26.5 h = normal MB = 5.0
	LmH	B	06 00.7	D = 44.00 Az = 45.2 (NEIC)
	LmV	B	00.7	PV A 1.4s 27.9nm M = 4.8
3.	ePKIKP	A	09 04 47	New Britain Region 6.22 S 151.44 E
				H = 08 45 50.6 h = 38 km MB = 5.0
				D = 124.51 Az = 330.1 (NEIC)
3.	LmH	B	18 00.4	Bismarck Sea 3.5 S 148.9 E
	LmV	B	00.4	H = 16 50 12.0 h = 9 km MB = 4.7 (NEIC)
				D = 120.9
				LmH B 20s 0.64/um M = 5.3
				LmV B 20 0.74/um 5.3
3.	eP	A	21 07 35.5	North Atlantic Ridge 28.41 N 39.25 W
				H = 20 59 29.7 h = normal MB = 4.7
				D = 44.03 Az = 45.3 (NEIC)
4.	+iP	AB	00 32 25.5	Hindu Kush Region 36.35 N 70.72 E
	esP	B	33 30	H = 00 24 36.5 h = 206 km MB = 5.6
	esPP	B	35 15	D = 44.03 Az = 308.2 (NEIC)
	ei	A	35 20.5	PV A 1.8s 392.0nm M = 5.6
	iS	B	38 41	PV B 9 1.42/um 5.5
	eisS	B	39 56	SH B 10 2.33/um 5.9
	eSS	C	41 48	
	e	B	42 15	
	e	C	42 49	

Day	Phase		h m s	Remarks
4.	eP	A	02 06 52	Hindu Kush Region 36.41 N 70.79 E
	e	A	12 04.5	H = 01 59 03.1 h = 206 km MB = 5.0
				D = 44.04 Az = 308.1 (NEIC)
4.	eP	A	14 15 29	North Atlantic Ridge 46.56 N 27.41 W
	LmH	B	23.5	H = 14 09 58.0 h = normal MB = 5.1
				D = 25.91 Az = 66.5 (NEIC)
				PV A 1.1s 16.1nm M = 4.5
				LmH B 12 0.4/um 4.2
4.	eP	A	18 35 58.5	Fox Islands, Aleutian Is.
				52.39 N 168.60 W
				H = 18 24 05.4 h = 33 km MB = 4.3
				D = 77.34 Az = 359.9 (NEIC)
				PV A 1.4s 14.0nm M = 4.8
4.	ePKP2	A	19 27 27	Balleny Islands Region 62.75 S 156.98 E
	LmH	B	20 45.5	H = 19 06 52.88 h = normal MB = 4.8 (NEIC)
	LmV	B	45.5	D = 157.8
				LmH B 19s 0.4/um M = 5.2
				LmV B 18 0.6/um 5.4
4.	eP	A	21 10 06	Near East Coast of Kamchatka
				55.41 N 161.90 E
				H = 20 58 52.9 h = 90 km MB = 5.1
				D = 71.45 Az = 340.6 (NEIC)
				PV A 0.9s 21.4nm M = 5.0
5.	eiP1	AB	02 08 48.5	Central Mid Atlantic Ridge
	-iP2	A	08 54.5	0.85 S 22.13 W
	Pm	AB	09 08	H = 01 58 51.7 h = normal MB = 6.3
	iS	B	16 56	D = 58.81 Az = 24.4 (NEIC)
	LQ	B	24.8	P1V A 2.4s 442.0nm M = 6.1
	LR	B	27.6	P2V A 2.4 1060.0nm 6.5
	LmH	B	35.9	PmV A 1.6 571.0nm 6.4
	eP'P'	A	38 46	PmV B 6 5.88/um 6.9
	LmV	B	40.7	SH B 23.5 131.0/um 7.3
				LmH B 16.5 77.9/um 6.9
				LmV B 14 60.0/um 6.9
				P'P'V A 1.7 24.2nm

Day	Phase		h m s	Remarks
5.	ePP	A	05 08 52	<u>Celebes</u> 2.69 S 119.92 E H = 04 50 31.10 h = normal MB = 5.6 (NEIC) D = 103.7 PPV A 2.0s 34.2nm M = 5.5
5.	eP	A	10 50 28	<u>Central Mid-Atlantic Ridge</u> 1.14 S 23.94 W H = 10 40 23.5 h = normal MB = 4.8 D = 59.82 Az = 25.4 (NEIC) PV A 1.2s 14.2nm M = 5.0
5.	eP	A	11 07 47.5	<u>Kurile Islands</u> 45.03 N 151.72 E H = 10 55 50.8 h = 60 km MB = 4.5 D = 78.62 Az = 335.4 (NEIC) PV A 1.0s 13.8nm M = 4.8
5.	eP	A	13 36 33.5	<u>Andaman Islands Region</u> 12.62 N 95.73 E H = 13 24 44.9 h = normal D = 76.64 Az = 319.4 (NEIC) PV A 1.5s 25.2nm M = 5.1
5.	eP	A	14 02 43	<u>Fox Islands, Aleutian Is.</u>
	e	A	03 26	55.65 N 165.01 W
	LmV	B	44.0	H = 13 51 08.3 h = normal MB = 5.2
	LmH	B	44.5	D = 74.04 Az = 2.2 (NEIC) PV A 1.3s 24.0nm M = 5.1 LmH B 17 0.5/um 4.9 LmV B 16 0.2/um 4.6
5.	eP	A	17 30 45.5	<u>Central Mid-Atlantic Ridge</u> 0.73 S 21.87 W H = 17 20 49.3 h = normal MB = 4.6 D = 58 59 Az = 24.3 (NEIC) PV A 1.2s 8.1nm M = 4.7
5.	e	A	20 21 34	

Day	Phase		h m s	Remarks
5.	eP	A	22 48 56	Andaman Islands Region 12.57 N 94.77 E
	epP	A	49 04	H = 22 37 10.9 h = 31 km MB = 5.0
	esP	A	49 07	D = 76.06 Az = 319.4 (NEIC) h = 30 km
	LmV	B	23 27.5	PV A 1.2s 16.3nm M = 4.9
	LmH	B	27.9	pPV A 1.6 38.5nm
				sPV A 1.4 32.6nm
				LmH B 19 1.0/um 5.1
				LmV B 21 0.5/um 4.8
5.	eP	A	23 10 48	Costa Rica 9.64 N 84.07 W
	eX	A	10 55.5	H = 22 58 08.3 h = normal MB = 4.9
				D = 86.21 Az = 39.4 (NEIC)
				PV A 2.0s 59.9nm M = 5.5
				XV A 1.6 30.2nm
5.	e	A	23 18 34.5	Upper Silesia, Poland 50.30 N 18.86 E
	e	A	18 39	H = 23 16 07.8 M = 3.4 (WAR)
	eSg	A	18 41	D = 4.64
6.	e	A	01 35 44	Upper Silesia, Poland 50.27 N 18.93 E
	eSg	A	35 53.5	H = 01 33 14.4 M = 2.7 (WAR)
				D = 4.68
6.	e	A	05 31 38	Switzerland 46.9 N 8.3 E
	eSg	A	31 40	H = 05 29 27 h = 0 km
				D = 4.36 Az = 29 (ISC)
6.	ePKIKP	A	19 02 21	New Ireland Region 4.52 S 153.24 E
				H = 18 43 29.2 h = 68 km MB = 4.5
				D = 123.91 Az = 331.6 (NEIC)
				PKIKPV A 1.2s 12.02nm
6.	eiPKIKP	AB	23 10 59	Fiji Islands Region 20.20 S 177.94 W
	eiPKHKP	AB	11 04	H = 22 52 15.3 h = 530 km MB = 5.3
	iPKP2	A	11 10	D = 148.69 Az = 348.3 (NEIC)
	epPKP	AB	13 07	PKIKPV A 1.5s 52.8nm
				PKHKPV A 1.4 263.0nm
				PKHKPV B 2.5 0.49/um
				PKP2V A 1.6 110.0nm

Day	Phase		h m s	Remarks
7.	e	A	04 22 20	
7.	e(Pg)	A	05 51 49	
	e(Sg)	A	52 25	
7.	ePKP	A	06 11 24.5	Samoa Islands Region 15.73 S 172.88 W
				H = 05 51 50.6 h = 38 km MB = 3.8
				D = 144.99 Az = 355.0 (NEIC)
				PKPV A 0.8s 7.7nm
7.	ePKIKP1	A	07 12 23	New Hebrides Islands 13.94 S 167.18 E
	ePKIKP2	AB	12 32.5	H = 06 53 27.0 h = 178 km MB = 5.4
	eSKP1	A	15 54	D = 138.37 Az = 336.6 (NEIC)
	eSKP2	A	16 02.5	PKIKPV A 1.4s 51.2nm
7.	e	A	09 14 01	XV A 0.9s 11.7nm
7.	e	A	10 28 11	
	e	A	28 14	
7.	eP	A	15 30 37	Kashmir-Sinkiang Border Region
	ePP	A	32 24	36.10 N 77.66 E
	eScS	B	40 24	H = 15 21 52.5 h = normal MB = 4.8
	e	B	41 22	D = 48.57 Az = 309.1 (NEIC)
	LmH	B	50.7	PV A 1.9s 53.0nm M = 5.3
	LmV	B	51.9	LmH B 13 0.4/um 4.5
				LmV B 16 0.3/um 4.4
7.	eP	A	17 11 30	Turkey 38.87 N 29.90 E
	LmH	B	17.4	H = 17 07 24.2 h = 20 km MB = 4.5
	LmV	B	20.4	D = 17.47 Az = 318.3 (NEIC)
				PV A 2.0s 68.4nm M = 4.4
				LmH B 16 0.6/um 3.9
				LmV B 12 0.3/um 3.9
7.	e(P)	A	18 53 53.5	Poland 50.30 N 18.88 E
				H = 18 51 19.2 (WAR)
				D = 4.7
				PV A 0.9s 9.7nm

Day	Phase		h m s	Remarks
7.	LmH	C	21 05.3	New Britain Region 5.76 S 152.28 E H = 19 46 28.4 h = 25 km MB = 4.9 (NEIC) D = 124.6 traces
7.	ePKIKP	A	23 29 26	New Britain Region 6.68 S 148.52 E
	LmH	B	24 25.5	H = 23 10 31.4 h = normal MB = 4.6
	LmV	B	26.4	D = 123.43 Az = 328.6 (NEIC) LmH B 20s 0.2/um M = 4.8 LmV B 16 0.2/um 4.9
8.	eP	AB	02 52 06.5	Carlsberg Ridge 8.23 N 58.36 E H = 02 42 20.5 h = normal MB = 5.1 D = 57.22 Az = 326.5 (NEIC)
8.	eP	A	04 01 20	Central Mid-Atlantic Ridge 0.68 S 22.19 W H = 03 51 22.9 h = normal MB = 4.7 D = 58.67 Az = 24.5 (NEIC) PV A 1.6s 16.5nm M = 4.9
8.	ePKIKP	A	06 51 04	Kermadec Islands 30.57 S 178.12 W
	e	A	51 39.5	H = 06 31 11.8 h = 51 km MB = 5.2 D = 158.66 Az = 342.8 (NEIC) PKIKPV A 2.0s 25.6nm
8.	LmH	B	17 25.5	
	LmV	B	26.1	
8.	eP	A	19 42 45	Greece 38.57 N 21.78 E
	LmH	B	48.9	H = 19 39 30.5 h = 56 km MB = 4.4
	LmV	B	48.9	D = 14.06 Az = 332.5 (NEIC) LmH B 13s 0.5/um M = 3.8 LmV B 12 0.3/um
9.	eP1	A	03 01 08.5	Iran 36.25 N 52.72 E
	eP2	AB	01 17	H = 02 54 36.8 h = 27 km MB = 5.2
	ePP	A	02 20	D = 32.67 Az = 309.2 (NEIC)
	eS	B	06 27	P2V A 2.4s 124.4nm M = 5.4

Day	Phase		h m s	Remarks
cont. 9.	ei(sS)	B	03 06 40	P2V B 10s 0.65 µm M = 5.5
	LmH	B	18.7	PPV A 1.5 55.3nm 5.2
	LmV	B	20.2	LmH B 17 2.7 µm 5.0
				LmV B 17 1.8 µm 5.0
9.	e	A	04 45 15.5	Turkey 37.55 N 29.76 E
	LmH	B	52.5	H = 04 40 46.5 h = 11 km MB = 4.8
				D = 18.40 Az = 321.1 (NEIC)
9.	ePKIKP	AB	12 30 57	Solomon Islands 5.89 S 154.31 E
	eiPP	AB	32 50	H = 12 12 01.7 h = 63 km MB = 6.0
	ePPP	B	35 24	D = 125.62 Az = 331.7 (NEIC)
	e	BC	40 41	PKIKPV A 1.3s 17.5nm
	e	BC	41 52	PPV A 1.6 85.2nm M = 5.6
	ePS	BC	42 46	PPV B 10 1.54 µm 6.1
	ePPS	BC	44 08	LmH B 21 5.0 µm 6.2
	LmH	B	13 18.2	LmV B 22 4.9 µm 6.1
	LmV	B	26.3	
9.	ePKIKP	AB	20 20 38	New Britain Region 5.79 S 152.20 E
	ePP	AB	22 22	H = 20 01 38.2 h = 24 km MB = 5.9 MS=6.2
	eSKKS	B	29 10	D = 124.51 Az = 330.6 (NEIC)
	ePS	B	32 20	PKIKPV A 1.6s 24.7nm
	ePPS	B	33 48	PPV A 2.0 47.0nm M = 5.6
	eSS	B	39 23	PPV B 10 1.56 µm 6.5
	LmV	B	21 17.2	
	LmH	B	20.6	
10.	ePKIKP	A	04 42 54	New Britain Region 5.70 S 152.09 E
	LmV	B	05 39.3	H = 04 23 56.0 h = normal MB=5.9 MS=5.4
	LmH	B	41.3	D = 124.39 Az = 330.6 (NEIC)
				LmV B 17s 0.8 µm M = 5.5
				LmH B 18 1.1 µm 5.6
10.	eX	A	04 59 08.5	XV A 1.5s 20.1nm
10.	ePKIKP	A	14 56 53	New Ireland Region 5.33 S 153.46 E
	e	A	58 34	H = 14 37 54.0 h = 38 km MB = 5.2

Day	Phase		h m s	Remarks
cont. 10.	ePP	AB	14 58 44	D = 124.73 Az = 331.4 (NEIC)
	eSKKS	B	05 41	PKIKPV A 1.2s 12.2nm
	ePS	B	08 40	PPV A 1.7 36.4nm M = 5.5
	ePPS	B	10 08	LmH B 20 4.0/um 6.1
	eSS	B	15 50	LmV B 16 2.2/um 5.9
	LmH	B	43.6	
	LmV	B	54.2	
10.	eP	A	15 11 18	<u>Baja California</u> 27.64 N 112.39 W
				H = 14 58 29.5 h = normal MB = 5.2
				D = 87.71 Az = 31.9 (NEIC)
10.	eX	A	23 24 06.5	XV A 2.0s 68.4nm
11.	ePKIKP	A	00 19 59	<u>Fiji Islands Region</u> 18.64 S 175.96 E
	iPKHKP	AB	20 02	H = 00 00 22.4 h = normal MB = 5.5
	LmH	B	01 25.4	D = 145.72 Az = 342.2 (NEIC)
	LmV	B	32.3	PKIKPV A 1.7s 103.0nm
				PKHKPV A 2.2 428.0nm
				LmH B 20 0.5/um M = 5.3
				LmV B 18 0.3/um 5.1
11.	e(P)	A	05 07 25.5	<u>Molucca Passage</u> 1.18 N 126.12 E
	eX	A	10 44	H = 04 53 19.7 h = 56 km MB = 5.6
				D = 104.37 Az = 323.3 (NEIC)
				PV A 1.4s 11.6nm M = 5.5
				XV A 1.4 16.3nm
11.	eP	A	05 41 11	<u>Southern Greece</u> 36.96 N 23.91 E
	LmH	B	47.0	H = 05 37 28.8 h = 120 km MB = 4.9
				D = 16.27 Az = 331.1 (NEIC)
11.	LmH	B	13 03.4	<u>South of Africa</u> 53.16 S 25.83 E
	LmV	B	08.5	H = 12 16 40.0 h = normal MB = 4.1
				D = 104.12 Az = 350.7 (NEIC)
				LmH B 16s 0.8/um M = 5.4
				LmV B 14 0.3/um 5.0

Day	Phase		h m s	Remarks
11.	-eP	A	13 26 09	South of Honshu, Japan 32.12 N 137.79 E
	epP	A	27 40	H = 13 14 15.8 h = 386 km MB = 5.4
				D = 84.88 Az = 328.9 (NEIC) h = 399 km
				PV A 1.4s 30.2nm M = 5.0
				pPV A 2.1 47.9nm
11.	e	A	13 57 28	XV A 1.4s 37.2nm
11.	ePKIKP	AB	14 43 24	Balleny Islands Region 62.75 S 155.71 E
	ePKP2	AB	43 58.5	H = 14 23 31.6 h = normal MB = 5.4
	eSS	C	15 07 35	D = 157.22 Az = 254.6 (NEIC)
	LmV	B	10.9	PKIKPV A 2.0s 17.1nm
	LmH	B	11.3	PKP2V A 2.0 72.6nm
				LmH B 18.5 3.7 µm M = 6.1
				LmV B 19 4.8 µm 6.3
11.	e	A	17 57(36)	Balleny Islands Region 62.63 S 155.25 E
	ePKP2	AB	57 48	H = 17 37 25.3 h = normal MB = 5.0
	LmH	B	19 23.6	D = 157.04 Az = 255.4 (NEIC)
	LmV	B	23.8	PKP2V A 2.0s 51.3nm
				LmH B 18.5 0.8 µm M = 5.5
				LmV B 18 0.8 µm 5.6
11.	eP	A	19 14 25.5	Near S. Coast of Honshu, Japan
				35.83 N 139.92 E
				H = 19 02 07.5 h = 57 km MB = 4.5
				D = 82.60 Az = 329.7 (NEIC)
				PV A 1.2s 14.2nm M = 4.8
12.	LmV	B	01 53.5	North Atlantic Ridge 23.9 N 45.7 W
	LmH	B	53.7	H = 01 25 53.4 h = normal MB = 4.2 (NEIC)
				D = 51.3
12.	+iP1	A	04 28 52.5	Andaman Islands Region 12.56 N 95.15 E
	eiP2	A	28 58	H = 04 17 05.6 h = 40 km MB = 5.3
	LmH	B	05 07.7	D = 76.31 Az = 319.4 (NEIC)
	LmV	B	07.7	P1V A 1.3s 35.0nm M = 5.2
				P2V A 1.5 70.4nm 5.4

Day	Phase		h m s	Remarks
cont. 12.				LmH B 19s 0.9/um M = 5.1
				LmV B 20 0.7/um 5.0
12.	eP	AB	14 33 48	Off Coast of Jalisco, Mexico
	ePP	AB	37 22	19.67 N 105.98 W
	eSKS	B	44 20	H = 14 20 43.0 h = normal MB = 5.6
	ePS	B	45 48	D = 91.14 Az = 34.4 (NEIC)
	eSS	C	50 48	PV A 2.6s 86.7nm M = 5.6
	LmH	B	15 14.7	PPV A 2.5 92.2nm 5.8
	LmV	B	14.7	PPV B 10 0.43/um 5.8
				LmH B 17 4.3/um 6.0
				LmV B 17.5 4.0/um 5.9
12.	eP	A	21 09 39	Kurile Islands 49.66 N 156.02 E
	LmH	B	47.0	H = 20 57 57.8 h = 45 km MB = 4.5
	LmV	B	47.0	D = 75.54 Az = 337.5 (NEIC)
12.	eP	A	12 20 30	North of Svalbard 82.75 N 6.40 W
				H = 23 13 57.6 h = normal MB = 4.4
				D = 32.67 Az = 158.6 (NEIC)
13.	eX	A	02 42 17.5	Mascarene Islands Region
				19.96 S 66.60 E
				H = 02 29 35.4 h = normal MB = 5.0
				D = 85.30 Az = 328.5 (NEIC)
				XV A 1.2s 10.2nm
13.	ePKP2	AB	15 03 40	South Island, New Zealand
	e	A	03 50	42.17 S 172.06 E
	ePP	B	07 24	H = 14 42 42.08 h = 18 km MB = 5.3 (NEIC)
	LmV	B	16 18.3	D = 164.1
	LmH	B	18.4	LmH B 20.5s 1.0/um M = 5.5
				LmV B 21 0.8/um 5.5
13.	ePKIKP	A	17 07 49	Solomon Islands 6.10 S 154.09 E
	e	A	09 36	H = 16 48 46.5 h = 21 km MB = 6.0
	ePP	AB	09 40	D = 125.70 Az = 331.5 (NEIC)
	ePPP	C	12 20	LmH B 20s 3.5/um M = 6.0

Day	Phase		h m s	Remarks
cont. 13.	ePS	B	17 19 36	LmV B 20s 3.7/um M = 6.1
	ePPS	B	21 00	
	e(SKKP)	A	21 38	
	eSS	B	26 35	
	LmH	B	18 05.5	
	LmV	B	09.4	
14.	ePKIKP	A	00 34 23	New Hebrides Islands 14.85 S 167.22 E
	eX	A	34 32	H = 00 15 17.9 h = 120 km MB = 5.5
	epPKIKP	A	34 48	D = 139.21 Az = 336.3 (NEIC) h = 90 km
	ePP	B	37 25	PKIKPV A 1.0s 57.4nm
	epPP	B	37 55	XV A 1.5 55.3nm
	eSKKS	B	44 10	pPKIKPV A 1.5 45.2nm
	e	C	51 20	LmH B 20 0.97/um
	eSS	C	55 35	LmV B 22 1.3/um
	LmH	B	01 34.4	
	LmV	B	36.5	
14.	ePKP	AB	09 21 10	Fiji Islands Region 15.70 S 178.57 W
	eX	A	21 35	H = 09 01 31.3 h = normal MB = 5.7
	eSS	C	43 06	D = 144.19 Az = 348.9 (NEIC)
				PKPV A 1.8s 20.3nm
				XV A 2.2 109.0nm
14.	ePKP	AB	09 22 28	Fiji Islands Region 15.20 S 178.42 W
	eSS	BC	44 24	H = 09 02 52.6 h = normal MB=5.6
	LmH	B	10 30.0	MS = 6.6 (NEIC)
	LmV	B	30.1	D = 143.8
				PKPV A 2.2s 81.8nm
				LmH B 20 15.4/um M = 6.7
				LmV B 19.5 13.4/um 6.7
14.	ePKP	AB	11 43 05	Fiji Islands Region 15.67 S 178.50 W
	eX	A	43 10	H = 11 23 28.6 h = normal MB = 5.2
	LmH	B	12 45.4	D = 144.17 Az = 349.0 (NEIC)
	LmV	B	50.8	PKPV A 1.5s 12.6nm
				XV A 2.0 42.7nm
				LmH B 22 1.9/um M = 5.8
				LmV B 20 1.33/um 5.7

Day	Phase		h m s	Remarks
14.	eP	A	18 37 34.5	Atlantic-Indian Rise 27.91 S 63.89 E H = 18 24 33.5 h = normal MB = 5.3 D = 90.82 Az = 329.8 (NEIC) PV A 1.6s 16.5nm M = 5.1
14.				In the intervall 20h 12m - 21h 34m the short-period seismographs (type A) have been out of operation.
14.	eP	A	22 24 15	Western Pakistan 26.6 N 64.2 E
	e	A	24 30	H = 22 15 44 MB = 4.9 D = 46.4 (ANUSSR)
15.	e	A	00 38 46.5	France 44.77 N 6.67 E
	e	A	38 48	H = 00 36 43.3 h = 0 km (ISC)
	ePg	A	38 55.5	D = 6.8
	e	A	40 04	LmH B 8s 0.28 µm M = 3.2
	eSg	A	40 19.5	
	LmH	B	41.0	
	LmV	B	41.0	
15.	eP	A	06 24 42	North Atlantic Ocean 47.44 N 49.41 W H = 06 17 15.6 h = normal MB = 4.6 D = 39.14 Az = 61.9 (NEIC)
15.	LmH	B	07 35.0	Kyushu, Japan 31.8 N 128.8 E
	LmV	B	40.0	H = 06 48 02 M = 5.2 (ANUSSR) D = 81.0 LmH B 15s 0.41 µm M = 4.9 LmV B 15 0.34 µm 4.9
15.	eP	A	12 27 05	Taiwan Region 21.88 N 121.76 E
	Pm	A	27 10	H = 12 14 29.3 h = 28 km MB = 4.8
	LmV	B	13 10.2	D = 85.30 Az = 323.1 (NEIC)
	LmH	C	10.3	PV A 1.8s 23.6nm M = 5.1 PmV A 2.0 51.3nm 5.4 LmH B 16 1.18 µm 5.4 LmV B 18 1.02 µm 5.3

Day	Phase		h m s	Remarks
15.	ePKP	A	15 12 27	Fiji Islands Region 14.96 S 176.98 W
	e	A	12 34	H = 14 52 59.7 h = normal MB = 5.1
	LmV	C	16 17.7	D = 143.73 Az = 350.7 (NEIC)
	LmH	C	18.0	LmH C 19s 0.50 µm M = 5.3
				LmV C 22 0.83 µm 5.5
15.	ePP	A	19 23 29	Chile-Bolivia Border Region
	eSKS	B	29 55	20.72 S 68.95 W
	eS	B	30 44	H = 19 05 45.3 h = 80 km MB = 5.3 (NEIC)
	ePS	B	32 20	D = 100.0
	eSS	B	37 40	PPV A 1.6s 27.5nm M = 5.6
	LmH	B	20 02.0	SH B 8.5 0.79 µm 6.2
	LmV	B	02.0	
15.	e	A	21 17 10	XV A 1.4s 11.6nm
16.	e	A	01 17 51	
16.	eP	AB	05 09 06.5	Szechwan Province, China
	eS	B	18 15	28.89 N 103.68 E
	ePS	B	19 10	H = 04 58 00.3 h = normal MB = 5.5
	eSS	B	22 40	D = 69.49 Az = 317.2 (NEIC)
	LmH	B	36.2	PV A 1.3s 63.4nm M = 5.5
	LmV	B	41.2	PV B 8 0.49 µm 5.6
				SH B 8.5 0.59 µm 5.7
				LmH B 20 3.7 µm 5.6
				LmV B 17 1.8 µm 5.5
16.	eP	A	05 39 24	Taiwan Region 21.73 N 121.80 E
	eX	A	39 34.5	H = 05 26 41.5 h = normal MB = 4.9
	LmH	B	06 22.7	D = 85.43 Az = 323.2 (NEIC)
	LmV	B	22.7	PV A 1.6s 22.0nm M = 5.1
				XV A 2.6 147.0nm
				LmH B 15.5 1.9 µm 5.6
				LmV B 16 2.3 µm 5.7
16.	eP	A	13 40 33.5	Szechwan Province, China
				28.83 N 103.73 E
				H = 13 29 24.8 h = normal MB = 4.8

Day	Phase		h m s	Remarks
cont. 16.				D = 69.55 Az = 317.2 (NEIC)
				PV A 1.6s 16.5nm M = 4.8
16.	eP	A	15 51 35	Off East Coast of Honshu, Japan
				40.75 N 143.89 E
				H = 15 39 27.9 h = 32 km MB = 4.7
				D = 79.86 Az = 331.4 (NEIC)
				PV A 1.8s 33.8nm M = 5.0
16.	eP	A	15 53 25	Off East Coast of Honshu, Japan
	eS	B	16 03 39	40.84 N 143.79 E
	LmH	B	27.7	H = 15 41 18.6 h = normal MB = 5.1
	LmV	B	33.0	D = 79.74 Az = 331.3 (NEIC)
				PV A 1.2s 32.5nm M = 5.2
				LmH B 17.5 4.0 µm 5.6
				LmV B 17.5 1.4 µm 5.2
16.	+iP	AB	19 05 02.7	Szechwan Province, China
	eS	B	14 10	28.88 N 103.74 E
	e	B	15 06	H = 18 53 54.7 h = normal MB = 5.3
	LmH	B	32.1	D = 69.52 Az = 317.2 (NEIC)
	LmV	B	37.1	PV A 1.7s 48.5nm M = 5.3
				SH B 8.5 0.61 µm 6.7
				LmH B 21 5.5 µm 5.8
				LmV B 17 2.6 µm 5.6
16.	eP	AB	22 48 40.5	Szechwan Province, China
	eX	A	51 34	28.84 N 103.59 E
	LmH	B	23 15.6	H = 22 37 33.6 h = normal MB = 5.4
	LmV	B	20.8	D = 69.47 Az = 317.2 (NEIC)
				PV A 2.0s 68.4nm M = 5.4
				XV A 1.6 27.5nm
				LmH B 21 0.86 µm 5.0
				LmV B 16 0.70 µm 5.0
17.	eP	AB	04 34 31	Turkey 37.10 N 36.82 E
	eiS	B	38 40	H = 04 29 33.1 h = normal MB = 5.0
	LmH	B	43.0	D = 22.52 Az = 314.9 (NEIC)

Day	Phase		h m s	Remarks
cont. 17.	LmV	B	04 46.1	PV A 1.5s 50.3nm M = 4.8
				SH B 8 1.7 µm 5.4
				LmH B 17.5 0.91 µm 4.3
				LmV B 16 1.05 µm 4.5
17.	eP	A	09 47 22	Szechwan Province, China
	LmH	B	10 14.5	28.92 N 103.72 E
	LmV	B	21.2	H = 09 36 15.5 h = normal MB = 4.9
				D = 69.48 Az = 317.2 (NEIC)
				LmH B 21s 0.60 µm M = 4.8
17.	eP	A	17 18 48	Szechwan Province, China
	Pm	A	18 51	28.85 N 103.68 E
				H = 17 07 40.4 h = normal MB = 4.9
				D = 69.51 Az = 317.2 (NEIC)
				PmV A 1.5s 20.1nm M = 5.0
17.	eiPKP	AB	24 06 11	Tonga Islands 15.52 S 173.45 W
	epPKP	A	06 28.5	H = 23 46 41.4 h = 70 km MB = 5.5
				D = 144.72 Az = 354.4 (NEIC)
				PKPV A 1.4s 107.0nm
18.	LmH	B	03 14.0	LmH B 18s 0.34 µm
	LmV	B	15.0	LmV B 16 0.35 µm
18.	eP	A	14 12 19	Southern Nevada 37.06 N 116.04 W
				H = 14 00 00.0 h = 0 km MB = 5.4
				D = 81.28 Az = 30.7
				Nevada test site "Algodones" (NEIC)
				PV A 1.2s 20.3nm M = 5.1
18.	LmH	C	15 48.5	LmV C 18s 0.33 µm
	LmV	C	48.5	
18.	eP	A	19 20 38	Peru-Ecuador Border Region
				3.90 S 80.90 W
				H = 19 07 17.4 h = 30 km MB = 5.1
				D = 94.59 Az = 39.6 (NEIC)

Day	Phase		h m s	Remarks
18.	eP	A	20 10 17.5	Kurile Islands 43.70 N 147.46 E
				H = 19 58 21.0 h = 60 km MB = 4.7
				D = 78.48 Az = 333.1 (NEIC)
				PV A 1.5s 15.1nm M = 4.7
19.	LmH	B	00 25.7	LmH B 17s 0.3 /um
	LmV	B	25.7	LmV B 17 0.4 /um
19.	e	A	01 33 52	
19.	eP1	A	08 41 21	Taiwan 23.92 N 121.87 E
	eP2	A	41 27	H = 08 28 53.1 h = 23 km MB = 5.4
	eP3	AB	41 36	D = 82.41 Az = 321.6 (NEIC)
	ePP	A	44 37	P1V A 1.5s 15.1nm M = 4.9
	eS	B	51 44	P2V A 1.7 15.2nm 4.9
	ei	B	51 58	P3V A 1.5 50.2nm 5.4
	LmH	B	09 23.7	LmH B 16.5 5.9 /um 6.0
	LmV	B	23.8	LmV B 16 6.8 /um 6.1
19.	eP	A	11 25 21	Taiwan Region 21.87 N 121.80 E
	LmV	B	12 08.5	H = 11 12 40.7 h = 15 km MB = 4.8
	LmH	B	08.6	D = 85.33 Az = 323.2 (NEIC)
				PV A 1.4s 32.6nm M = 5.4
				LmH B 16 2.2 /um 5.6
				LmV B 16 2.7 /um 5.8
19.	eP	A	13 46 38	South of Honshu, Japan 32.50 N 137.29 E
				H = 13 34 47.1 h = 389 km MB = 4.8 (NEIC)
				D = 84.3
19.	eX	A	17 08 29	XV A 1.4s 16.3nm
19.	+iP1	AB	22 27 21.5	Kurile Islands 49.29 N 155.45 E
	-iP2	A(B)	27 23	H = 22 15 37.7 h = normal MB = 6.0
	i	B	27 47	D = 75.74 Az = 337.2 (NEIC)
	eiS	B	36 56	P1V A 1.1s 90.7nm M = 5.7
	eiSKS	B	37 24	P2V A 1.0 516.0nm 6.5

Day	Phase	h m s	Remarks
cont. 19.	LmV B	23 05.7	P2V B 5s 2.83 µm M = 6.5
	LmH B	05.8	SH B 10 1.83 µm 6.2
			LmH B 17.5 15.2 µm 6.4
			LmV B 18 15.8 µm 6.4.
20.	ePKHKP AB	02 16 21	Tonga Islands 20.59 S 174.11 W
	e A	16 31	H = 01 56 35.5 h = 52 km MB = 5.2
	LmH B	03 33.0	D = 149.67 Az = 352.8 (NEIC)
	LmV B	33.0	PKHKPV A 1.6s 71.4nm
			LmH B 18 0.4 µm M = 5.2
			LmV B 16 0.4 µm 5.2
20.	eP A	11 32 41.5	Mindoro, Philippine Islands
			13.88 N 120.69 E
			H = 11 19 51.6 h = 139 km MB = 4.9
			D = 91.05 Az = 323.0 (NEIC)
20.	eP A	19 09 17	Southern Norway 61.79 N 5.00 E
	e(Sg) A	12 59	H = 19 06 28.6 h = 35 km MB = 4.5
	LmH B	14.3	D = 11.76 Az = 158.9 (NEIC)
20.	eP AB	21 49 01	Off Coast of Chiapas, Mexico
	eX A	49 13	13.38 N 92.37 W
	ePP B	52(29)	H = 21 36 09.6 h = normal MB = 5.8
	eSKS B	59 29	D = 88.43 Az = 38.2 (NEIC)
	eS B	59 42	PV A 2.2s 174.0nm M = 6.0
	eSS B	22 05 30	PV B 12 1.73 µm 6.3
	LmH B	31.2	XV A 2.0 188.0nm
	LmV B	31.4	LmH B 18 4.2 µm 5.9
			LmV B 17.5 4.0 µm 5.9
21.	ePg A	01 47 03	Northern Italy 45.5 N 9.3 E
	eSg A	48 11	H = 01 45 28 h = 0 km
			D = 5.42 Az = 16 (ISC)
21.	ePg A	03 29 53	Northern Italy 44.5 N 7.4 E
	e(Sn) A	30 27.5	H = 03 27 55 h = 0 km
	eSg A	31 15	D = 6.80 Az = 24 (ISC)

Day	Phase		h m s	Remarks
21.	+iP	A	04 01 34.3	<u>Tyrrhenian Sea</u> 39.63 N 12.83 E H = 03 59 04.0 h = 484 km MB = 4.3 D = 11.05 Az = 356.0 (NEIC) PV A 1.0s 51.2nm M = 4.9
21.	ePKP	A	10 42 45	<u>New Hebrides Islands Region</u>
	epPKP	A	42 57	19.28 S 167.92 E
	esPKP	A	43 03	H = 10 23 11.8 h = 20 km MB = 5.0
	ePP	AC	46 02	D = 143.50 Az = 334.5 (NEIC) h = 43 km PKPV A 1.5s 20.1nm pPKPV A 1.3 15.3nm sPKPV A 1.2 12.2nm PPV A 2.0 51.3nm M = 5.4
21.	ePKP	A	12 10 30	<u>New Hebrides Islands Region</u>
	e(pPKP)	A	10 41	19.20 S 167.77 E H = 11 50 59.8 h = normal MB = 4.7 D = 143.36 Az = 334.5 (NEIC) PKPV A 1.6s 24.7nm
21.	LmH	B	13 49.0	<u>Near coast of Northern Chile</u>
	LmV	B	49.3	21.77 S 70.21 W H = 12 47 44.9 h = 17 km MB = 5.8 (NEIC) D = 101.6 LmH B 18s 0.2 µm M = 4.7 LmV B 18 0.4 µm 5.0
21.	ePKP	AB	16 28 06.5	<u>New Hebrides Islands Region</u>
	e	AB	28 12	19.68 S 173.61 E
	LmH	B	17 25.0	H = 16 08 27.7 h = normal MB = 5.1
	LmV	B	25.0	D = 145.97 Az = 339.4 (NEIC) PKPV A 1.8s 47.3nm LmH B 16.5 0.9 µm M = 5.6 LmV B 17 1.0 µm 5.6 It is not possible to decide whether the surface waves belong to the New Hebrides Islands earthquake or to the following (Azores Islands) quake.

Day	Phase		h m s	Remarks
21.	eP	A	17 14 21.5	Azores Islands Region 42.08 N 29.27 W H = 17 08 21.5 h = normal MB = 4.7 D = 29.17 Az = 58.8 (NEIC) PV A 1.6s 27.5nm M = 4.7 LmH B 16.5 0.9 µm 4.5 LmV B 17 1.0 µm 4.6 It is not possible to decide whether the surface waves belong to the Azores Islands earthquake or to the foregoing(New Hebrides Islands) quake.
21.	ePKP	A	17 28 54.5	New Hebrides Islands Region 19.24 S 167.74 E H = 17 09 23.5 h = normal MB = 4.5 D = 143.39 Az = 334.4 (NEIC)
21.	eP	A	19 42 17	East of Severnaya Zemlya 81.92 N 118.88 E H = 19 34 23.2 h = normal MB = 4.6 D = 42.55 Az = 296.0 (NEIC) PV A 1.2s 24.4nm M = 4.8
21.	ePKP	A	20 56 39	New Ireland Region 4.62 S 153.12 E H = 20 37 46.8 h = 75 km MB = 4.8 D = 123.94 Az = 331.5 (NEIC) PKPV A 1.4s 18.6nm
21.	ePKHKP ePKP2	A A	22 36 10 36 19	Tonga Islands 20.25 S 174.08 W H = 22 16 23.0 h = normal MB = 4.7 D = 149.34 Az = 352.9 (NEIC)
21.	+eP	A	22 55 20	Alaska Peninsula 54.28 N 162.49 W H = 22 43 38.4 h = normal MB = 5.2 D = 75.33 Az = 3.9 (NEIC) PV A 1.3s 72.0nm M = 5.5
21.	LmV	B	23 06.7	LmV B 20s 0.52 µm

Day	Phase		h m s	Remarks
22.	ePKP	A	09 35 52.5	Fiji Islands Region 17.65 S 178.93 W H = 09 17 14.5 h = 563 km MB = 4.5 D = 146.01 Az = 348.0 (NEIC) PKPV A 1.2s 12.2nm
22.	eP	A	11 13 54.5	North of Svalbard 82.98 N 6.47 W H = 11 07 21.7 h = normal MB = 4.6 D = 32.89 Az = 158.7 (NEIC) PV A 1.2s 14.2nm M = 4.8
22.	LmV	B	11 57.0	Near coast of Guatemala 13.76 N 91.40 W H = 11 01 59.0 h = 69 km MB = 4.8 (NEIC) D = 87.5
22.	eP	A	18 01 11	Iran 30.08 N 50.73 E
	eS	B	06 51	H = 17 54 14.6 h = normal MB = 5.1
	LmH	B	17.7	D = 35.64 Az = 316.4 (NEIC)
	LmV	B	17.7	PV A 2.0s 42.7nm M = 5.0
				LmH B 17 0.7 μm 4.5
				LmV B 16 0.6 μm 4.5
23.	ePKP	A	04 27 00	East New Guinea Region 3.97 S 146.06 E
	e	A	27 15	H = 04 08 02.0 h = normal MB = 5.2
	ePP	B	28 20	D = 119.86 Az = 328.4 (NEIC)
	eS diff.	C	36 00	LmH B 20s 19.0 μm M = 6.7
	ePS	C	38 00	LmV B 20 19.2 μm 6.7
	e	C	40 00	
	eSS	C	44 40	
	iPSPS	C	45 18	
	eSSS	C	49 00	
	LmH	B	05 19.4	
	LmV	B	20.0	
23.	eP	A	05 47 20.5	Szechwan Province, China 28.81 N 103.75 E H = 05 36 11.4 h = normal MB = 5.2 D = 69.58 Az = 317.2 (NEIC) PV A 1.5s 35.2nm M = 5.2

Day	Phase		h m s	Remarks
23.	eP	A	05 57 33	Szechwan Province, China 28.86 N 103.66 E
				H = 05 46 25.0 h = normal MB = 5.2
				D = 69.49 Az = 317.2 (NEIC)
				PV A 1.4s 11.6nm M = 4.7
23.	e(Sg)	A	09 12 19.5	Poland 50.38 N 18.89 E
				H = 09 09 47.9 M = 3.3 (WAR)
				D = 4.66
23.	eP	A	14 46 01.5	Dodekanese Islands 36.83 N 28.5 E
				H = 14 41 46.0 h = 32 km
				D = 18.35 Az = 324 (ISC)
				PV A 1.0s 96.6nm M = 4.0
23.	-iPKHKP	A	15 59 27	Fiji Islands Region 20.76 S 178.57 W
	ePKP2	A	59 34	H = 15 40 45.4 h = 615 km MB = 4.8
				D = 149.11 Az = 347.3 (NEIC)
				PKHKPV A 1.2s 36.6nm
23.	ePg	A	16 12 10	Northern Italy 44.69 N 10.44 E
	eSg	A	13 21	H = 16 10 04.2 h = 0 km (ISC)
				D = 6.03
23.	ePP	A	18 45 28.5	Southwest of Sumatra 9.6 S 107.5 E
	ePPP	A	47 39	H = 18 27 31 MB = 5.4 (ANUSSR)
	e	A	47 48	D = 101.2
	e	A	47 55	PPV A 1.8s 16.9nm M = 5.2
	LmH	B	19 49.4	LmV B 20 0.30 µm 4.8
	LmV	B	53.0	
23.	+iP	AB	22 07 13	Kurile Islands 45.62 N 150.96 E
	e	B	07 33	H = 21 55 17.7 h = 34 km MB = 5.7
	eS	B	16 56	D = 77.86 Az = 334.9 (NEIC)
	ePS	C	17 42	PV A 1.2s 122.0nm M = 5.8
	eSS	C	23 38	PV B 9 0.87 µm 5.8
	eSSS	C	25 46	LmH B 19 6.9 µm 6.0
	LmH	B	44.7	LmV B 19 7.9 µm 6.1
	LmV	B	45.1	

Day	Phase		h m s	Remarks
24.	ePKHKP	A	01 46 14	South of Fiji Islands 22.42 S 178.10 W H = 01 27 03.4 h = 354 km MB = 4.5 D = 150.82 Az = 347.3 (NEIC)
24.	LmH	B	04 47.7	Mid- Indian Rise 38.6 S 78.6 E
	LmV	B	48.0	H = 03 33 15.7 h = normal MB = 4.7 MS = 5.3 (NEIC) D = 106.4 LmH B 20s 0.48 µm M = 5.0 LmV B 18 0.77 µm 5.3
24.	ePKP	A	06 04 35	New Hebrides Islands Region 19.21 S 167.81 E H = 05 45 03.8 h = normal MB = 4.6 D = 143.38 Az = 334.5 (NEIC) PKPV A 1.6s 11.0nm
24.	LmH	B	08 55.4	Near Coast of Nicaragua 11.5 N 86.3 W
	LmV	B	09 01.0	H = 08 09 07.5 h = 33 km MB = 5.1 MS = 4.2 (NEIC) LmH B 20s 0.6 µm M' = 5.0 LmV B 18 0.6 µm 5.1
24.	eP	A	10 04 48.5	Kurile Islands 45.35 N 151.27 E H = 09 52 52.0 h = normal MB = 4.7 D = 78.20 Az = 335.1 (NEIC) PV A 1.2s 10.2nm M = 4.7
24.	ePKIKP	A	13 48 11.5	New Britain Region 5.53 S 152.85 E
	e	A	48 14	H = 13 29 12.8 h = 36 km MB = 5.2
	ePP	B	50 00	D = 124.61 Az = 331.0 (NEIC)
	ePPP	B	52 36	LmH B 14s 0.3 µm M = 5.1
	ePS	B	59 50	LmV B 16 0.4 µm 5.1
	LmH	B	14 49.5	
	LmV	B	53.6	
24.	ePKHKP	A	15 56 53	South of Fiji Islands 24.15 S 177.17 W H = 15 37 16.7 h = 180 km MB = 5.1 D = 152.69 Az = 347.8 (NEIC)

Day	Phase		h m s	Remarks
24.	eP1	A	16 41 55	Central Russia 52.21 N 91.43 E
	eP2	AB	41 59.5	H = 16 33 22.7 h = normal MB = 5.2
	+ePP	AB	43 51	D = 47.38 Az = 301.6 (NEIC)
	e	A	46 49	P2V A 1.3s 140.0nm M = 5.8
	eS	B	48 49	PPV A 1.4 55.8nm 5.3
	eSS	B	52 16	LmH B 12.5 6.1 μm 5.8
	LmV	B	17 03.3	LmV B 13 5.6 μm 5.8
	LmH	B	03.4	
24.	eiPKP	A	22 33 23	New Hebrides Islands 19.44 S 167.96 E
	e	A	35 24.5	H = 22 13 52.8 h = normal MB = 5.3
				D = 143.65 Az = 334.5 (NEIC)
				PKPV A 1.5s 40.2nm
25.	ePKP	A	07 31 19	New Hebrides Islands Region
	e	A	31 29.5	19.29 S 167.80 E
				H = 07 11 48.3 h = normal MB = 4.8
				D = 143.45 Az = 334.4 (NEIC)
				PKPV A 1.6s 24.7nm
25.	e	A	08 26 37	New Hebrides Islands Region
	e	A	26 45	19.11 S 167.79 E
				H = 08 06 55.2 h = normal
				D = 143.29 Az = 334.5 (NEIC)
25.	eP	A	09 11 54	Turkey 38.42 N 40.81 E
				H = 09 06 40.6 h = 41 km
				D = 23.96 Az = 310 (ISC)
25.	LmH	B	18 56.5	Ryukyu Islands 28.4 N 128.1 E
	LmV	B	19 02.8	H = 18 08 44.6 h = 40 km MB = 4.4 (NEIC)
				D = 83.4
				LmH B 17.5s 0.5 μm M = 4.9
				LmV B 16 0.4 μm 4.9
26.	ePKIKP	A	03 29 43	New Ireland Region 5.67 S 153.02 E
				H = 03 10 41.4 h = 23 km MB = 4.9
				D = 124.81 Az = 331.1 (NEIC)

Day	Phase		h m s	Remarks
26.	eP	A	06 37 33.5	<u>Hokkaido, Japan Region</u> 40.99 N 143.75 E
	LmH	B	07 16.5	H = 06 25 28.3 h = 35 km MB = 4.9
	LmV	B	16.5	D = 79.60 Az = 331.3 (NEIC)
				PV A 1.2s 10.2nm M = 4.7
				LmH B 18 0.5 µm 4.8
				LmV B 16 0.5 µm 4.9
26.	eP	A	06 54 10	<u>Central Mid-Atlantic Ridge</u> 8.70 N 39.80 W
	e	A	54 17	H = 06 44 07.0 h = normal MB = 4.8
				D = 59.45 Az = 35.3 (NEIC)
26.	eP	A	07 02(04)	<u>Iran</u> 29.99 N 50.72 E
	e	A	02 23	H = 06 55 08.7 h = 45 km MB = 4.8
				D = 35.70 Az = 316.5 (NEIC)
26.	ePKP	A	09 51 34.5	<u>Tonga Islands</u> 14.99 S 175.19 W
				H = 09 32 32.5 h = 275 km MB = 5.2
				D = 144.02 Az = 352.6 (NEIC)
				PV A 1.4s 44.2nm
26.	LmH	B	11 47.8	<u>South Sandwich Islands</u> 55.7 S 26.1 W
	LmV	B	52.0	H = 10 50 40.2 h = normal
				MB = 5.9 MS = 5.5 (NEIC)
				D = 110.4
26.	e	A	11 57(34)	<u>Northern Italy</u> 44.49 N 9.61 E
	ePg	A	57 55	H = 11 55 51.0 h = normal MB = 4.0
	eSn	A	58 33.5	D = 6.30 Az = 11.7 (NEIC)
	eSg	A	59 15	
26.	eSn	A	13 34(28)	<u>Northern Italy</u> 44.58 N 9.52 E
	eSg	A	35(09)	H = 13 31 47.5 (BCIS)
				D = 6.24 Az = 12
				station clock out of operation
26.	+iPKP	AB	17 53 21.5	<u>Loyalty Islands</u> 20.06 S 168.66 E
	+i	A	53 24.5	H = 17 33 45.7 h = 16 km MB = 5.0
	LmH	B	19 05.5	D = 144.49 Az = 334.7 (NEIC)

Day	Phase		h m s	Remarks
cont. 26.	LmV	B	19 05.8	PKPV A 1.9s 90.9nm
				PKPV B 5 0.55/um
				LmH B 18 0.5/um M = 5.3
				LmV B 18 0.5/um 5.3
26.	ePKP	A	18 12 40.5	<u>Loyalty Islands</u> 20.05 S 168.69 E
	e	A	12 55.5	H = 17 53 07.3 h = 27 km MB = 4.9
	LmH	B	19 24.6	D = 144.49 Az = 334.7 (NEIC)
	LmV	B	24.6	PKPV A 1.4s 20.9nm
				LmH B 16 0.38/um M = 5.2
				LmV B 18 0.51/um 5.3
26.	ePKP	A	23 59 12	<u>Loyalty Islands</u> 20.12 S 168.62 E
	e	A	59 26	H = 23 39 36.5 h = 18 km MB = 5.0
				D = 144.53 Az = 334.6 (NEIC)
				PKPV A 1.2s 8.1nm
26.	-ePKP	A	24 12 29.5	<u>Loyalty Islands</u> 20.12 S 168.89 E
				H = 23 52 56.1 h = normal MB = 4.3
				D = 144.64 Az = 334.8 (NEIC)
				PKPV A 1.6s 27.5nm
27.	eP	A	02 21 27.5	<u>Greece</u> 38.0 N 20.2 E
				H = 02 18 05 h = 0 km
				D = 14.02 Az = 337 (ISC)
				PV A 0.7s 17.2nm
27.	ePKP2	A	04 03 10	<u>Kermadec Islands Region</u> 31.3 S 179.7 W
				H = 03 43 29.8 h = 566 km MB = 5.0 (NEIC)
				D = 158.9
				PKP2V A 0.8s 9.6nm
27.	eP	A	05 27 09	<u>Iran</u> 30.17 N 50.74 E
	LmV	B	45.3	H = 05 20 15.1 h = 54 km MB = 5.0
	LmH	B	45.4	D = 35.58 Az = 316.3 (NEIC)
				PV A 1.2s 20.4nm M = 4.9
				LmH B 16 0.2/um
				LmV B 16 0.2/um

Day	Phase		h m s	Remarks
27.	ePKP	A	05 37 53.5	New Hebrides Islands Region
	e	A	38 49.5	19.33 S 167.90 E
	e	A	38 53	H = 05 18 19.8 h = 12 km MB = 4.6
				D = 143.53 Az = 334.5 (NEIC)
				PKPV A 1.4s 14.0nm
27.	LmH	B	07 26.0	North Atlantic Ocean 19.2 N 68.1 W
	LmV	B	26.0	H = 06 37 53.2 h = normal MB = 4.7 (NEIC)
				D = 68.9
27.	eX	A	09 34 53.5	Off Coast of Central America
				11.84 N 88.88 W
				H = 09 21 57.5 h = normal MB = 5.0
				D = 87.50 Az = 38.8 (NEIC)
				XV A 0.8s 7.7nm
27.	ePKP2	A	09 45 47.5	Kermadec Islands Region 31.0 S 176.9 W
				H = 09 25 14.4 h = 37 km MB = 4.6 (NEIC)
				D = 159.4
27.	ePg	A	11 17 54	Northern Italy 44.6 N 9.6 E
	e(Sn)	A	18 29	H = 11 15 51 (BCIS)
	e(Sg)	A	19 17	D = 6.3
27.	eP	A	14 02 06	Off East Coast of Honshu, Japan
				40.85 N 143.80 E
				H = 13 49 59.4 h = normal MB = 4.9
				D = 79.74 Az = 331.4 (NEIC)
				PV A 1.2s 14.2nm M = 4.9
28.	ePKIKP	A	03 24 16	Fiji Islands Region 21.10 S 178.74 W
	ePKHKP	A	24 21.5	H = 03 05 30.3 h = 521 km MB = 5.0
	ePKP2	A	24 27.5	D = 149.40 Az = 347.0 (NEIC)
				PKHKP A 1.2s 30.5nm
28.	e	A	04 12 26.5	

Day	Phase		h m s	Remarks
28.	-iPKIKP	AB	04 28 33.5	<u>Tonga Islands</u> 18.81 S 174.70 W
	+iPKHKP	A	28 36.5	H = 04 09 07.4 h = 153 km MB = 5.4
	epPKIKP	A	29 13	D = 147.84 Az = 352.4 (NEIC) h = 151 km
	epPKHKP	A	29 16	PKHKPV A 1.5s 173.0nm
	e(sPKHKP)	AB	29 35	
28.	e	A	06 21 21.5	<u>Probably Poland</u>
				H = 06 18 39.5 (WAR)
28.	eiP	AB	16 10 20.5	<u>Ryukyu Islands</u> 28.26 N 130.71 E
	e	A	10 25	H = 15 57 47.7 h = 35 km MB = 5.7
	eS	B	20 49	D = 84.83 Az = 326.0 (NEIC)
	LmH	B	48.6	PV A 2.1s 105.0nm M = 5.7
	LmV	B	54.8	SH B 8 0.89/um 5.9
				LmH B 15.5 2.2/um 5.6
				LmV B 14 2.1/um 5.7
28.	eP	A	16 41 25.5	<u>Iran - USSR Border Region</u>
				37.64 N 55.82 E
				H = 16 34 44.4 h = normal MB = 4.8
				D = 33.77 Az = 307.0 (NEIC)
				PV A 0.7s 13.4nm M = 5.0
29.	eP	A	03 49 00.5	<u>Philippine Islands Region</u>
	e	A	49 15.5	6.32 N 127.38 E
	LmV	B	04 41.6	H = 03 35 15.1 h = 58 km MB = 5.4
	LmH	B	44.4	D = 100.99 Az = 324.3 (NEIC)
				LmH B 16s 0.5/um
				LmV B 18 0.5/um
29.	LmH	B	09 13.2	LmH B 19s 0.7/um
	LmV	B	15.1	LmV B 20 1.0/um
29.	eP1	AB	11 01 16	<u>Iceland Region</u> 67.58 N 18.84 W
	eP2	A	01 22.5	H = 10 56 17.2 h = normal MB = 4.7
	eS	C	05 30	D = 22.68 Az = 123.2 (NEIC)
	eS	B	05 35	P1V A 1.5s 37.7nm M = 4.6

Day	Phase		h m s	Remarks
cont. 29.	LmH	B	11 10.4	P1V B 5s 0.99/um M = 5.5
	LmV	B	13.3	P2V A 2.0 205.0nm 5.2
				SH B 8 0.97/um 5.2
				LmH B 17 2.0/um 4.6
				LmV B 12 1.4/um 4.8
29.	ePKP	A	12 21 03	New Hebrides Islands Region
				19.37 S 167.84 E
				H = 12 01 28.9 h = 13 km MB = 4.6
				D = 143.54 Az = 334.4 (NEIC)
30.	eP	A	02 38 04	Off East Coast of Honshu, Japan
				35.14 N 143.51 E
				H = 02 25 31.8 h = normal MB = 4.9
				D = 84.64 Az = 331.6 (NEIC)
				PV A 1.4s 9.3nm M = 4.8
30.	eP	A	08 50 00	North of Severnaya Zemlya
				85.8 N 83.0 E
				H = 08 42 35 MB = 4.9 (ANUSSR)
				D = 38.2
30.	ePKIKP	A	21 32 07.5	New Hebrides Islands 14.81 S 167.26 E
				H = 21 12 52.4 h = 120 km MB = 5.0
				D = 139.19 Az = 336.3 (NEIC)
				PKIKPV A 1.6s 19.2nm
30.	ePKIKP	A	22 56 50.5	Southern Pacific Ocean 36.43 S 98.42 W
	eSS	B	16 15	H = 22 37 40.0 h = normal MB = 5.3
	eSSS	B	21 00	MS = 5.3
	LmH	B	23 55.2	D = 129.18 Az = 50.5 (NEIC)
	LmV	B	55.6	PKIKP A 1.8s 23.6nm
				LmH B 18 0.6/um M = 5.3
				LmV B 18 0.6/um 5.4
31.	ePKIKP	A	00 25 08.5	New Ireland Region 4.87 S 153.29 E
				H = 00 06 15.3 h = 60 km MB = 4.5
				D = 124.24 Az = 331.5 (NEIC)

Day	Phase		h m s	Remarks
31.	e	A	12 18 01	
31.	LmH	B	16 18.5	<u>Azores Islands Region</u> 42.5 N 29.3 W
	LmV	B	18.6	H = 16 02 08.6 h = normal MB = 4.6 (NEIC)
				D = 29.0
				LmH B 18s 0.2 μm M = 3.8
				LmV B 18 0.3 μm 4.0
31.	ePKHKP	A	21 33 11.5	<u>Tonga Islands</u> 20.83 S 173.78 W
				H = 21 13 22.6 h = normal MB = 5.0
				D = 149.94 Az = 353.1 (NEIC)
				PKHKPV A 1.5s 37.7nm

Day	Phase		h m s	Remarks
1.	LmV	B	01 37.3	<u>South Sandwich Islands Region</u>
	LmH	B	38.6	59.3 S 27.0 W
				H = 00 32 08.6 h = normal
				MB = 5.4 MS = 5.6 (NEIC)
				D = 113.8
				LmH B 18s 0.4 µm M = 5.1
				LmV B 18 0.4 µm 5.1
1.	LmV	B	03 32.5	LmV B 18s 0.38 µm
	LmH	B	33.0	
1.	eSg	A	03 39 56	<u>Poland</u> 50.25 N 18.92 E
				H = 03 37 26.7 M = 3.2 (WAR)
				D = 4.69
1.	e	A	07 09 24	
1.	eP	A	11 05 50	<u>Kurile Islands</u> 48.35 N 154.87 E
				H = 10 54 04.6 h = 50 km MB = 4.9
				D = 76.46 Az = 336.9 (NEIC)
1.	LmH	C	13 46.0	<u>New Britain Region</u> 5.5 S 151.7 E
				H = 12 30 11.2 h = 64 km MB = 5.2 (NEIC)
				D = 124.0
1.	ePKIKP	A	20 36 35.5	<u>New Britain Region</u> 5.54 S 152.89 E
	eX	A	36 40	H = 20 17 37.4 h = 34 km MB = 5.2
	LmH	C	21 22.0	D = 124.63 Az = 331.0 (NEIC)
				XV A 1.1s 20.2nm
1.	ePKP	A	23 05 42.5	<u>New Britain Region</u> 5.56 S 152.85 E
				H = 22 46 37.3 h = 19 km MB = 4.8
				D = 124.63 Az = 331.0 (NEIC)
1.	ePKP	A	23 51 11.5	<u>Tonga Islands</u> 15.47 S 175.09 W
				H = 23 32 10.6 h = 300 km MB = 4.9
				D = 144.49 Az = 352.6 (NEIC)
				PKPV A 1.3s 21.8nm

Day	Phase		h m s	Remarks
2.	ePKIKP	A	06 52 20.5	Santa Cruz Islands 11.07 S 166.30 E
				H = 06 33 20.7 h = 175 km MB = 5.4
				D = 135.41 Az = 337.2 (NEIC)
2.	-eP	A	18 31 43	Iran 30.10 N 50.78 E
	ePcP	A	34 20	H = 18 24 47.3 h = 45 km MB = 5.1
	e	A	34 26	D = 35.65 Az = 316.4 (NEIC)
	LmH	B	48.0	PV A 1.1s 20.2nm M = 5.0
	LmV	B	48.0	
2.	LmH	C	22 43.0	Iran 30.08 N 50.76 E
				H = 22 21 39.0 h = 39 km MB = 5.0 (NEIC)
				D = 37.1
2.	ePKP	A	23 29 02	New Hebrides Islands 20.71 S 169.37 E
	epPKP	A	29 12	H = 23 09 27.9 h = 50 km MB = 4.8
				D = 145.36 Az = 334.9 (NEIC)
				PKPV A 1.2s 34.5nm
2.	ePKIKP	AB	24 07 41	South of Tonga Islands 24.36 S 175.19 W
	ePKHKP	A	07 48	H = 23 47 52.2 h = normal MB = 5.7
	ePKP2	A	08 00	D = 153.24 Az = 350.4 (NEIC)
	LmH	B	25(20)	PKIKPV A 2.0s 59.8nm
	LmV	B	22.5	PKHKP A 1.8 74.3nm
				PKP2 A 2.0 94.0nm
				LmV B 18 0.64 µm
3.	e	A	11 32 58	
3.	LmH	B	13 01.3	LmH B 15s 0.54 µm
	LmV	B	01.3	LmV B 15 0.68 µm
3.	ePn	A	21 34 11	France 48.22 N 6.53 E
	iPg	A	34 24	H = 21 33 08.5 h = 26 km
	iSg	A	35 17	D = 4.12 Az = 51.9 (NEIC)
3.	ePKP	A	23 48 29	Tonga Islands 15.33 S 173.03 W
				H = 23 28 51.4 h = 37 km MB = 4.8

Day	Phase		h m s	Remarks
cont. 3.	epPKP	A	23 48 36	D = 144.57 Az = 354.9 h = 25 km (NEIC) pPKPV A 1.6s 33.0nm
3.	LmH	C	24 52.0	South Sandwich Islands Region
	LmV	C	53.5	56.3 S 26.6 W H = 23 37 29.4 h = normal MB = 5.3 (NEIC) D = 111.0 LmH C 22s 0.33 µm M = 4.9 LmV C 20 0.58 µm 4.2
4.	eP	A	13 49 17	Iran 30.08 N 50.86 E H = 13 42 19.5 h = 30 km MB = 4.6 D = 35.71 Az = 316.4 (NEIC)
4.	eSg	A	15 17 43	Pyrenees 42.3 N 1.9 E H = 15 11 54 h = 0 km (ISC) D = 11.6
4.	eP	A	15 22 11.5	Kurile Islands 45.58 N 151.42 E H = 15 10 17.4 h = 60 km MB = 4.5 D = 78.03 Az = 335.2 (NEIC)
4.	+iP	AB	16 04 54.5	Unimak Island Region 54.98 N 163.36 W H = 15 53 25.4 h = 107 km MB = 5.8 D = 74.66 Az = 3.3 (NEIC) PV A 1.5s 314.0nm M = 5.9 PV B 2 0.77 µm 6.2
5.	e	A	03 19 18	New Ireland Region 5.2 S 153.0 E
	LmV	B	04 18.5	H = 03 00 12.7 h = 45 km MB = 5.1 (NEIC)
	LmH	B	18.7	D = 124.5 LmV B 20s 0.46 µm M = 5.2
5.	LmH	C	10 10.1	Guerrero, Mexico 17.1 N 99.8 W
	LmV	C	11.3	H = 09 13 15.3 h = 50 km MB = 5.2 MS = 5.0 (NEIC) D = 89.8

Day	Phase		h m s	Remarks
5.	eP	A	12 24 18.5	Turkey 37.26 N 30.32 E
	LmH	B	30.9	H = 12 19 56.9 h = 20 km MB = 4.5
	LmV	B	32.7	D = 18.91 Az = 320.9 (NEIC)
				LmH B 14s 1.1 μm M = 4.3
				LmV B 12 0.6 μm 4.3
5.	eP	A	15 06 40	Komandorsky Islands Region
	e	A	06 47.5	55.95 N 165.09 E
				H = 14 55 21.0 h = normal MB = 5.1
				D = 71.51 Az = 342.6 (NEIC)
5.	+eP1	AB	18 47 01.5	Sakhalin Island 46.50 N 141.17 E
	eiP2	A	47 04.3	H = 18 35 25.0 h = 9 km MB = 6.3
	eiP3	AB	47 08.5	D = 73.89 Az = 329.3 (NEIC)
	+iPP	B	49 37	P1V A 1.2s 32.5nm M = 5.3
	-iPP3	B	49 55	P2V A 2.0 368.0nm 6.1
	eiPPP	B	51 29	P3V A 2.3 1670.0nm 6.7
	iS	CB	56 35	P3V B 10 27.2 μm 7.3
	ePS	B	57 20	PPV B 13 15.1 μm 7.1
	eSS	B	21 01 24	SH B 14 33.9 μm 7.2
	e	B	02 30	LmH B 16 625.0 μm 8.0
	eSSS	B	04 28	LmV B 18.5 377.0 μm 7.8
	ei	B	05 16	
	LmH	B	18.3	
	LmV	B	22.5	
5.	+eP	A	19 45 23	Sakhalin Island 46.41 N 141.17 E
				H = 19 33 47.2 h = 15 km MB = 5.5
				D = 73.97 Az = 329.3 (NEIC)
				PV A 1.6s 55.0nm M = 5.3
5.	eP	A	20 48 33	Sakhalin Island 46.69 N 141.19 E
	eX	A	48 42	H = 20 37 00.8 h = normal MB = 5.0
				D = 73.73 Az = 329.3 (NEIC)
				PV A 2.0s 25.7nm M = 4.9
				XV A 1.7 36.4nm

Day	Phase		h m s	Remarks
5.	eP	A	20 54 30	Sakhalin Islands 46.65 N 141.02 E
				H = 20 42 58.0 h = normal MB = 4.4
				D = 73.71 Az = 329.2 (NEIC)
5.	ePP	A	22 53 16	Northern Easter I. Cordillera
				8.9 S 108.4 W
				H = 22 33 16.6 h = normal
				MB = 5.2 MS = 5.4 (NEIC)
				D = 115.6
6.	eP	A	00 41 46	Afghanistan 33.21 N 69.90 E
	e(PP)	A	43 53	H = 00 33 25.9 h = 37 km MB = 4.9
				D = 45.50 Az = 310.6 (NEIC)
				PV A 1.7s 31.5nm M = 4.9
6.	eP	A	01 22 43	South of Panama 2.32 N 79.00 W
	eX	A	22 50	H = 01 09 51.6 h = normal MB = 5.0
				D = 88.61 Az = 39.6 (NEIC)
				PV A 1.4s 9.3nm M = 4.6
				XV A 2.0 25.7nm
6.	+eP	AB	03 11 11.5	Sakhalin Island 46.71 N 141.21 E
	LmH	B	43.0	H = 02 59 39.2 h = normal MB = 5.4
	LmV	B	45.5	D = 73.72 Az = 329.3 (NEIC)
				PV A 1.0s 29.5nm M = 5.2
				LmH B 13.5 1.6 μm 5.5
				LmV B 14 1.0 μm 5.3
6.	e(P)	A	04 33 23	Sakhalin Island 46.78 N 141.41 E
				H = 04 21 43.9 h = 26 km MB = 4.8
				D = 73.73 Az = 329.4 (NEIC)
				PV A 1.2s 12.2nm M = 4.8
6.	eP	AB	06 57 35	Sakhalin Island 46.37 N 141.13 E
	+iX	A	57 40	H = 06 45 59.0 h = 16 km MB = 5.7
	e	A	07 00 08	D = 73.99 Az = 329.3 (NEIC)
	eS	B	07 05	PV A 1.1s 20.2nm M = 5.1
	LmH	B	28.7	XV A 1.8 81.0nm
	LmV	B	33.0	SH B 8 0.38 μm 5.6

Day	Phase		h m s	Remarks
6.	LmH	B	11 35.8	Probably <u>Sakhalin Island</u>
	LmV	B	36.0	46.42 N 141.37 E
				H = 10 48 32.8 h = 41 km MB = 4.7 (NEIC)
				D = 74.2
				LmH B 14s 1.34/um M = 5.4
				LmV B 15 1.03/um 5.3
6.	+iP	AB	13 48 44	<u>Sakhalin Island</u> 46.75 N 141.36 E
	iX	AB	48 50	H = 13 37 10.8 h = 29 km MB = 6.1
	eS	B	58 13	D = 73.74 Az = 329.3 (NEIC)
	eSS	C	14 03 00	PV A 1.1s 181.5nm M = 6.0
	LmH	B	20.0	XV A 1.6 269.0nm
	LmV	B	27.5	SH B 8 0.91/um 6.1
				LmH B 15.5 7.3/um 6.1
				LmV B 12 7.2/um 6.4
6.	ePKIKP	A	20 23 05.5	<u>Fiji Islands Region</u> 21.10 S 177.78 W
	ePKHKP	A	23 10	H = 20 04 02.2 h = 356 km MB = 4.7
	ePKP2	A	23 16.5	D = 149.60 Az = 348.1 (NEIC)
				PKHKPV A 1.0s 27.5nm
6.	e	A	23 27 29.5	
6.	LmH	C	24 37	<u>East New Guinea Region</u> 5.8 S 146.7 E
	LmV	C	45.8	H = 23 35 32.1 h = 116 km MB = 5.2 (NEIC)
				D = 121.8
7.	ePn	A	04 03 32.8	<u>Northern Italy</u> 46.10 N 12.45 E
	i	A	03 33.2	H = 04 02 24.3 h = 25 km MB = 4.2
	iPg	AB	03 52.2	D = 4.59 Az = 353.4 (NEIC)
	iSn	A	04 24.5	LmH B 5s 0.6/um M = 3.5
	iSg	AB	04 49.7	LmV B 6 0.5/um
	LmV	B	05.6	
	LmH	B	05.7	
7.	ePKHKP	A	06 54 37	<u>South of Tonga Islands</u> 24.40 S 175.38 W
				H = 06 34 42.1 h = 61 km MB = 4.3
				D = 153.24 Az = 350.1 (NEIC)

Day	Phase		h m s	Remarks
7.	e	A	07 58 16	France 43.16 N 0.20 E
	e	A	08 00 14	H = 07 54 28.7 h = normal MB = 4.0
				D = 10.81 Az = 42.2 (NEIC)
8.	eP	A	03 29 00	Sakhalin Island 46.73 N 141.34 E
	eX	A	29 05	H = 03 17 25.5 h = 16 km MB = 5.0
	e	A	29 21	D = 73.75 Az = 329.3 (NEIC)
	ePP	A	31 52.5	XV A 1.4s 32.5nm
	eS	B	38 32	LmH B 14.5 5.8 µm M = 6.0
	LmH	B	04 01	LmV B 15 2.5 µm 5.7
	LmV	B	03.4	
8.	+iP	AB	07 37 30	Near East Coast of Honshu, Japan
	e	AB	37 42.5	37.16 N 141.31 E
	e	A	37 56	H = 07 25 14.5 h = 56 km MB = 5.5
	ePP	A	40 38.5	D = 82.01 Az = 330.3 (NEIC)
	e(S)	C	48 00	PV A 1.2s 40.6nm M = 5.3
	LmH	B	11.2	LmH B 15 0.5 µm
	LmV	B	18.3	LmV B 17 0.6 µm
8.	eP	AB	12 00 01.5	Sakhalin Island 46.38 N 141.15 E
	e	A	02 54	H = 11 48 23.5 h = 6 km MB = 5.9
	ePP	B	02 48	D = 73.98 Az = 329.3 (NEIC)
	ePPP	B	04 31	PV A 4.5s 2030.0nm M = 6.5
	eiS	B	09 32	PV B 8 3.21 µm 6.5
	e	B	10 15	PPV B 11 1.73 µm 6.2
	eSS	C	15 12	SH B 14 4.30 µm 6.3
	eSSS	C	17 30	LmH B 13.5 75.0 µm 7.1
	LmH	B	32.2	LmV B 12 37.4 µm 6.9
	LmV	B	37.3	
8.	eP	A	12 50 31	Sakhalin Islands 46.40 N 141.05 E
				H = 12 38 55.4 h = 19 km MB = 5.2
				D = 73.93 Az = 329.2 (NEIC)
8.	+iP	A	13 01 27.0	Southern Iran 29.18 N 60.01 E
	e	A	01 34	H = 12 53 34.0 h = 12 km MB = 5.4
	e	A	01 40	D = 41.95 Az = 314.6 (NEIC)
	e	A	01 49	PV A 1.5s 50.2nm M = 5.0

Day	Phase		h m s	Remarks
8.	e(P)	A	15 13 50.5	Sakhalin Island 46.59 N 141.14 E
				H = 15 02 14.5 h = normal MB = 4.7
				D = 73.80 Az = 329.2 (NEIC)
				PV A 1.8s 27.0nm M = 4.9
8.	eP	A	17 05 32	Turkey 37.28 N 30.25 E
	eS	C	09 11	H = 17 01 09.2 h = 6 km MB = 4.8
				D = 18.86 Az = 321.0 (NEIC)
				PV A 1.2s 16.3nm M = 4.0
8.	eiP	AB	17 11 29	Near E. Coast of Eastern Russia
	Pm	A	11 38	46.25 N 140.94 E
	ePP	B	14 20	H = 16 59 52.6 h = 16 km MB = 5.9
	ePPP	B	16 00	D = 74.02 Az = 329.2 (NEIC)
	eS	B	21 00	PV A 1.8s 148.5nm M = 5.8
	e	C	26 40	PV B 6 2.72 µm 6.5
	eSSS	C	29 00	PmV A 1.8 338.0 µm 6.1
	LmH	B	44.0	LmH B 12.5 27.4 µm 6.7
	LmV	B	48.7	LmV B 12.5 21.2 µm 6.7
8.	eP	A	19 33 51	Near E. Coast of Eastern Russia
				46.41 N 140.93 E
				H = 19 22 15.4 h = 19 km MB = 5.3
				D = 73.88 Az = 329.2 (NEIC)
				PV A 1.6s 27.5nm M = 5.0
8.	eP	AB	20 38 24	North Atlantic Ocean 53.83 N 35.31 W
	eS	C	43 10	H = 20 32 29.9 h = normal MB = 4.9
	LmH	B	51.4	D = 28.50 Az = 77.0 (NEIC)
	LmV	B	49.8	SH C 9s 0.56 µm M = 5.1
				LmH B 14 2.0 µm 4.9
				LmV B 15 2.3 µm 5.0
8.	ePKIKP	A	22 27 30.5	Banda Sea 6.48 S 130.57 E
				H = 22 08 55.6 h = normal MB = 5.5
				D = 113.12 Az = 322.7 (NEIC)
				PKIKPV A 0.8s 15.4nm

Day	Phase		h m s	Remarks
8.	eP	A	22 40(31)	<u>Turkey-USSR Border Region</u>
	e	A	40 35	41.11 N 43.77 E
	eX	A	40 46.5	H = 22 35 15.8 h = 33 km MB = 4.8
	LmH	B	50.5	D = 24.16 Az = 304.1 (NEIC)
	LmV	B	51.9	XV A 1.8s 33.8nm
				LmV B 16 0.62 µm M = 4.2
				LmH B 13 0.44 µm 4.3
9.	e	A	04 10 49	<u>Near West Coast of Colombia</u>
				2.28 N 78.89 W
				H = 03 57 47.2 h = 14 km MB = 4.8
				D = 88.58 Az = 39.6 (NEIC)
9.	eP	A	06 54 29	<u>Greece</u> 38.20 N 20.14 E
	e	A	54 35	H = 06 51 08.8 h = 5 km MB = 4.3
	LmH	B	59.6	D = 13.84 Az = 336.8 (NEIC)
	LmV	B	07 01.0	PV A 0.9s 17.5nm M = 4.2
9.	-eP1	AB	15 14 24	<u>Turkey</u> 37.30 N 30.24 E
	+iP2	A	14 30.5	H = 15 10 03.4 h = 23 km MB = 5.3
	eS	B	18 00	D = 18.84 Az = 321.0 (NEIC)
	LmH	B	21.2	P1V A 1.6s 66.0nm M = 4.6
	LmV	B	23.7	P2V A 1.7 157.5nm 4.9
				SH B 7 1.44 µm 5.3
				LmH B 14 5.2 µm 5.2
				LmV B 10 2.0 µm 4.9
9.	e	A	15 22 47.3	
9.	-eP1	AB	23 13 12.5	<u>Kurile Islands</u> 44.44 N 150.86 E
	-iP2	AB	13 15.5	H = 23 01 06.8 h = 7 km MB = 6.0
	ei	B	13 43	D = 78.90 Az = 334.9 (NEIC)
	eiS	B	23 03	P1V A 1.5s 105.5nm M = 5.7
	eSS	C	28.5	P2V A 1.6 366.0nm 6.2
	LmH	B	52.9	P2V B 5 2.97 µm 6.6
	LmV	B	52.9	SH B 8 3.66 µm 6.5
				LmH B 16.5 8.8 µm 6.2
				LmV B 17 11.6 µm 6.3

Day	Phase		h m s	Remarks
10.	-iPKIKP	AB	06 48 35.3	<u>Tonga Islands</u> 20.44 S 174.21 W
	+iPKHKP	AB	48 39.3	H = 06 28 51.2 h = normal MB = 5.8
	-iX	A	48 53	D = 149.50 Az = 352.7 (NEIC)
	ePP	B	52 20	PKHKPV A 1.9s 265.0nm
	ePPP	B	55 30	XV A 1.7 382.0nm
	eSKSP	B	07 02 30	LmH B 22 2.1 µm M = 5.8
	LmH	B	53.6	LmV B 19 2.08 µm 5.9
	LmV	B	08 03.6	
10.	e	A	10 30 16	
10.	ePKIKP	A	14 11 10.5	<u>Fiji Islands Region</u> 18.14 S 179.59 W
	-eiPKHKP	A	11 12	H = 13 52 41.8 h = 671 km MB = 5.2
	eiPKP2	A	11 14.5	D = 146.35 Az = 347.1 (NEIC)
				PKHKPV A 1.0s 31.5nm
11.	eX	A	02 06 40	<u>Greece</u> 38.92 N 22.07 E
	LmH	B	12.5	H = 02 03 09.9 h = 4 km MB = 4.5
	LmV	B	12.5	D = 13.86 Az = 331.2 (NEIC)
				XV A 0.8s 11.5nm
				LmH B 11 1.1 µm M = 4.2
				LmV B 10 1.0 µm
11.	eP	A	06 01 05	<u>South Burma</u> 15.24 N 96.18 E
				H = 05 49 24.4 h = normal MB = 4.9
				D = 74.95 Az = 319.0 (NEIC)
				PV A 1.0s 15.7nm M = 5.0
11.	eP	A	06 39 53	<u>South Burma</u> 15.14 N 96.29 E
	iX	A	39 57.5	H = 06 28 11.1 h = 28 km MB = 5.3
	LmH	C	07 14.0	D = 75.10 Az = 319.0 (NEIC)
	LmV	C	19.0	PV A 0.8s 13.5nm M = 5.0
				XV A 1.0 23.6nm
				LmH C 22 0.33 µm M=4.6
				LmV C 18 0.55 µm 4.9

Day	Phase		h m s	Remarks
11.	eP	A	15 06 20	<u>South Burma</u> 15.30 N 96.34 E
	LmH	C	40.0	H = 14 54 39.4 h = 33 km MB = 4.6
	LmV	C	45.0	D = 75.01 Az = 319.0 (NEIC)
				LmV C 18s 0.33 µm M = 4.7
11.	ePn	A	23 19 35.3	<u>Northern Italy</u> 44.8 N 10.4 E
	iPn	A	19 36.0	H = 23 18 13 h = 0 km
	ePg	A	20 02	D = 5.91 Az = 8 (ISC)
	iSn	A	20 39.5	
12.	e	A	05 54 35	
12.	+iPKIKP	AB	08 26 36.5	<u>South of Fiji Islands</u> 26.70 S 177.14 W
	eiPKHKP	AB	26 46	H = 08 06 54.1 h = 101 km MB = 5.8
	eiPKP2	A	27 01	D = 155.17 Az = 346.7 (NEIC)
	LmH	B	09 35.5	PKIKPV A 1.6s 49.5nm
	LmV	B	35.5	PKHKPV A 1.4 60.3nm
				PKP2V A 1.4 69.8nm
				LmH B 20 0.31 µm
				LmV B 20 0.37 µm
12.	ePKP2	A	17 01 52.5	<u>South of Fiji Islands</u> 26.58 S 177.73 W
				H = 16 41 50.20 h = 130 km MB = 4.8 (NEIC)
				D = 154.8
				PKPV A 1.6s 16.5nm
12.	ePP	A	19 25 26	<u>Mariana Islands</u> 18.25 N 146.34 E
	epPP	A	25 40	H = 19 07 37.1 h = 77 km MB = 4.9 (NEIC)
	LmH	B	20 10.5	D = 100.5
	LmV	B	10.5	PPV A 1.7s 24.3nm M = 5.6
				pPPV A 1.8 30.4nm
				LmH B 18 0.3 µm
				LmV B 19 0.5 µm
12.	ePKIKP	A	20 32 59	<u>New Hebrides Islands</u> 17.51 S 168.66 E
	e	A	34 36	H = 20 13 57.5 h = 229 km MB = 5.0
	ePP	A	36 14	D = 142.19 Az = 336.1 (NEIC)
				PKIKPV A 1.3s 21.8nm
				PPV A 2.0 42.7nm M = 5.1

Day	Phase		h m s	Remarks
13.	-iP	A	01 21 57	<u>South of Honshu, Japan</u> 33.92 N 138.32 E
	eiPP	A	25 13	H = 01 09 55.8 h = 249 km MB = 5.2
				D = 83.58 Az = 329.1 (NEIC)
				PV A 1.1s 68.6nm M = 5.4
				PPV A 1.6 49.4nm 5.5
13.	LmH	B	02 54.7	<u>Near Coast of Northern Chile</u>
	LmV	B	54.7	24.4 S 70.2 W
				H = 01 55 43.1 h = normal MB = 5.0 (NEIC)
				D = 103.6
13.	+iP	AB	04 29 16.5	<u>Dominican Republic Region</u>
	e	A	29 36	17.93 N 69.73 W
	eS	B	38 25	H = 04 18 01.9 h = 48 km MB = 5.7
	LmH	B	55.5	D = 70.90 Az = 41.8 (NEIC)
	LmV	B	55.5	PV A 1.6s 71.4nm M = 5.4
				SH B 11 0.35 μm 5.3
				LmH B 21 1.2 μm 5.1
				LmV B 21 1.3 μm 5.2
13.	LmH	C	23 58.0	<u>South Sandwich Islands Region</u>
	LmV	C	58.5	59.3 S 25.9 W
				H = 22 58 47.2 h = normal
				MB = 5.1 MS = 5.2 (NEIC)
				D = 113.6
14.	eiPKHKP	AB	03 22 28	<u>Solomon Islands</u> 7.37 S 155.98 E
	ePKIKP	A	22 41.5	H = 03 03 39.2 h = 46 km MB = 5.7 (NEIC)
	ePP	AB	24 41	D = 127.8
	e	B	31 50	PKHKPV A 1.5s 85.5nm
	eSKKS	B	36 40	PKIKPV A 1.4 58.2nm
	ePS	B	39 50	LmH B 16 4.8 μm M = 6.3
	LmH	B	56.6	LmV B 16 4.5 μm 6.2
	LmV	B	59.0	
14.	-ePKIKP	AB	05 39 27	<u>New Britain Region</u> 6.48 S 151.52 E
	ePP	AB	41 16	H = 05 20 29.3 h = normal MB = 6.1
	ePPP	B	43 50	D = 124.78 Az = 330.1 (NEIC)
	eiPS	B	51 12	PKIKPV A 2.0s 196.7nm

Day	Phase		h m s	Remarks
cont. 14.	ePPS	B	05 52 36	PKIKPV B 7s 0.87/um
	ePcPPKP	A	53 07	PPV B 9 3.75/um M = 6.8
	LmH	B	06 33.6	LmH B 17 5.2/um 6.3
	LmV	B	33.6	LmV B 17 6.0/um 6.3
14.	eP	A	07 08 31	<u>Hokkaido, Japan Region</u> 42.36 N 144.74 E
				H = 06 56 31.8 h = 39 km MB = 4.9
				D = 78.75 Az = 331.7 (NEIC)
				PV A 1.2s 10.2nm M = 4.7
14.	ePKIKP	A	10 56 56.5	<u>Tonga Islands</u> 15.16 S 174.20 W
				H = 10 37 34.5 h = 123 km MB = 4.5 (NEIC)
				D = 144.3
14.	ePKIKP	A	14 27 38	<u>Tonga Islands</u> 20.53 S 174.30 W
	iPKHKP	AB	27 43	H = 14 07 55.1 h = 33 km MB = 5.2
	e(PKP2)	A	27 52.5	D = 149.58 Az = 352.6 (NEIC)
	LmH	B	15 39.0	PKHKPV A 1.8s 74.2nm
	LmV	B	43.0	PKP2V A 1.6 49.5nm
14.	eP	AB	20 03 22	<u>Central Mid-Atlantic Ridge</u>
	e	A	03 31.5	0.82 N 28.98 W
	e	A	03 39.5	H = 19 53 14.3 h = normal MB = 5.2
	e	A	03 46	D = 60.38 Az = 28.5 (NEIC)
	e	A	03 49.5	PV A 1.2s 32.5nm M = 5.3
	eX	B	07 18	PV B 5 0.50/um 5.9
	eS	B	11 30	XV B 11 0.54/um
	LmH	B	26.1	SH B 16 2.26/um 5.8
	LmV	B	27.5	LmH B 15 1.5/um 5.2
				LmV B 15 1.2/um 5.2
14.	eP	A	20 18 19.5	<u>Eastern Caucasus</u> 42.8 N 45.3 E
				H = 20 13 17 h = 154 km MB = 4.4
				D = 24.26 Az = 301 (ISC)
				PV 1.3s 10.9nm M = 4.2
15.	e	A	09 18 47	<u>Chagos Archipelago Region</u> 7.05 S 67.95 E
				H = 09 06 58.1 h = normal MB = 4.7
				D = 75.16 Az = 326.8 (NEIC)

Day	Phase		h m s	Remarks
15.	eP	A	10 11 32	Hokkaido, Japan Region 41.89 N 142.26 E
				H = 09 59 38.5 h = 74 km MB = 5.0
				D = 78.27 Az = 330.4 (NEIC)
				PV A 0.7s 17.2nm M = 5.1
15.	eP	AB	13 42 11	South of Honshu, Japan 31.50 N 139.99 E
	e	A	42 16	H = 13 29 34.4 h = 72 km MB = 5.2
	ePP	AB	45 32	D = 86.37 Az = 330.0 (NEIC)
	eS	B	52 40	PV A 2.0s 42.7nm M = 5.2
	eSS	C	58 25	PV B 7 0.93 µm 6.0
	LmH	B	14 21.4	PPV A 2.0 42.7nm 5.5
	LmV	B	25.1	PPV B 7 0.6 µm 6.1
				SH B 9 1.14 µm 5.9
				LmH B 15 2.7 µm
				LmV B 18 2.9 µm
15.	+iP	AB	15 07 21.1	Off East Coast of Honshu, Japan
	iX	A	07 30.5	39.11 N 143.45 E
	iY	A	07 36.8	H = 14 55 05.3 h = 17 km MB = 5.8
	ePP	AB	10 23	D = 81.14 Az = 331.3 (NEIC)
	e	A	10 28	PV A 1.8s 236.1nm M = 5.9
	e	A	10 37	PV B 8 2.62 µm 6.3
	e	A	10 51	XV A 1.8 230.0nm
	eiS	B	17 28	YV A 1.8 192.1nm
	LmH	B	42.7	PPV B 8.5 1.69 µm 6.3
	LmV	B	49.8	SH B 14 4.99 µm 6.4
				LmH B 17.5 26.3 µm 6.7
				LmV B 15 18.5 µm 6.6
15.	e(X)	A	15 48 07	XV A 1.0s 9.8nm
15.	eSg	A	21 55 56	Poland 50.29 N 18.88 E
				H = 21 53 26.2 (WAR)
				D = 4.67
16.	eP diff	A	06 37 08.5	Banda Sea 5.92 S 130.65 E
	e	A	38 30	H = 06 22 37.6 h = 115 km MB = 6.2 (NEIC)
	ePKIKP	AB	41 06	D = 112.7
	ePP	AB	41 51	P diffV A 1.5s 30.2nm

Day	Phase		h m s	Remarks
cont. 16.	eSKS	B	06 47 29	PKIKPV A 2.0s 111.0nm
	eS diff	B	49 18	PPV A 1.8 257.0nm M = 5.3
	eSP	AB	51 08	PPV B 6 1.09/um 6.4
	e	B	51 43	LmH B 19 1.23/um
	e	B	52 35	LmV B 18 1.27/um
	LmH	B	07 31.2	station clock out of operation
	LmV	B	32	
16.	eP	A	19 03 57	Off East Coast of Honshu, Japan
	epP	A	04 06.5	39.04 N 143.53 E
	eS	B	14 05	H = 18 51 44.1 h = 43 km MB = 5.2
	LmH	B	38.5	D = 81.22 Az = 331.4 h = 34 km (NEIC)
	LmV	B	43.4	PV A 1.2s 28.4nm M = 5.1
				pPV A 1.8 57.5nm
				LmH B 18 1.1/um 5.3
				LmV B 17 0.92/um 5.2
17.	e(pP)	A	09 27 25	Off East Coast of Honshu, Japan
	LmH	B	10 05.2	38.96 N 143.66 E
	LmV	B	06.6	H = 09 15 00.9 h = 35 km MB = 4.4
				D = 81.34 Az = 331.4 (NEIC)
				(pP)V A 1.7s 15.1nm
				LmH B 14.5 0.37/um
				LmV B 16 0.29/um
17.	ePKHKP	A	14 40 37	Tonga Islands 21.57 S 174.54 W
				H = 14 20 47.9 h = normal MB = 5.2
				D = 150.58 Az = 352.0 (NEIC)
				PKHKPV A 1.5s 35.2nm
18.	ePKP	A	04 02 15.5	Tonga Islands 16.12 S 173.09 W
				H = 03 42 38.8 h = normal MB = 4.7
				D = 145.35 Az = 354.7 (NEIC)
				PKPV A 1.4s 16.3nm
18.	eP	A	17 24 32	Near Coast of Guatemala 13.53 N 90.03 W
				H = 17 11 54.7 h = 82 km MB = 4.8
				D = 86.89 Az = 38.6 (NEIC)
				PV A 1.1s 12.1nm M = 4.9

Day	Phase		h m s	Remarks
18.	eP	A	18 24 52	Off East Coast of Honshu, Japan
				39.20 N 143.30 E
				H = 18 12 38.9 h = normal MB = 4.4
				D = 81.00 Az = 331.2 (NEIC)
19.	ePKIKP	AB	10 19 03	Fiji Islands Region 21.30 S 178.49 W
	-iPKHKP	AB	19 08.5	H = 10 00 19.3 h = 545 km MB = 5.4
	eiPKP2	A	19 15.5	D = 149.65 Az = 347.2 (NEIC) h = 564 km
	epPKIKP	A	21 11.5	PKIKPV A 2.0s 59.8nm
	epPKHKP	A	21 17	PKHKPV A 1.6 173.0nm
	epPKP2	A	21 25	PKHKPV B 3 0.77/um
				PKP2V A 1.6 82.5nm
19.	eP	A	11 04 24	Western Russia 57.78 N 41.10 E
				H = 11 00 06.8 h = normal MB = 4.5
				D = 18.55 Az = 260.1 (NEIC)
19.	e	A	22 59 28	Kurile Islands 45.68 N 151.73 E
				H = 22 47 13.6 h = 3 km MB = 4.2
				D = 78.03 Az = 335 (ISC)
20.	eP	A	02 07 30	Off East Coast of Honshu, Japan
				39.38 N 143.31 E
				H = 01 55 18.1 h = 30 km MB = 4.6
				D = 80.85 Az = 331.2 (NEIC)
20.	eP	A	02 55 38	Mid-Indian Rise 12.68 S 66.12 E
	eX	A	55 42	H = 02 43 35.8 h = normal MB = 5.0
				D = 78.89 Az = 328.1 (NEIC)
				PV A 1.1s 16.1nm M = 5.0
				XV A 1.3 17.4nm
20.	e(pP)	A	03 34 38	Off East Coast of Honshu, Japan
	LmH	B	04 13.5	39.28 N 143.47 E
	LmV	B	15.7	H = 03 22 16.5 h = 33 km MB = 4.4
				D = 81.00 Az = 331.3 (NEIC)
				pPV A 1.3s 13.1nm
				LmH B 14 0.3/um M = 4.8
				LmV B 15 0.3/um 4.9

Day	Phase		h m s	Remarks
20.	e	A	05 27 38	Off East Coast of Honshu, Japan
	e	A	27 42.5	39.34 N 143.44 E
				H = 05 15 19 h = 31 km MB = 4.3
				D = 80.93 Az = 331 (ISC)
20.	e	A	06 13 09	Off East Coast of Honshu, Japan
				39.29 N 143.3 E
				H = 06 00 40 h = 49 km MB = 3.9 (ISC)
				D = 80.8
20.	LmH	B	06 32	LmH B 12s 0.5 µm
	LmV	B	32.2	LmV B 12 0.5 µm
20.	eP	AB	06 55 39.5	Kodiak Island Region 56.37 N 153.00 W
	e	A	55 47	H = 06 44 14.2 h = normal MB = 5.2 (NEIC)
	eS	B	07 05 05	D = 72.6
	eSS	B	10 05	PV A 1.6s 33.0nm M = 5.1
	LmH	B	32.0	LmH B 18 0.8 µm 5.0
	LmV	B	35.0	LmV B 16 0.6 µm 5.0
20.	eP	A	21 05 40.5	Kodiak Island Region 56.63 N 153.0 W
	e	A	05 51.5	H = 20 54 14.4 h = 37 km MB = 4.7
				D = 72.34 Az = 10 (ISC)
20.	LmH	B	21 42.0	Kodiak Islands Region 56.4 N 153.4 W
	LmV	B	45.0	H = 20 54 13.9 h = 37 km
				MB = 4.8 MS = 4.8 (NEIC)
				D = 72.6
				LmH B 18s 0.28 µm M = 4.6
				LmV B 16 0.24 µm 4.6
20.	e(Sg)	A	21 44 46	Probably Poland
				H = 21 42 00.7 (WAR)
21.	e(P)	A	01 09 58	Turkey 39.2 N 44.3 E
	LmH	C	21.0	H = 01 04 20 MB = 4.3 MS = 4.2
	LmV	C	21.0	D = 25.6 (ANUSSR)

Day	Phase		h m s	Remarks
21.	ePKP2	A	05 23 32.5	<u>South of Fiji Islands</u> 26.56 S 177.49 W
				H = 05 03 20.5 h = 77 km MB = 5.0
				D = 154.96 Az = 346.2 (NEIC)
				PKP2V A 1.2s 20.3nm
21.	-iP	AB	08 55 25.5	<u>Near West Coast of Honshu, Japan</u>
	epP	A	56 13	37.30 N 138.60 E
	ePP	A	58 13	H = 08 43 31.9 h = 186 km MB = 5.4 (NEIC)
	LmH	B	09 47.5	D = 80.3 h = 198 km
	LmV	B	51.3	PV A 1.4s 88.4nm M = 5.3
				LmH B 19 0.4/um
				LmV B 16 0.5/um
21.	e(Pn)	A	09 18 21.5	<u>Austria</u> 47.6 N 15.7 E
	eSg	A	19 32	H = 09 17 22 h = 0 km
				D = 4.09 Az = 321 (ISC)
21.	ePn	A	13 02 54	<u>Swabian Jura, Federal Rep. of Germany</u>
	iPg	A	03 02.4	48°19' N 9°04' E
	iSn	A	03 25	H = 13 02 12 (BCIS)
	iSg	A	03 36	D = 2.85
21.	e	A	13 12 40.5	<u>Swabian Jura, Federal Rep. of Germany</u>
	ePg	A	12 45	48°19' N 9°04' E
	eSn	A	13 08	H = 13 11 56 (BCIS)
	eSg	A	13 19	D = 2.85
21.	eP	AB	16 53 10	<u>Turkey</u> 37.27 N 30.17 E
	eX	A	53 16	H = 16 48 51.6 h = 37 km MB = 4.8 (NEIC)
	eS	B	56 48	D = 18.83
	LmH	B	59.9	PV A 1.6s 38.5nm M = 4.3
	LmV	B	17 01.1	XV A 1.3 56.7nm
				SH B 10.5 0.83/um 4.8
				LmH B 12.5 2.9/um 4.8
				LmV B 16 1.2/um 4.4

Day	Phase		h m s	Remarks
21.	e	A	19 04 07	<u>South of Kermadec Islands</u> 33.4 S 178.1 W
				H = 18 43 25 h = 43 km
				D = 161.36 Az = 340 (ISC)
22.	eP	A	06 37 51	<u>North Atlantic Ridge</u> 31.06 N 41.56 W
	eX	A	37 57	H = 06 29 49.0 h = normal MB = 4.8
				D = 43.67 Az = 47.6 (NEIC)
				XV A 1.3s 17.5nm
22.	eP	A	08 52 33	<u>Azores Islands Region</u> 42.28 N 29.38 W
	LmH	B	09 03.0	H = 08 46 33.5 h = normal MB = 4.3
	LmV	B	03.0	D = 29.13 Az = 59.1 (NEIC)
				LmH B 18s 0.56 µm M = 4.2
				LmV B 18 0.76 µm 4.5
22.	eP	AB	14 27 08.5	<u>Near Coast of Chiapas, Mexico</u>
	e	A	27 10.5	15.06 N 93.70 W
	e	A	27 14.5	H = 14 14 20.5 h = normal MB = 5.2
	ePP	C	30 30	D = 87.90 Az = 37.9 (NEIC)
	eSKS	BC	37 43	PV A 2.0s 51.3nm M = 5.5
	eS	B	37 58	PV B 12 0.98 µm 6.0
	eSP	C	38 46	LmH B 19 4.9 µm 5.9
	LmV	B	15 07.6	LmV B 19 5.4 µm 6.0
	LmH	B	07.8	
22.	iPn	A	18 36 34	<u>Swabian Jura, Federal Rep. of Germany</u>
	iPg	A	36 43.5	48.4 N 8.8 E
	iSn	A	37 08.5	H = 18 35 55 h = 20 km
	eSg	A	37 20	D = 2.86 Az = 38 (ISC)
22.	LmH	C	23 07.5	<u>Near Coast of Peru</u> 10.9 S 78.1 W
	LmV	C	07.5	H = 22 14 59.9 h = 59 km MB = 5.3 (NEIC)
				D = 98.2
23.	eP	A	01 02 27.5	<u>Near E. Coast of Eastern Russia</u>
	ePP	A	04 55	46.24 N 137.15 E
				H = 00 51 30.6 h = 291 km MB = 4.8
				D = 72.64 Az = 327.1 (NEIC)
				PV A 1.5s 30.2nm M = 4.8

Day	Phase		h m s	Remarks
23.	eP	A	03 05 24	Near East Coast of Honshu, Japan 40.26 N 142.57 E H = 02 53 20.6 h = 60 km MB = 4.4 D = 79.80 Az = 330.8 (NEIC)
23.	ePn	A	07 59 07	Czechoslovakia 50°45.7' N 14°25.5' E
	ePg	A	59 12.5	explosion yield 14.6 t (PRU)
	iSg	A	59 32	D = 1.77
23.				In the intervall $08^h\ 59^m - 15^h\ 19^m$ the short-period seismographs (type A) have been out of operation.
23.	e	A	23 08 53.5	
23.	ePn	A	23 12 55.5	Swabian Jura, Federal Rep. of Germany
	ePg	A	13 06	48°19' N 9°04' E
	eSn	A	13 30	H = 23 12 09 h = 20 km (BCIS)
	e	A	13 40.5	D = 2.85
	iSg	A	13 43	
24.	ePKHKP	A	00 22 41.5	Fiji Islands Region 19.63 S 177.71 W H = 00 04 00.1 h = 598 km MB = 4.1 D = 148.18 Az = 348.7 (NEIC)
24.	+iP	AB	01 22 14.5	Off East Coast of Honshu, Japan
	eX	B	24 34	39.43 N 143.25 E
	ePP	B	25 18	H = 01 10 00.4 h = 22 km MB = 5.6
	eS	B	32 20	D = 80.78 Az = 331.2 (NEIC)
	LmH	B	02 01.2	PV B 10s 1.29 µm M = 5.9
	LmV	B	03.4	XV A 14 1.22 µm
				PPV B 14 1.16 µm 5.9
				SH B 18 2.8 µm 6.0
				LmH B 16 11.4 µm 6.3
				LmV B 15.5 11.7 µm 6.4
24.	LmV	B	05 29.3	Near Coast of Peru 16.4 S 73.7 W
	LmH	B	29.8	H = 04 32 55.0 h = 28 km

Day	Phase		h m s	Remarks
cont. 24.				MB = 5.2 MS = 5.6 (NEIC)
				D = 99.6
				LmH B 18.5s 2.3/um M = 5.7
				LmV B 19 3.0/um 5.8
24.	LmH	B	16 44.4	LmH B 18.5s 0.6/um
	LmV	B	53.0	
24.	iPKHKP	A	19 39 15	South of Fiji Islands 22.72 S 178.45 W
				H = 19 20 09.5 h = 426 km MB = 5.0
				D = 151.03 Az = 346.7 (NEIC)
				PKHKPV A 1.0s 21.7nm
24.	eP	A	19 51 08	Off East Coast of Honshu, Japan
				39.86 N 143.15 E
				H = 19 38 55.9 h = 20 km MB = 4.5
				D = 80.36 Az = 331.1 (NEIC)
				traces
25.	ePKP	A	03 09 36	Fiji Islands Region 17.91 S 178.76 W
				H = 02 51 01.3 h = 616 km MB = 4.6
				D = 146.31 Az = 348.1 (NEIC)
25.	ePKP2	A	03 43 09	Kermadec Islands 29.55 S 177.05 W
				H = 03 22 43.6 h = 43 km MB = 5.1
				D = 157.94 Az = 345.2 (NEIC)
				PKP2V A 1.2s 12.2nm
25.	-iPKIKP	AB	04 54 57.3	East New Guinea Region 6.54 S 146.58 E
	ePP	B	56 35	H = 04 36 14.0 h = 115 km MB = 5.3
	esPP	B	57 15	D = 122.29 Az = 327.8 (NEIC)
	eiPPP	B	59 10	PKIKPV A 1.4s 265.0nm
	e(PKKP)	B	05 05 25	PPV A 10 2.91/um M = 6.2
	e	B	06 28	LmH B 22 8.8/um
	e	B	06 50	LmV B 20 5.3/um
	eSS	B	13 08	
	ePSPS	BC	13 50	
	LmH	B	39.6	
	LmV	B	43.9	

Day	Phase		h m s	Remarks
25.	ePn	A	10 35 44	Northern Italy 44.17 N 8.64 E
	eSn	A	36 57	H = 10 34 04.3 h = 19 km MB = 4.1
	eSg	A	37 45	D = 6.79 Az = 16.3 (NEIC)
	LmH	B	38.6	LmH B 11.5s 1.5 /um M = 3.8
	LmV	B	38.9	LmV B 11 1.0 /um
25.	LmH	C	14 10.0	Off Coast of Central Chile
	LmV	B	10.0	32.4 S 73.0 W
				H = 13 05 59.6 h = normal MB = 5.5 (NEIC)
				D = 111.0
25.	LmH	B	15 47.5	LmH B 14s 0.89 /um
	LmV	B	48.0	LmV B 16 0.82 /um
25.	LmH	B	19 14.0	West New Guinea 3.3 S 139.7 E
	LmV	B	14.0	H = 18 01 31.9 h = normal MB = 5.1 (NEIC)
				D = 115.8
				LmH B 20s 0.52 /um M = 5.2
				LmV B 20 0.59 /um 5.2
25.	LmH	B	22 30.0	LmH B 20s 0.36 /um
	LmV	B	30.5	LmV B 19 0.41 /um
25.	ePKIKP	A	22 37 06	Fiji Islands Region 19.50 S 176.80 W
	iPKHKP	A	37 09.8	H = 22 18 05.2 h = 373 km MB = 5.0
	ePKP2	A	37 13.4	D = 148.21 Az = 349.8 (NEIC)
26.	ePKP2	A	01 52 33	South of Fiji Islands 26.08 S 177.31 W
				H = 01 32 27.9 h = 93 km MB = 5.3
				D = 154.53 Az = 346.7 (NEIC)
26.	iPKHKP	A	05 28 27.5	Fiji Islands Region 19.97 S 178.46 W
				H = 05 09 48.0 h = 623 km MB = 4.2
				D = 148.37 Az = 347.7 (NEIC)
26.	eP	A	05 48 04	Southern Greece 37.94 N 22.25 E
	LmH	B	54.1	H = 05 44 34.9 h = 107 km MB = 4.1
	LmV	B	54.4	D = 14.79 Az = 332.6 (NEIC)
				LmH B 16s 0.84 /um
				LmV B 12 0.65 /um

Day	Phase		h m s	Remarks
26.	ePKIKP	A	11 21 13	South Sandwich Islands Region
				56.66 S 27.43 W
				H = 11 02 59.1 h = 181 km MB = 5.6
				D = 111.70 Az = 25.6 (NEIC)
26.	e	A	15 39 19	Talaud Islands 3.22 N 125.79 E
				H = 15 25 18.0 h = 48 km MB = 5.3
				D = 102.55 Az = 323.5 (NEIC)
26.	eP	AB	16 47 03.5	Molucca Sea 0.00 S 124.86 E
	ePP	A	51 16	H = 16 33 04.7 h = 71 km MB = 5.8
	e	B	51 24	D = 104.56 Az = 322.8 (NEIC)
	e	B	17 01 08	LmH B 25s 2.6 µm
	eSPP	BC	01 28	LmV B 22 1.8 µm
	e	B	06 56	
	LmH	B	34.3	
	LmV	B	38.4	
27.	+iP	AB	06 06 00.8	Novaya Zemlya 73.39 N 55.10 E
	iPn	B	06 28.5	H = 05 59 55.2 h = 0 km MB = 6.4 MS=5.2
	ePP	B	06 50	D = 29.37 Az = 243.3 (NEIC)
	iPcP	B	09 07	PH A 1.4s 651.0nm M = 6.4
	eS	B	10 56	PV B 4 2.74 µm 6.4
	eiSn	B	11 32	LmH B 7.5 21.6 µm 6.2
	LmH	B	21.8	LmV B 6 15.0 µm 6.3
	LmV	B	21.8	underground explosion
27.	ePKHKP	A	07 17 16	South of Fiji Islands 25.25 S 177.24 W
				H = 09 58 05.3 h = 124 km MB = 4.7
				D = 153.74 Az = 347.2 (NEIC)
27.	ePKHKP	A	10 17 50	South of Fiji Islands 25.2 S 177.2 W
	ePKP2	A	18 03.5	H = 09 58 05.3 h = 124 km MB = 4.7 (NEIC)
				D = 153.7
27.	-eX	A	12 31 16.5	South of Kermadec Islands
				32.24 S 177.86 W
				H = 12 10 39 h = 28 km MB = 4.8
				D = 160.31 Az = 342 (ISC)
				XV A 1.5s 35.2nm

Day	Phase		h m s	Remarks
27.	ePKIKP	A	15 14 53	<u>East New Guinea Region</u> 5.71 S 145.44 E
				H = 14 56 10.9 h = 103 km MB = 5.4
				D = 120.99 Az = 327.6 (NEIC)
				PKIKPV A 1.6s 19.2nm
27.	eP	A	15 49 07.5	<u>Kurile Islands</u> 49.39 N 155.56 E
	LmH	B	16 27.5	H = 15 37 23.2 h = normal MB = 5.3
	LmV	B	27.5	D = 75.68 Az = 337.2 (NEIC)
				PV A 1.2s 26.4nm M = 5.1
27.	eP	AB	19 13 21	<u>Sakhalin Island</u> 46.39 N 141.11 E
	ePP	B	16 05	H = 19 01 45.4 h = 21 km MB = 5.9
	ePPP	B	17 52	D = 73.97 Az = 329.3 (NEIC)
	iS	B	22 54	PV A 2.2s 246.0nm M = 5.9
	eSS	C	27 10	PV B 6 1.55/um 6.3
	eSSS	C	30 52	PPV B 12 0.76/um 5.8
	LmH	B	44.5	SH B 14 2.83/um 6.3
	LmV	B	48.8	LmH B 16.5 41.9/um 6.8
				LmV B 18.5 20.2/um 6.5
27.	ePKIKP	A	22 38(55)	<u>Bismarck Sea</u> 3.21 S 148.14 E
	e	A	39 10	H = 22 20 03.2 h = normal MB = 5.5
	ePP	BC	40 16	D = 120.28 Az = 329.5 (NEIC)
	e	A	40 26.5	LmH B 21s 9.6/um M = 6.4
	e	B	40 36	LmV B 21 12.7/um 6.5
	e	A	40 39.5	
	e	A	43 14	
	ePS	CB	50 00	
	eSS	C	57 10	
	eSSS	B	23 01 40	
	LmV	B	31.7	
	LmH	B	31.8	
28.	eP	AB	05 14 46	<u>Turkey</u> 37.12 N 30.11 E
	eS	B	18 20	H = 05 10 24.8 h = 36 km MB = 4.7
	LmH	C	21.4	D = 18.92 Az = 321.5 (NEIC)
	LmV	C	23.1	PV B 3.5s 0.50/um M = 5.1
				LmH C 16 0.78/um 4.1
				LmV C 15 2.23/um 4.8

Day	Phase		h m s	Remarks
28.	ePn	A	18 11 39.5	Federal Rep. of Germany 50.2 N 7.6 E
	ePg	A	11 48.5	H = 18 10 59 h = 0 km
	eSn	A	12(13)	D = 2.61 Az = 78 (ISC)
	iSg	A	12 23	
28.	ePKIKP	A	20 04 24	Solomon Islands 8.96 S 157.36 E
	e	A	04 29.5	H = 19 45 15.8 h = normal MB = 5.5
	LmH	B	21 03.2	D = 129.73 Az = 332.2 (NEIC)
	LmV	B	05.0	LmH B 18s 0.8 /um M = 5.4
				LmV B 16 0.6 /um 5.4
28.	ePKIKP	A	22 41 50	New Britain Region 5.40 S 152.81 E
	LmH	C	23 44.0	H = 22 22 52.9 h = 43 km MB = 4.8
	LmV	C	44.0	D = 124.48 Az = 331.1 (NEIC)
				LmH C 16s 0.40 /um M = 5.2
				LmV C 18 0.55 /um 5.3
28.	ePKIKP	A	23 34 20	Chile-Argentina Border Region
				32.03 S 69.97 W
				H = 23 16 01.8 h = 110 km MB = 5.6
				D = 109.14 Az = 41.8 (NEIC)
				PKIKPV A 1.5s 20.1nm
28.	eX	A	23 45 39.5	XV A 1.2s 14.2nm
29.	iPn	A	07 19 50	Switzerland 47.09 N 8.97 E
	iPg	A	20 05.5	H = 07 18 51.6 h = 24 km MB = 4.5
	i	B	20 07	D = 3.96 Az = 25.1 (NEIC)
	i	B	20 15	LmH B 8s 7.5 /um M = 4.3
	iSn	AB	20 38.5	LmV B 5 2.9 /um
	eSg	AB	20 56	
	LmH	B	21.3	
	LmV	B	21.5	
29.	LmH	B	14 55.7	Near North Coast of New Guinea
	LmV	B	15 02.7	2.9 S 142.0 E
				H = 13 50 02.5 h = 16 km
				MB = 5.5 MS = 5.3 (NEIC)
				D = 116.8

Day	Phase		h m s	Remarks
cont. 29.				LmH B 17.5s 0.5/um M = 5.2
				LmV B 17.5 0.4/um 5.1
29.	eP	A	15 59 19	Off East Coast of Kamchatka
				55.43 N 163.61 E
				H = 15 47 58.3 h = normal MB = 5.0
				D = 71.75 Az = 341.7 (NEIC)
				PV A 1.0s 19.7nm M = 5.1
29.	eP	A	21 06 19	Southern Greece 37.12 N 23.04 E
	LmH	B	13.9	H = 21 02 33.0 h = 39 km MB = 4.4
	LmV	B	13.9	D = 15.80 Az = 332.4 (NEIC)
				LmH B 10s 0.30/um M = 3.8
				LmV B 10 0.32/um 4.0
30.	eP1	AB	08 30 48	Gulf of California 26.77 N 110.78 W
	eP2	A	30 53	H = 08 17 54.6 h = normal MB = 5.6
	eSKS	B	41 23	D = 87.68 Az = 32.5 (NEIC)
	-iS	B	41 35	P1V A 1.8s 43.9nm M = 5.5
	iSS	C	47 08	P2V A 1.7 94.0nm 5.8
	eSSS	C	50 58	SH B 18 8.16/um 6.5
	LmH	B	09 08.3	LmH B 17 45.1/um 7.0
	LmV	B	08.5	LmV B 17.5 54.5/um 7.0
				NEIC: Appears to be a complex multiple event. Arrival times used pertain to first event. Second, larger event occurred about 6.5 sec. after first. Magnitude result for mb is probably due to a mixture of data from both events.
30.	e	A	08 50 20	Turkey 37.64 N 30.13 E
	e	A	50 30.5	H = 08 45 56 h = 16 km MB = 4.4
				D = 18.52 Az = 320 (ISC)
30.	ePKP	A	14 10 39	Loyalty Islands 21.6 S 169.6 E
				H = 13 51 02 h = 33 km
				D = 146.30 Az = 335 (ISC)

Day	Phase		h m s	Remarks
30.	LmH	B	16 28.5	Near Coast of Jaliso, Mexico
	LmV	B	28.5	18.3 N 104.9 W
				H = 15 33 29.4 h = normal MB = 4.7 (NEIC)
				D = 91.7
				LmH B 18.5s 0.4/um M = 4.9
				LmV B 16 0.4/um 4.9
30.	+iP1	AB	21 33 25	South Atlantic Ocean 0.46 S 4.84 W
	eiP2	AB	33 30	H = 21 24 11.0 h = normal MB = 6.0
	ePP	B	35 24	D = 52.82 Az = 13.1 (NEIC)
	iS	B	40 54	P1V A 1.5s 332.0nm M = 6.1
	iSS	B	44 32	P1V B 5 1.54/um 6.2
	eSSS	B	46 10	P2V A 2.4 787.0nm 6.2
	LmV	B	56.7	P2V B 6 2.65/um 6.3
	LmH	B	56.9	SH B 10.5 2.96/um 6.1
				LmH B 17 6.3/um 6.3
				LmV B 18 6.2/um 5.8
30.	-iPKHKP	A	21 38 12.5	South of Fiji Islands 22.40 S 179.60 E
	iPKP2	A	38 21.5	H = 21 19 25.6 h = 585 km MB = 4.9 (NEIC)
				D = 150.2
				PKHKPV A 1.1s 42.4nm
				PKP2V A 1.2 26.4nm
30.	eiP	A	21 41 36.5	Eastern Siberia 61.61 N 140.30 E
				H = 21 31 25.9 h = normal MB = 5.4
				D = 60.85 Az = 325.3 (NEIC)
				PV A 1.9s 110.0nm M = 5.7

Day	Phase		h m s	Remarks
1.	e	A	00 48 02	Easter Siberia 61.67 N 140.07 E H = 00 37 46.9 h = 0 km MB = 4.6 D = 60.74 Az = 325 (ISC) PV A 1.0s 7.9nm M = 4.8
1.	ePKIKP	A	05 35 50.5	New Britain Region 4.34 S 152.87 E H = 05 16 56.2 h = 45 km MB = 5.4 (NEIC) D = 123.6 PKIKPV A 1.2s 20.3nm LmH B 18 0.5/um M = 5.2 LmV B 16 0.5/um 5.2
	LmH	B	34.4	
	LmV	B	34.5	
1.	+ePKP	AB	11 48 02	Loyalty Islands Region 21.44 S 169.72 E H = 11 28 23.4 h = 32 km MB = 5.2 D = 146.16 Az = 334.8 (NEIC) PKPV A 1.8s 77.7nm PKPV B 6 0.6/um
	e	A	48 16.5	
	LmH	C	12 58.0	
	LmV	C	58.0	
1.	ePKHKP	A	16 29 37	Fiji Islands Region 19.90 S 177.90 W H = 16 10 55.4 h = 590 km MB = 4.6 D = 148.41 Az = 348.4 (NEIC) PKHKPV A 1.2s 26.4nm
1.	eP	A	16 35 37	Tadzhik SSR 38.60 N 39.75 E H = 16 27 47.7 h = 36 km MB = 4.9 D = 42.06 Az = 306.2 (NEIC) PV A 2.0s 42.7nm M = 4.8 LmH B 20 0.3/um 4.1 LmV B 14 0.4/um 4.5
	LmH	B	55.5	
	LmV	B	56.5	
1.	LmV	C	22 07.5	New Hebrides Islands 16.7 S 167.2 E H = 21 53 04.0 h = 17 km MB = 4.9 (NEIC) D = 140.8
2.	LmH	C	00 35.5	LmH C 20s 0.2/um LmV C 16 0.4/um C : EW-component out of operation
	LmV	C	43.5	

Day	Phase		h m s	Remarks
2.	+iP	A	03 28 16	Northwest Territories, Canada
	-ipP	A	28 24.5	64.36 N 86.50 W
	LmV	C	46.5	H = 03 19 28.2 h = 16 km MB = 5.0
	LmH	C	49.5	D = 49.07 Az = 56.5 (NEIC)
				h = 37 km
				PV A 1.2s 38.6nm M = 5.3
				pPV A 1.5 45.2nm
				LmV C 25s 0.4/um 4.4
				C : EW-component out of operation
2.	ePKHKP	A	09 48 27	Tonga Islands 17.30 S 174.84 W
	epPKHKP	A	49 16.5	H = 09 29 07.9 h = 200 km MB = 4.4
				D = 146.33 Az = 352.6 (NEIC) h = 191 km
2.	e(P)	A	11 01 35.5	Southwestern Russia 45.3 N 42.5 E
	e	A	06 29.5	H = 10 56 40 MS = 4.0
	e	A	06 47	D = 21.3 (ANUSSR)
	LmH	B	10.0	LmH B 14s 0.2/um M = 3.7
	LmV	B	11.8	LmV B 12 0.3/um 4.0
2.	eP	A	19 55 31.5	Kurile Islands 46.87 N 151.93 E
				H = 19 43 48.6 h = 100 km MB = 4.7
				D = 77.01 Az = 335.3 (NEIC)
				PV A 1.2s 14.5nm M = 4.7
3.	eP	A	07 48 32	Turkey 38.95 N 29.87 E
	LmH	B	54.5	H = 07 44 26.9 h = 23 km MB = 4.7
	LmV	B	57.4	D = 17.40 Az = 318.2 (NEIC)
				PV A 2.0s 51.3nm M = 4.3
				LmH B 16 0.6/um 4.0
3.	ePKIKP	A	08 55 02.5	Solomon Islands 6.61 S 155.00 E
	LmH	B	09 54.0	H = 08 36 04.8 h = 58 km MB = 5.9
	LmV	B	54.0	D = 126.57 Az = 331.8 (NEIC)
				LmH B 19s 0.7/um
				LmV B 20 1.0/um

Day	Phase		h m s	Remarks
3.	ePKIKP	A	13 44 13	New Hebrides Islands Region
	e	B	44 18	14.70 S 171.67 E
	ePP	B	47 05	H = 13 24 37.05 h = 10 km
	e	B	47 32	MB = 5.7 MS = 6.4 (NEIC)
	e(PPS)	C	59 35	D = 140.7
	e	C	14 00 38	LmH B 23s 22.1nm M = 6.8
	eiSS	B	05 38	LmV B 23.5 23.6nm 6.9
	eiSSS	B	10 45	
	LmH	B	42.7	
	LmV	B	42.8	
3.	eP	A	17 13 44.5	Gulf of Campeche 18.09 N 94.14 W
	epP	A	13 59	H = 17 01 06.8 h = 38 km MB = 5.0
	e	A	14 15	D = 85.79 Az = 37.9 (NEIC)
				PV A 1.6s 30.2nm M = 5.3
3.	eP	A	17 23 18.5	Turkey 36.76 N 29.93 E
	e	A	23 23	H = 17 18 53.7 h = 27 km MB = 4.4
	LmH	B	31.6	D = 19.11 Az = 322.3 (NEIC)
	LmV	B	37.7	PV A 1.8s 33.8nm M = 4.3
				LmH B 12 0.15 µm 3.6
				LmV B 13 0.25 µm 3.9
3.	eP	A	22 05 41.5	Near East Coast of Kamchatka
				55.66 N 162.30 E
				H = 21 54 09 h = 7 km MB = 4.8 (ISC)
				D = 71.2
3.	eP	A	23 24 09	Crete 34.33 N 26.15 E
	e	A	24 14.5	H = 23 19 42.4 h = 40 km MB = 4.7
	eS	B	27 44	D = 19.46 Az = 331.3 (NEIC)
	e	B	28 00	LmH B 11s 0.9 µm M = 4.4
	LmH	B	32.3	LmV B 11.5 1.0 µm 4.5
	LmV	B	33.7	
4.	eP diff	B	01 46 05	Solomon Islands 5.85 S 154.18 E
	ePKIKP	B	49 28	H = 01 30 34.1 h = 74 km MB = 5.7
	e	A	49 33	D = 125.52 Az = 331.6 (NEIC)

Day	Phase		h m s	Remarks
cont. 4.	ePP	B	01 51 24	P diff V B 20s 0.59/um
	e	A	51 26	PKIKPV B 14 1.11/um
	e	A	51 33	PPV B 16 4.0/um M = 6.3
	e	A	51 45.5	LmH B 21.5 20.0/um
	ePPP	C	54 00	LmV B 20 21.0/um
	eiSKSP	C	02 01 16	
	ePS	B	01 28	
	ePKKS	C	02 54	
	e	A	02 56	
	eSS	C	08 48	
	eSSS	C	13 35	
	LmH	B	43.5	
	LmV	B	47.0	
4.	eP	A	10 05 02	<u>Western Russia</u> 61.61 N 47.12 E
	e(Sn)	A	09 16	H = 10 00 02.1 h = 13 km MB = 5.1
				D = 22.32 Az = 256.8 (NEIC)
				PV A 1.4s 37.3nm M = 4.5
4.	ePKHKP	A	16 10 49	<u>Loyalty Islands Region</u> 21.97 S 169.59 E
				H = 15 51 09.1 h = 35 km MB = 4.4
				D = 146.58 Az = 334.3 (NEIC)
4.	eP	A	16 39 39	<u>Crete</u> 34.37 N 26.12 E
	e	A	39 44	H = 16 35 12.9 h = 33 km MB = 4.8
				D = 19.41 Az = 331.3 (NEIC)
4.	ePn	A	16 45 25	<u>Central Italy</u> 42.87 N 13.10 E
	e	A	45 34	H = 16 43 31.8 h = normal MB = 5.0 (NEIC)
	ePg	A	46 07.5	D = 7.90
	eiSn	A	46 51.5	PnV A 1.3s 32.7nm
	eSg	A	47 49.5	LmH B 8.5 3.4/um M = 4.4
	LmH	B	48.5	LmV B 14 3.3/um
	LmV	B	49.4	
4.	ePn	A	22 23 50.5	<u>Central Italy</u> 43.06 N 12.90 E
	ePg	A	24 31	H = 22 21 57.2 h = 43 km MB = 4.4 (NEIC)

Day	Phase		h m s	Remarks
cont. 4.	eSn	A	22 25 18	D = 7.67
	LmH	B	27.7	LmH B 15s 0.35/um M = 3.2
	LmV	B	27.8	
5.	ePg	A	00 06 57	Central Italy 42.82 N 13.10 E H = 00 04 18.3 h = 0 km (ISC) D = 7.90
5.	ePg	A	00 07 51	Central Italy 42.8 N 13.2 E H = 00 05 15 (BCIS) D = 7.90
5.	eiSn	A	00 10 33	Central Italy 42.79 N 13.0 E H = 00 07 07.3 h = 0 km (ISC) D = 7.90
5.	eP	A	01 18 08.5	Crete 34.06 N 26.24 E H = 01 13 30.4 h = 0 km D = 19.73 Az = 332 (ISC)
5.	eP	A	01 51 04.5	Near N. Coast of Eastern Siberia
	LmH	B	02 21.0	67.45 N 172.76 W
	LmV	B	27.7	H = 01 40 41.1 h = normal MB = 5.3 (NEIC) D = 62.2 PV A 1.4s 23.2nm M = 4.9 LmH B 19 0.35/um 4.6 LmV B 18 0.40/um 4.7
5.	ePn	A	04 30 03	Central Italy 42.2 N 14.8 E
	eSg	A	32 30	H = 04 27 54 h = 0 km D = 8.71 Az = 346 (ISC)
5.	eP	A	17 30 56.5	Near East Coast of Kamchatka 56.11 N 161.19 E H = 17 19 47.7 h = 80 km MB = 4.8 D = 70.66 Az = 340.0 (NEIC) PV A 0.6s 11.5nm M = 5.0

Day	Phase		h m s	Remarks
5.	eP	A	18 38 57	Southern Iran 27.19 N 55.79 E
	LmH	C	56.0	H = 18 31 17.7 h = 39 km MB = 5.1
	LmV	C	58.0	D = 40.78 Az = 317.2 (NEIC)
				PV A 2.0s 42.8nm M = 4.8
				LmH C 22 0.30 μm 4.1
6.	eP	A	01 50 51	Turkey 38.31 N 30.15 E
	eS	A	54 20	H = 01 46 38.3 h = 19 km MB = 4.6
	LmH	B	59.0	D = 18.02 Az = 319.1 (NEIC)
	LmV	B	02 00.9	PV A 1.5s 35.2nm M = 4.3
				LmH B 10.5 0.8 μm 4.3
				LmV B 12 0.8 μm 4.4
6.	eP	A	02 22 40	Kamchatka 55.89 N 160.34 E
				H = 02 11 41.2 h = 170 km MB = 4.4
				D = 70.70 Az = 339.5 (NEIC)
				PV A 1.2s 16.3nm M = 4.7
6.	ePKHKP	A	04 07 41.5	Fiji Islands Region 20.26 S 177.97 W
				H = 03 48 53.7 h = 540 km MB = 4.6
				D = 148.75 Az = 348.2 (NEIC)
				PKHKPV A 1.0s 23.6nm
6.	ePKIKP	A	07 34 17	Fiji Islands Region 20.65 S 178.45 W
	ePKHKP	A	34 22	H = 07 15 41.7 h = 625 km MB = 4.9
	ePKP2	A	34 28	D = 149.03 Az = 347.5 (NEIC)
				PKIKPV A 1.0s 11.8nm
				PKHKPV A 1.2 44.7nm
				PKP2V A 1.0 23.6nm
6.	ePKP	A	10 46 27	Tonga Islands 17.43 S 173.56 W
	eX	A	46 47	H = 10 26 44.4 h = 7 km MB = 4.7
				D = 146.60 Az = 354.0 (NEIC)
				XV A 1.6s 53.7nm
6.	ePKP	A	11 51 24	New Hebrides Islands Region
	e	A	51 27	19.77 S 172.05 E

Day	Phase		h m s	Remarks
cont. 6.	e	A	11 51 31	H = 11 31 46.7 h = 21 km MB = 5.2
				D = 145.52 Az = 337.9 (NEIC)
				PKPV A 1.2s 40.6nm
6.	eP	A	23 20 07	Turkey 38.06 N 27.27 E
	LmH	B	25.9	H = 23 16 04.8 h = 0 km
	LmV	B	26.7	D = 16.80 Az = 324 (ISC)
				PV A 1.1s 20.2nm M = 4.2
				LmH B 12 0.6 μm 4.0
6.	iX	A	23 25 14.8	XV A 0.8s 25.0nm
	e	A	25 21.5	
6.	eP	A	23 51 30	Tadzhik SSR 37.47 N 72.11 E
	LmV	B	24 11.5	H = 23 43 28.9 h = 107 km MB = 4.8
	LmH	B	11.6	D = 44.22 Az = 307.4 (NEIC)
				LmH B 12s 0.4 μm
				LmV B 14 0.3 μm
8.	ePKHKP	A	04 10(03)	South of Fiji Islands 25.96 S 177.25 W
	ePKP2	A	10 17.5	H = 03 50 12.9 h = 92 km MB = 4.8
				D = 154.43 Az = 346.9 (NEIC)
				PKP2V A 1.1s 10.1nm
8.	ePKIKP	A	14 34 30	New Guinea 6.85 S 144.26 E
				H = 14 15 35.9 h = 18 km MB = 5.6
				D = 121.31 Az = 326.8 (NEIC)
8.	-iPKIKP	AB	16 23 23.8	Fiji Islands Region 18.00 S 178.32 W
	iPKHKP	AB	23 26.7	H = 16 04 45.4 h = 548 km MB = 5.7
				D = 146.48 Az = 348.5 (NEIC)
				PKIKPV A 1.5s 226.0nm
				PKIKPV B 3 1.87 μm
				PKHKPV A 1.7 1060.0nm
9.	ePg	A	02 32 33.5	Marl-Hüls, Fed. Rep. of Germany
	eSn	A	32 58	51°39' N 7°04' E
	eiSg	A	33 07	H = 02 31 35 (Bensberg, BCIS)
				D = 3.00

Day	Phase		h m s	Remarks
9.	+iP	A	06 10 43	<u>Eastern Kazakh SSR</u> 50.00 N 77.70 E
	ePn	A	12 15	H = 06 02 57.1 h = 0 km MB = 5.4
				D = 40.92 Az = 297.3 (NEIC)
				underground explosion (UPP)
				PV A 0.7s 42.2nm M = 5.3
9.	+eP	A	11 16 53.5	<u>Hokkaido, Japan Region</u> 43.05 N 143.77 E
	epP	A	17 22.5	H = 11 05 08.1 h = 122 km MB = 5.3
				D = 77.80 Az = 331.1 (NEIC)
				h = 118 km
				PV A 1.3s 32.8nm M = 5.0
9.	+eP	AB	13 27 53.5	<u>Taiwan Region</u> 24.89 N 122.06 E
	eS	B	38 00	H = 13 15 36.4 h = 85 km MB = 5.7
	LmV	B	14 10.3	D = 83.06 Az = 323.1 (NEIC)
	LmH	B	10.4	PV A 1.6s 66.0nm M = 5.3
				LmH B 15 0.6µm
				LmV B 16 0.7µm
9.	eP	A	22 33 38.5	<u>Turkey</u> 37.28 N 30.24 E
	LmH	B	40.2	H = 22 29 16.0 h = 33 km MB = 4.5
	LmV	B	42.7	D = 18.86 Az = 321.0 (NEIC)
				LmH B 13s 0.4µm M = 3.9
				LmV B 12 0.3µm 4.0
10.	eP	A	02 15 15.5	<u>Turkey</u> 37.17 N 30.13 E
				H = 02 10 51.5 h = 26 km MB = 4.3
				D = 18.88 Az = 321.3 (NEIC)
10.	e	A	06 49 07	
10.	eX	A	18 36 30	<u>Burma</u> 23.01 N 95.96 E
	eS	C	45 20	H = 18 25 14.6 h = 33 km MB = 5.1
	eSS	C	50 40	D = 69.05 Az = 317.3 (NEIC)
	LmH	B	19 05.1	XV A 1.4s 18.6nm
	LmV	B	11.0	LmH B 18.5 2.1µm M = 5.4
				LmV B 17 1.6µm 5.4

Day	Phase		h m s	Remarks
10.	eP	A	18 42 08.5	Southern Greece 37.63 N 21.36 E H = 18 38 38.7 h = 98 km MB = 4.0 D = 14.75 Az = 335.0 (NEIC)
10.	eP	A	19 13 25	Burma 24.97 N 95.29 E H = 19 02 44.0 h = 133 km MB = 4.7 (ISC) D = 67.2
11.	eiP	A	01 41 29	Kurile Islands 45.40 N 150.67 E
	LmH	C	02 18.6	H = 01 29 36.2 h = 60 km MB = 5.2
	LmV	C	19.0	D = 77.97 Az = 334.7 (NEIC) PV A 1.3s 28.4nm M = 5.1 LmH C 18 0.46 µm LmV C 20 0.85 µm
11.	ePKP	A	05 10 00	Tonga Islands 17.60 S 173.95 W
	epPKP	A	10 36.5	H = 04 50 36.7 h = 184 km MB = 4.9 D = 146.73 Az = 353.5 (NEIC) h = 136 km
11.	eP	A	10 28 35.5	Near East Coast of Honshu, Japan
	epP	A	28 49	35.82 N 140.42 E
	LmH	B	11 05.8	H = 10 16 14.7 h = 47 km MB = 5.1
	LmV	B	08.6	D = 82.82 Az = 330.0 (NEIC) h = 50 km PV A 1.1s 10.1nm M = 4.8 pPV A 2.0 68.5nm LmH B 18 0.4 µm 4.9 LmV B 12 0.4 µm 5.1
12.	e	A	00 20 20	
12.	ePKHKP	A	08 09 27.5	South of Fiji Islands 22.10 S 175.98 W
	epPKP	A	10 05	H = 07 49 54.5 h = 180 km MB = 3.6 D = 150.90 Az = 350.0 (NEIC) h = 139 km

Day	Phase		h m s	Remarks
12.	+eP	AB	09 57 36.5	Mexico-Guatemala Border Region
	+i	A	57 41.4	15.84 N 91.17 W
	eS	B	10 08 07	H = 09 44 59.3 h = 36 km MB = 5.7
	LmH	B	41.3	D = 85.78 Az = 38.5 (NEIC)
	LmV	B	41.3	PV A 1.3s 188.0nm M = 6.2
				LmH B 19 1.0/um 5.2
				LmV B 20 1.2/um 5.3
12.	ePn	A	11 46 12.5	Northern Italy 44.54 N 10.96 E
	e	A	46 14	H = 11 44 42.2 h = 25 km MB = 4.6
	ei	A	46 15	D = 6.13 Az = 3.9 (NEIC)
	e	A	46 37.5	
	ePg	A	46 45	
	eiSn	A	47 19.5	
12.	ePKP	A	18 12 12.5	Tonga Islands 15.14 S 173.36 W
	eX1	A	12 26	H = 17 52 36.4 h = 15 km MB = 5.2
	eX2	A	12 40	D = 144.35 Az = 354.6 (NEIC)
				X1V A 1.4s 16.3nm
				X2V A 1.4 18.6nm
12.	eP	A	18 16 25	Kodiak Island Region 56.75 N 152.01 W
				H = 18 05 02.1 h = 33 km MB = 4.7
				D = 72.12 Az = 10.9 (NEIC)
13.	eP	A	03 30 53	Crete 34.26 N 26.09 E
	eS	AB	34 44	H = 03 26 27.2 h = 27 km MB = 5.1
	LmV	B	40.4	D = 19.49 Az = 331.5 (NEIC)
	LmH	B	40.5	PV A 1.6s 79.7nm M = 4.7
				SH A 2.1 368.0nm 5.1
				SH B 5.0 1.4/um 4.4
				LmH B 12.5 2.3/um 4.7
				LmV B 13 2.7/um 4.9
13.	LmH	C	10 29.7	West New Guinea Region 2.9 N 132.8 E
	LmV	C	36.6	H = 09 31 00.5 h = 41 km MB = 5.3 (NEIC)
				D = 106.8
				LmH C 20s 0.6/um M = 5.2
				LmV C 20 0.9/um 5.5

Day	Phase		h m s	Remarks
13.	e	C	12 24 44	<u>West New Guinea Region</u> 2.9 N 132.8 E
	LmH	B	49.5	H = 11 50 34.2 h = normal MB = 5.3 (NEIC)
	LmV	B	56.2	D = 106.8
				LmH B 18.5s 1.0 µm M = 5.4
				LmV B 18 0.5 µm 5.3
13.	eP	A	14 13 33	<u>Andreanof Islands, Aleutian Is.</u>
				51.95 N 179.60 W
				H = 14 01 47.4 h = 95 km MB = 5.3
				D = 77.35 Az = 352.7 (NEIC)
				PV A 1.0s 15.7nm M = 4.8
13.	eP	A	22 49 39	<u>Chile-Bolivia Border Region</u>
	ePP	A	53 39.5	21.67 S 68.18 W
	eX	A	53 43.5	H = 22 36 01.8 h = 92 km MB = 5.8
	e	A	53 57	D = 100.26 Az = 39.6 (NEIC)
	eSKS	BC	23 00 10	PV A 1.5s 17.6nm M = 5.5
	eS	BC	01 00	PPV A 1.5 15.1nm 5.4
	e	BC	01 10	XV A 2.0 59.9nm
	esS	BC	02 00	LmH C 25 0.4 µm
	e	BC	03 20	LmV C 22 0.6 µm
	eSS	C	08 00	
	LmH	C	30.0	
	LmV	C	30.8	
14.	eP	A	13 06 23.5	<u>Burma</u> 23.07 N 95.80 E
	e	A	06 28	H = 12 55 23.3 h = 63 km MB = 5.2
	e	A	06 37	D = 68.91 Az = 317.2 (NEIC)
	eS	B	15 22	PV A 1.3s 13.1nm M = 4.7
	eSS	C	20 45	LmH B 20 3.1 µm
	eSSS	C	23 20	LmV B 17 2.1 µm
	LmH	B	35.4	
	LmV	B	41.1	
14.	-iP	A	22 03 56	<u>Afghanistan-USSR Border Region</u>
	epP	A	04 28	36.38 N 71.09 E
				H = 21 55 55.9 h = 112 km MB = 5.1
				D = 44.24 Az = 308.2 (NEIC) h = 151 km
				PV A 1.2s 36.6nm M = 5.0

Day	Phase		h m s	Remarks
14.	e(P)	C	22 50 44	Bouvet Island Region 54.35 S 1.84 E
	e	BC	52 24	H = 22 36 28.6 h = normal MB = 5.3
	e	BC	54 05	D = 104.94 Az = 6.4 (NEIC)
	ePP	BC	54 52	LmH B 18.5s 15.9/um M = 6.6
	e	A	55 05	LmV B 19 15.5/um 6.6
	eSKS	C	23 01 25	
	e	C	02 46	
	ePS	BC	04 12	
	eSS	BC	09 58	
	LmH	B	37.4	
	LmV	B	38.1	
15.	epP	A	02 46 11.5	Philippine Islands Region
	e	A	46 28	19.03 N 121.14 E
				H = 02 33 16.1 h = 60 km MB = 5.1
				D = 87.21 Az = 323.1 (NEIC)
				pPV A 1.8s 27.0nm
15.	epP	A	03 53 24.5	Southern Sumatra 2.21 S 102.22 E
				H = 03 39 54.1 h = 188 km MB = 4.7 (NEIC)
				D = 92.1
				pPV A 1.5s 20.1nm
15.	eP	A	08 23 06	Crete 34.66 N 25.49 E
	e	A	23 26.5	H = 08 18 39.9 h = normal MB = 4.3
				D = 18.90 Az = 331.9 (NEIC)
				PV A 1.5s 17.6nm M = 4.1
15.	+eP	A	10 47 19	Peru 14.09 S 73.31 W
	eSKS	C	58 04	H = 10 33 46.8 h = 54 km MB = 5.7
	eSSS	C	11 09.4	D = 97.58 Az = 39.8 (NEIC)
	LmV	B	30.9	PV A 1.8s 54.1nm M = 5.1
	LmH	B	31.9	LmH B 17 1.5/um
				LmV B 18 1.9/um

Day	Phase		h m s	Remarks
15.	-eP	A	16 30 09	<u>Afghanistan-USSR Border Region</u>
	epP	A	30 44	36.95 N 71.83 E
				H = 16 22 13.1 h = 167 km MB = 4.9
				D = 44.37 Az = 307.8 (NEIC)
				h = 165 km
				PV A 1.2s 30.5nm M = 4.8
				pP A 0.8 15.4nm
15.	e	A	17 14 11	<u>Eastern Caucasus</u> 41.43 N 48.57 E
				H = 17 08 06.3 h = normal MB = 4.9
				D = 27.01 Az = 302.6 (NEIC)
15.	eP	A	18 38 17.5	<u>Andreanof Islands, Aleutian Is.</u>
				50.45 N 176.69 W
				H = 18 26 12.8 h = 16 km MB = 4.9
				D = 79.03 Az = 354.6 (NEIC)
				PV A 1.7s 24.2nm M = 5.0
15.	eP	AB	22 08 36	<u>Central Mid-Atlantic Ridge</u>
	eX	A	09 12	7.72 N 37.34 W
	eS	B	16 44	H = 21 58 35.8 h = normal MB = 5.0
	eScS	B	18 36	D = 58.85 Az = 34.1 (NEIC)
	eSS	C	20 34	XV A 1.8s 60.9nm
	eiSSS	C	23 08	SH B 16 1.6/um M = 5.7
	LmV	B	29.1	LmH B 19 2.3/um 5.3
	LmH	B	29.4	LmV B 17 2.2/um 5.4
16.	ePKIKP	A	05 34 21	<u>Solomon Islands</u> 5.98 S 154.48 E
	ePP	B	36 14	H = 05 15 25.0 h = 76 km MB = 5.7
	ePPS	B	48 00	D = 125.78 Az = 331.7 (NEIC)
	e	B	51 17	LmH B 20.5s 2.9/um
	eSS	B	53 48	LmV B 20 2.4/um
	LmV	B	06 33.8	
	LmH	B	33.9	
16.	eP	A	06 00 21	<u>South Atlantic Ridge</u> 22.79 S 13.25 W
				H = 05 48 33.2 h = normal MB = 4.9
				D = 76.35 Az = 16.0 (NEIC)
				PV A 1.6s 24.7nm M = 5.0

Day	Phase		h m s	Remarks
16.	epP	A	06 20 01.5	Afghanistan-USSR Border Region 37.18 N 71.25 E H = 06 11 39.1 h = 105 km MB = 4.8 D = 43.86 Az = 307.6 (NEIC) h = 120 km
16.	ePKP	A	08 12 48	Fiji Islands Region 17.82 S 178.02 W H = 07 54 09.2 h = 570 km MB = 4.6 D = 146.36 Az = 348.9 (NEIC) PKPV A 1.6s 30.2nm
16.	eP e(pP)	A A	09 49 50 50 03	Dodecanese Islands 36.60 N 28.42 E H = 09 45 36.0 h = 61 km MB = 4.8 D = 18.51 Az = 324.6 (NEIC) PV A 1.0s 19.7nm M = 4.3 pPV A 1.4 55.8nm
17.	ePKP	A	07 11 30.5	Kermadec Islands 30.81 S 178.07 W H = 06 50 59.4 h = 45 km MB = 5.0 (ISC) D = 159.0
17.	e e	A A	14 24 19.5 24 31	
17.	eP	A	18 18 41.5	Southern Sumatra 4.92 S 102.70 E H = 18 05 21.8 h = normal MB = 5.5 D = 94.46 Az = 320.3 (NEIC)
17.	ePKP	A	20 44 29	Fiji Islands Region 18.01 S 178.05 W H = 20 25 51.8 h = 599 km MB = 4.3 D = 146.54 Az = 348.8 (NEIC)
18.	ePKIKP ePKHKP +iPKP2	A A A	04 04 27.5 04 40 05 04.3	Kermadec Islands Region 30.41 S 179.72 W H = 03 45 14.3 h = 364 km MB = 5.1 D = 158.07 Az = 340.4 (NEIC) PKP2V A 1.3s 89.5nm

Day	Phase		h m s	Remarks
18.	+iPKP	AB	10 19 13.3	Loyalty Islands Region 22.66 S 171.16 E
	LmV	B	27.7	H = 09 59 33.6 h = 59 km MB = 5.2
	LmH	B	28.1	D = 147.82 Az = 335.3 (NEIC)
				PKPV A 1.2s 50.8nm
				LmV B 18 0.8 /um
18.	eX	A	10 25 24	XV A 1.0s 13.8nm
18.	ePKP	A	12 25 53.5	Loyalty Islands 22.95 S 170.91 E
				H = 12 06 11.7 h = 33 km
				D = 147.99 Az = 335 (ISC)
				PKPV A 0.9s 9.7nm
18.	ePKP	A	12 50 18	Loyalty Islands Region 22.62 S 171.11 E
				H = 12 30 38.9 h = 63 km MB = 5.3
				D = 147.76 Az = 335.3 (NEIC)
				PKPV A 1.1s 46.4nm
18.	e	A	12 57 48.5	
18.	eiPKP	AB	13 15 14	Loyalty Islands Region 22.64 S 171.13 E
	LmH	B	14 21.1	H = 12 55 38.3 h = 65 km MB = 5.5
	LmV	B	24.6	D = 147.79 Az = 335.3 (NEIC)
				PKPV A 1.1s 175.0nm
				PKPV B 10 2.05 /um
				LmH B 18 1.7 /um
				LmV B 20 2.7 /um
19.	ePKP	A	13 23 18.5	Loyalty Islands Region 22.0 S 171.3 E
	eX	A	23 23	H = 13 03 50 h = 113 km
				D = 147.28 Az = 336 (ISC)
				PKPV A 1.1s 14.1nm
				XV A 1.3 24.0nm
19.	ePKP	A	02 28 38	Loyalty Islands Region 22.59 S 171.1 E
				H = 02 09 01 h = 70 km
				D = 147.75 Az = 335 (ISC)

Day	Phase		h m s	Remarks
19.	eiPKP	A	08 32 31	Loyalty Islands Region 22.72 S 171.06 E
				H = 08 12 53.3 h = 63 km MB = 4.9
				D = 147.84 Az = 335.2 (NEIC)
				PKPV A 1.1s 44.4nm
19.	ePKP	A	09 22 27.5	Loyalty Islands Region 22.8 S 170.8 E
	e	A	23 21.5	H = 09 02 50 h = 74 km MB = 4.8
				D = 147.82 Az = 335 (ISC)
				PV A 1.1s 16.1nm
19.	-eiP	AB	11 14 30.5	Fox Islands, Aleutian Is.
	eipP	A	14 40	52.66 N 166.97 W
	LmH	B	59.4	H = 11 02 37.7 h = 22 km MB = 5.6
	LmV	B	12 05.3	D = 77.07 Az = 0.9 (NEIC) h = 35 km
				PV A 1.7s 142.0nm M = 5.7
				pPV A 1.3 52.4nm
				LmH B 16 0.5/um 4.9
				LmV B 20 0.7/um 5.0
19.	e	A	20 07 39	
19.	-iPKP	A	21 28 34	Tonga Islands 16.56 S 173.69 W
				H = 21 08 55.8 h = 30 km MB = 4.9
				D = 145.73 Az = 354.0 (NEIC)
				PKPV A 1.9s 87.1nm
20.	+eiP	AB	08 52 52.5	Taiwan Region 21.93 N 121.35 E
	epP	A	53 05	H = 08 40 19.0 h = 35 km MB = 5.5
	e	A	55 51	D = 85.02 Az = 323.0 (NEIC) h = 55 km
	e	A	56 26	PV A 1.6s 137.0nm M = 5.9
	eS	BC	09 03 20	LmH B 16 4.96/um 6.0
	LmH	B	36	LmV B 16 6.81/um 6.2
	LmV	B	36.1	
20.	ePKIKP	A	14 44 38.5	New Britain Region 4.27 S 152.52 E
				H = 14 25 55.1 h = 131 km MB = 4.9
				D = 123.35 Az = 331.3 (NEIC)

Day	Phase		h m s	Remarks
20.	iPg	A	15 05 20	Hilders (Rhön), Fed. Rep. of Germany
	iSg	A	05 35	50°32.7' N 10°02.5' E
				explosion, yield 9.2 t (Hannover)
				H = 15 05 00
				D = 1.0
20.	LmH	B	20 12.5	Near Coast of Central Chile
	LmV	B	12.5	32.7 S 71.8 W
				H = 19 07 30.6 h = 51 km MB = 5.2 (NEIC)
				D = 110.7
21.	+iP	A	06 10 43.2	Eastern Kazakh SSR 49.99 N 77.65 E
	ePn	A	12 14	H = 06 02 57.3 h = 0 km MB = 5.6
				D = 40.89 Az = 297.3 (NEIC)
				PV A 0.9s 54.5nm M = 5.3
				underground explosion
21.	ePKP2	A	07 04 35	South of Kermadec Islands
				33.67 S 179.78 W
				H = 06 43 55.0 h = normal
				MB = 4.6 MS = 5.2 (NEIC)
				D = 161.1
				PKP2V A 1.8s 40.5nm
21.	eP	A	07 15 53	Turkey 38.01 N 30.40 E
				H = 07 11 36.9 h = normal MB = 4.6
				D = 18.38 Az = 319.4 (NEIC)
				PV A 1.5s 27.6nm M = 4.2
21.	+eP	A	08 35 15	Southwestern Ryukyu Islands
				24.80 N 125.28 E
				H = 08 22 45.7 h = 63 km MB = 5.2
				D = 84.87 Az = 324.2 (NEIC)
				PV A 1.3s 67.6nm M = 5.6
21.	iPg	A	15 05 30.2	Dorheim, Fed. Rep. of Germany
	eSg	A	05 50	explosion (GRF)
				D c. 1.5

Day	Phase		h m s	Remarks
21.	e	A	22 55 00	Crete 34.96 N 24.43 E H = 22 50 39.5 h = normal MB = 4.1 D = 18.24 Az = 333.2 (NEIC)
21.	eP	A	23 16 16	Central Russia 54.7 N 90.9 E H = 23 07 49 h = 10 km MB = 4.9 MS = 4.4 D = 45.9 (ANUSSR) PV A 1.1s 16.1nm M = 4.9
22.	+iP	A	05 05 42.2	Western Russia 51.58 N 54.54 E H = 05 00 00.5 h = 6 km MB = 5.3 D = 26.68 Az = 285.1 (NEIC) PV A 1.0s 33.4nm M = 5.0 underground explosion (UPP.)
22.	eX	A	18 01 47.5	Mexico-Guatemala Border Region 15.89 N 91.30 W H = 17 49 01.6 h = 20 km MB = 4.5 D = 85.82 Az = 38 (ISC) XV A 1.0s 11.8nm
22.	eiPKP eipPKP	A A	20 57 55.5 58 50.5	Tonga Islands 16.75 S 174.61 W H = 20 38 39.3 h = 204 km MB = 5.2 (NEIC) D = 145.8 h = 218 km PKPV A 2.0s 226.0nm pPKPV A 1.4 51.1nm
22.	ePKHKP	A	22 30 32.5	Tonga Islands 20.11 S 173.81 W H = 22 10 44.7 h = normal MB = 4.9 D = 149.23 Az = 353.2 (NEIC) PKHKPV A 1.3s 17.5nm
22.	ePKP	A	22 58 55	Fiji Islands Region 17.86 S 178.63 W H = 22 40 20.7 h = 597 km MB = 4.8 D = 146.28 Az = 348.2 (NEIC) PKPV A 1.2s 30.5nm

Day	Phase		h m s	Remarks
23.	+ePKHKP	A	02 01 08	West of Macquarie Island
	ePKP2	A	01 18	54.74 S 146.34 E
	eSS	C	24 28	H = 01 41 12.7 h = normal MB = 5.5
	eSSS	C	30 38	D = 152.63 Az = 280.3 (NEIC)
	LmH	B	03 10.6	PKHKPV A 1.5s 42.7nm
	LmV	B	20.5	PKP2V A 1.6 102.0nm
				LmH B 23 5.7 µm M = 6.2
				LmV B 18 3.3 µm 6.2
23.	e	A	11 56 30	Southern Iran 29.59 N 51.32 E
				H = 11 49 20.6 h = normal MB = 4.5
				D = 36.34 Az = 316.7 (NEIC)
23.	eP	C	22 47 35	South of Mariana Islands
	ePP	C	52 00	11.83 N 142.48 E
	eSKS	BC	58 12	H = 22 33 30.9 h = 28 km MB = 5.9
	eS	BC	59 24	D = 104.52 Az = 330.2 (NEIC)
	ePS	BC	23 00 00	LmH B 18s 28.8 µm M = 6.9
	eSS	BC	06 45	LmV B 19.5 34.3 µm 6.9
	eSSS	BC	10 48	
	LmV	B	38.5	
	LmH	B	40.3	
23.	e(PP)	A	23 20 40	South Sandwich Islands Region
				57.2 S 25.5 W
				H = 23 01 19.0 h = normal MB = 5.6 (NEIC)
				D = 111.5
24.	eP	AB	01 51 32	South of Mariana Islands
	ePP	B	55 56	11.84 N 142.29 E
	eSKS	B	02 02 07	H = 01 37 30.4 h = normal MB = 6.0
	eS	C	03 20	D = 104.42 Az = 330.1 (NEIC)
	ePS	BC	05 00	LmH B 18s 30.2 µm M = 6.9
	ePPS	C	05 55	LmV B 19 36.0 µm 6.9
	ePKKP	A	07 21	
	e	A	07 29	
	eSS	BC	10 55	
	eSSS	C	14 20	

Day	Phase		h m s	Remarks
cont. 24.	LmV	B	02 41.9	
	LmH	B	44.0	
24.	eP	A	02 35 05.5	South of Mariana Islands
	ePP	A	39 24	11.85 N 142.29 E
	LmV	B	03 25.4	H = 02 21 02.3 h = normal MB = 6.2
	LmH	B	27.5	D = 104.41 Az = 330.1 (NEIC)
				PV A 1.7s 36.4nm M = 6.0
				PPV A 2.0 94.0nm 6.0
				LmH B 19 9.2 µm 6.3
				LmV B 20 10.3 µm 6.4
24.	eP	A	09 09 07	Tibet 28.25 N 87.16 E
				H = 08 59 04.6 h = 44 km MB = 5.1
				D = 59.78 Az = 314.5 (NEIC)
				traces
24.	ePKP	A	14 40 30.5	Tonga Islands 15.14 S 173.89 W
	epPKP	A	40 54	H = 14 21 02.0 h = 66 km MB = 5.6
	esPKP	A	41 07	D = 144.30 Az = 354.0 (NEIC)
				h = 90 km
				PKPV A 1.6s 77.0nm
24.	LmV	B	24 10.5	South of Mariana Islands
	LmH	B	12.7	11.8 N 142.3 E
				H = 23 05 34.1 h = 37 km MB = 5.1 (NEIC)
				D = 104.4
				LmH B 16s 0.9 µm M = 5.4
				LmV B 19 1.3 µm 5.5
25.	eX	A	00 17 28.5	XV A 1.3s 21.8nm
25.	+iP	A	00 21 18	South of Honshu, Japan 29.96 N 137.07 E
	ePP	A	24 46.5	H = 00 09 29.8 h = 508 km MB = 5.3
				D = 86.40 Az = 328.7 (NEIC)
				PV A 1.3s 24.0nm M = 4.8
				PPV A 1.0 17.7nm 5.1

Day	Phase		h m s	Remarks
25.	ePKIKP	AB	04 05 58	New Hebrides Islands 13.09 S 166.45 E
	ePP	AB	08 36	H = 03 46 36.3 h = 34 km MB = 5.5
	ePS	BC	18 47	D = 137.31 Az = 336.5 (NEIC)
	ePPS	C	21 00	PKIKPV B 9s 1.7 /um
	eSS	C	26 52	PPV B 8 2.2 /um M = 6.5
	LmV	B	05 13.2	LmH B 18.5 4.8 /um 6.2
	LmH	B	13.9	LmV B 18 5.8 /um 6.4
26.	ePKIKP	A	06 31 18	Fiji Islands Region 17.91 S 178.39 W
	ePKHKP	A	31 20	H = 06 12 46.6 h = 618 km MB = 4.7
	ePKP2	A	31 22.5	D = 146.37 Az = 348.5 (NEIC)
				PKHKPV A 1.0s 27.6nm
26.	eP	A	13 04 04	Hokkaido, Japan Region
				42.25 N 142.97 E
				H = 12 52 10.0 h = 69 km MB = 4.9
				D = 78.22 Az = 330.8 (NEIC)
26.	LmV	B	16 18.5	Near Coast of Central Chile
	LmH	B	20.6	32.2 S 71.7 W
				H = 15 14 40.7 h = 44 km
				MB = 5.1 MS = 5.3 (NEIC)
				D = 110.2
				LmH B 18.5s 0.6 /um M = 5.2
				LmV B 20 1.0 /um 5.4
26.	LmV	B	20 09.1	South of Mariana Islands
	LmH	B	11.9	11.9 N 142.5 E
				H = 19 05 06.2 h = normal
				MB = 5.6 MS = 5.1 (NEIC)
				D = 104.5
				LmH B 17.5s 1.0 /um M = 5.4
				LmV B 18 1.3 /um 5.5
26.	LmV	B	22 06.7	LmH B 18s 0.3 /um
	LmH	B	07.4	LmV B 18 0.4 /um
26.	e	A	23 21 10	

Day	Phase		h m s	Remarks
27.	eP	AB	09 40 04	Luzon, Philippine Islands
	epP	A	40 15	17.43 N 120.12 E
	eSKS	B	50 29	H = 09 27 18.6 h = 51 km MB = 5.4
	eSS	C	51 48	D = 87.90 Az = 322.8 (NEIC)
	LmH	B	10 22.3	h = 40 km
	LmV	B	26.6	PV A 1.4s 23.2nm M = 5.2
				pPV A 2.6 104.0nm
				LmH B 15.5 1.6 µm
				LmV B 16 2.0 µm
27.	-eiP	A	10 53 56	Near East Coast of Kamchatka
				53.50 N 160.09 E
				H = 10 42 29.0 h = normal MB = 5.1
				D = 72.89 Az = 339.6 (NEIC)
				PV A 1.2s 36.6nm M = 5.3
27.	ePKP	A	12 31 17	New Hebrides Islands 19.60 S 168.85 E
				H = 12 11 47.8 h = 47 km MB = 4.6
				D = 144.15 Az = 335.1 (NEIC)
				PKPV A 1.0s 19.7nm
27.	eP diff	B	18 15 18	New Hebrides Islands 15.50 S 167.22 E
	+ePKIKP	B	17 52	H = 17 58 36.9 h = 40 km MB = 6.0
	e	A	17 57	D = 139.80 Az = 336.0 (NEIC)
	e	A	18 03	PKIKPV A 10s 6.36 µm
	ePP	AB	20 53	PPV B 16 22.2 µm M = 7.2
	eSKSP	B	31 05	LmH B 21.5 55.7 µm 7.2
	eSS	B	39 30	LmV B 21 53.0 µm 7.3
	LmH	B	19 22.0	
	LmV	B	24.7	
27.	eP	A	22 02 08	Guerrero, Mexico 18.35 N 100.31 W
	esP	A	03 34.5	H = 21 49 18.3 h = 72 km MB = 5.1
				D = 89.11 Az = 36.2 (NEIC)
				h = 71 km
				PV A 1.7s 30.3nm M = 5.3

Day	Phase		h m s	Remarks
28.	ePKIKP	A	01 13 18	**West New Guinea** 3.20 S 139.50 E
	epPKIKP	A	13 35	H = 01 56 00.7 h = normal MB = 5.4 (NEIC)
				D = 124.0
28.	ePKHKP	A	06 01 22	**Tonga Islands** 16.82 S 174.49 W
	ePKP2	A	01 25	H = 05 41 43.2 h = normal MB = 4.8
	e	A	01 49	D = 145.90 Az = 353.1 (NEIC)
				PKHKPV A 1.6s 30.2nm
				PKP2V A 1.6 38.5nm
28.	eP	A	06 43 48	**Peru-Ecuador Border Region**
				3.86 S 80.65 W
				H = 06 30 32.3 h = 51 km MB = 5.2
				D = 94.40 Az = 39.6 (NEIC)
				PV A 1.6s 16.5nm M = 5.2
28.	eP	A	07 42 00	**Kurile Islands** 47.6 N 152.9 E
				H = 07 30 15.1 h = 72 km
				D = 76.64 Az = 336 (ISC)
28.	ePg	A	11 45 45.5	**Czechoslovakia** 50°33.5' N 14°00.4' E
	iSg	A	46 06.2	explosion, yield 13.2 t (PRU)
				D = 1.52
28.	-iPKHKP	A	12 13 35.2	**Fiji Islands Region** 20.16 S 176.35 W
				H = 11 54 17.1 h = 260 km MB = 4.8
				D = 148.94 Az = 350.2 (NEIC)
				PKHKPV A 1.4s 41.9nm
28.	+iP1	A	13 38 47	**Kirgiz SSR** 41.86 N 72.37 E
	iP2	A	38 49.5	H = 13 30 57.1 h = 22 km MB = 5.5
	ePP	B	40 27	D = 41.84 Az = 303.6 (NEIC)
	eS	B	45 00	P1V A 1.4s 112.0nm M = 5.4
	eSa	B	48 05	P2V A 2.2 316.0nm 5.7
	e	B	48 24	PPV B 7 0.9 μm 5.7
	LmH	B	57.6	LmH B 14 6.0 μm 5.6
	LmV	B	57.6	LmV B 14 6.8 μm 5.8

Day	Phase		h m s	Remarks
28.	ePKP	A	14 35 35	<u>New Hebrides Islands</u> 15.56 S 167.35 E
				H = 14 16 07.6 h = 56 km MB = 4.8
				D = 139.91 Az = 336.0 (NEIC)
28.	ePKIKP	A	14 39 53.5	<u>New Hebrides Islands</u> 13.28 S 166.47 E
	ePP	B	42 32	H = 14 20 29.5 h = normal MB = 5.6
	e	A	42 38	D = 137.48 Az = 336.4 (NEIC)
				PKIKPV A 1.3s 17.5nm
				PPV B 9 0.8 μm M = 5.9
28.	e(PKIKP)	A	15 32 30.5	<u>New Ireland Region</u> 5.48 S 153.90 E
	e	A	32 51	H = 15 13 37.8 h = 120 km MB = 5.7
	ePP	B	34 18	D = 125.06 Az = 331.6 (NEIC)
	i	B	35 10	PKIKPV A 1.4s 14.0nm
	e	B	45 00	LmH B 20 12.8 μm M = 6.6
	e	B	47 07	LmV B 19.5 15.0 μm 6.7
	e	B	51 56	
	LmH	B	16 31.6	
	LmV	B	31.7	
28.	ePKIKP	AB	18 19 21.5	<u>New Hebrides Islands</u> 13.36 S 166.44 E
	ePP	B	22 10	H = 18 00 00.5 h = normal MB = 5.7
	ePS	B	32 13	D = 137.55 Az = 336.3 (NEIC)
	LmH	B	19 27.0	PKIKPV A 1.3s 32.8nm
	LmV	B	27.3	PKIKPV B 12 1.73 μm
				PPV B 15 4.1 μm M = 6.5
				LmH B 18 11.2 μm 6.6
				LmV B 18 12.7 μm 6.7
29.	ePKIKP	A	01 02 48.5	<u>Kermadec Islands Region</u>
	e(PKP2)	A	03 19	28.16 S 178.30 E
	e(pPKP2)	A	05 41	H = 00 44 08.5 h = 658km MB = 5.2
				D = 155.38 Az = 339.4 (NEIC)
				PKIKPV A 1.4s 25.6nm
29.	eP	A	07 19 44	<u>Kurile Islands</u> 44.36 N 149.12 E
				H = 07 07 47.4 h = 47 km
				D = 78.43 Az = 334 (ISC)

Day	Phase		h m s	Remarks
29.	+iP	A	13 28 01.5	PV A 0.9s 25.3nm
29.	eiP	AB	14 17 59	Kurile Islands 49.88 N 154.86 E
	e	BZ	18 40	H = 14 06 27.0 h = 95 km MB = 5.6
	e	A	19 30	D = 75.05 Az = 336.8 (NEIC)
	e	AB	21 15	PV A 2.2s 213.0nm M = 5.6
	e	B	23 19	LmH B 17.5 0.7/um
	e	B	28 00	LmV B 19 1.0/um
	LmV	B	55.1	
	LmH	B	55.2	
29.	eP	A	17 26 24.5	Tibet 34.12 N 86.28 E
				H = 17 16 52.1 h = normal MB = 5.0
				D = 55.23 Az = 311.6 (NEIC)
				PV A 1.4s 18.6nm M = 4.9
29.	ePKIKP	AB	20 26 36	New Hebrides Islands 13.52 S 166.26 E
	ePP	AB	29 22	H = 20 07 13.2 h = normal MB = 5.3
	LmH	B	21 35.0	D = 137.62 Az = 336.1 (NEIC)
	LmV	B	43.8	LmH B 16s 0.6/um M = 5.4
				LmV B 16 0.7/um 5.5
29.	eP	A	20 38 04	Near East Coast of Honshu, Japan
				37.36 N 141.55 E
				H = 20 25 47.7 h = 52 km MB = 4.7
				D = 81.93 Az = 330.5 (NEIC)
29.	ePKIKP	A	23 28 53	New Hebrides Islands 13.58 S 166.35 E
	LmH	B	24 37.3	H = 23 09 31.5 h = 29 km MB = 5.0
	LmV	B	37.4	D = 137.72 Az = 336.2 (NEIC)
				PKIKPV A 1.6s 16.5nm
				LmH B 18 0.4/um M = 5.1
				LmV B 18 0.3/um 5.1
30.	-iP	A	14 28 16.5	South of Honshu, Japan
	epP	A	29 49	32.09 N 137.68 E
	ePP	A	31 39.5	H = 14 16 23.6 h = 393 km MB = 5.6
	eS	B	38 10	D = 84.86 Az = 328.9 (NEIC)

Day	Phase		h m s	Remarks
cont. 30.	esS	B	14 40 44	h = 360 km
	eSS	B	43 48	PV A 1.2s 63.0nm M = 5.3
	esSS	B	46 25	pPV A 1.6 55.0nm
	eSSSS	C	50 35	PPV A 1.1 42.3nm 5.5
	LmH	B	15 04.3	SH B 8 2.8 µm 5.9
	LmV	B	10.1	LmH B 18.5 0.8 µm
				LmV B 16 0.6 µm
30.	ePKP	A	16 50 28.5	<u>New Hebrides Islands</u> 15.06 S 167.33 E
				H = 16 31 18.6 h = 121 km MB = 5.0
				D = 139.45 Az = 336.2 (NEIC)
30.	eP	AB	21 01 17	<u>Taiwan</u> 23.01 N 121.39 E
	eS	B	11 40	H = 20 48 48.0 h = 35 km MB = 5.3
	LmV	B	43.9	D = 84.19 Az = 323.0 (NEIC)
	LmH	B	44.0	PV A 1.3s 35.0nm M = 5.4
				PV B 12 0.8 µm 5.7
				LmH B 15.5 7.3 µm 6.2
				LmV B 15.5 11.0 µm 6.4
31.	ePKIKP	A	12 30 14	<u>North Island, New Zealand</u>
	ePKHKP	A	30 37	38.81 S 176.38 E
	ePKP2	A	31 07	H = 12 10 21.9 h = 92 km MB = 5.4 (NEIC)
				D = 164.1
				PKP2V A 1.1s 38.3nm
				PKIKP and PKHKP only traces
31.	eP	A	16 05 17	<u>Eastern India</u> 26.17 N 90.73 E
				H = 15 54 47.9 h = normal MB = 4.6
				D = 63.50 Az = 315.7 (NEIC)

Day	Phase		h m s	Remarks
1.	eP	A	05 38 37.5	Northern Sinkiang Province, China
	LmH	B	59.9	44.04 N 85.05 E
	LmV	B	59.9	H = 05 29 57.2 h = normal MB = 5.0
				D = 48.31 Az = 305.2 (NEIC)
				PV A 1.6s 38.5nm M = 5.2
				LmH B 12.5 1.6/um 5.2
				LmV B 13 2.3/um 5.4
1.	ePKP	A	07 42 39	New Hebrides Islands 19.15 S 169.51 E
				H = 07 23 34.4 h = 252 km MB = 5.0
				D = 144.00 Az = 335.9 (NEIC)
				PKPV A 0.9s 15.6nm
2.	LmV	C	23 20.4	West Caroline Islands 11.25 N 139.56 E
	LmH	C	20.5	H = 22 16 23.9 h = 35 km MB = 5.0 (NEIC)
				D = 103.2
				LmH C 18s 0.61/um M = 5.2
				LmV C 18 1.54/um 5.6
3.	eP	A	10 43 56.5	Kurile Islands 43.42 N 146.79 E
				H = 10 32 00.6 h = 60 km MB = 4.8
				D = 78.51 Az = 332.7 (NEIC)
				PV A 1.0s 19.7nm M = 5.0
3.	eP	A	14 55 56	Off East Coast of Kamchatka
				52.28 N 159.05 E
				H = 14 44 23.0 h = normal MB = 4.9
				D = 73.81 Az = 339.1 (NEIC)
				PV A 1.0s 19.7nm M = 5.1
3.	eSg	A	18 12 24	Poland 50.30 N 18.93 E
				H = 18 09 56.8 M = 3.0 (WAR)
				D = 4.7
3.	iPn	A	21 31 52	Northern Italy 46.62 N 12.65 E
	iSn	A	32 41	H = 21 30 50.5 h = normal
	iSg	A	33 04	D = 4.09 Az = 350.7 (NEIC)
				PnV A 0.7s 26.8nm

Day	Phase		h m s	Remarks
4.	eP	A	14 29 53	Mediterranean Sea 35.27 N 22.71 E
				H = 14 25 48.2 h = 32 km MB = 4.4
				D = 17.35 Az = 335.7 (NEIC)
				PV A 0.7s 11.5nm M = 4.1
4.	eP	A	20 23 27.5	Szechwan Province, China 28.84 N 103.68 E
				H = 20 12 20.5 h = 34 km MB = 5.0
				D = 69.51 Az = 317.2 (NEIC)
				PV A 1.6s 24.7nm M = 5.0
5.	e	A	00 18 53	Switzerland 46.03 N 6.5 E
	ePg	A	19 02	H = 00 17 21 h = 0 km
	eSg	A	20 07	D = 5.73 Az = 34 (ISC)
5.	eP	A	11 29 36	Central Mid Atlantic Ridge
	e	A	29 37.5	0.81 N 25.45 W
	e	A	33 22	H = 11 19 39.3 h = normal MB = 5.4
	eS	B	37 40	D = 58.75 Az = 26.7 (NEIC)
	LmH	B	12 01.6	PV A 1.9s 75.8nm M = 5.5
	LmV	B	01.6	SH B 11 0.96/um 5.5
				LmH B 15 1.6/um 5.3
				LmV B 16 1.8/um 5.3
5.	eP	A	15 04 20.5	Near Coast of West Pakistan
				24.68 N 63.27 E
				H = 14 55 48.8 h = normal MB = 5.1
				D = 47.19 Az = 317.1 (NEIC)
				PV A 0.8s 15.4nm M = 5.0
5.	eP1	AB	22 23 02.5	Andaman Islands Region 10.18 N 92.98 E
	+iP2	AB	23 03.7	H = 22 11 15.5 h = 55 km MB = 5.7
	e(PP)	B	26 12	D = 76.73 Az = 319.7 (NEIC) h = 55 km
	epP	AB	23 23	P1V A 0.7s 23.0nm M = 5.3
	eS	B	32 42	P2V A 1.5 171.0nm 5.8
	LmV	B	23 01.0	LmH B 20 0.7/um 4.9
	LmH	B	01.4	LmV B 20 0.5/um 4.9

Day	Phase		h m s	Remarks
5.	eP	A	24 10 29.5	<u>Southern Sumatra</u> 0.07 S 100.19 E
	e	A	10 43.5	H = 23 58 05.5 h = 290 km MB = 5.0
				D = 89.14 Az = 320.5 (NEIC)
				PV A 1.4s 32.6nm M = 5.1
6.	LmH	B	17 45.5	<u>Off East Coast of Honshu, Japan</u>
	LmV	B	49.5	40.0 N 144.0 E
				H = 16 52 58.0 h = 65 km MB = 4.3 (NEIC)
				D = 80.5
				LmH B 15s 0.83/um
				LmV B 12 0.54/um
6.	+eP	AB	19 47 53.5	<u>Turkey</u> 39.02 N 29.73 E
	eS	B	51 14	H = 19 43 46.9 h = 14 km MB = 5.0
	LmH	B	53.8	D = 17.27 Az = 318.2 (NEIC)
	LmV	B	55.7	PV A 1.5s 50.3nm M = 4.4
				LmH B 16 3.4/um 4.7
				LmV B 12 3.3/um 4.9
6.	+iP	AB	22 11 55.2	<u>Rat Islands, Aleutian Is.</u>
	iPP	B	14 50	51.47 N 179.11 E
	eS	C	22 00	H = 22 00 00.1 h = 2 km MB = 6.8
	e(PS)	C	22 36	D = 77.71 Az = 351.9 (NEIC)
	e	C	22 55	underground explosion "Cannikin" (USAEC)
	ePKKP	A	30 56.5	PV A 1.8s 875.0nm M = 6.6
	ePKPPKP	A	38 46	PV B 2.0 1.76/um 6.8
	LmH	B	50.8	PKKPV A 1.4 18.6nm
	LmV	B	54.3	LmH B 17.5 4.0/um 5.8
				LmV B 16 2.9/um 5.4
6.	ePg	A	24 00 58	<u>France</u> 46.0 N 6.3 E
	eSg	A	02 10	H = 23 59 15 h = 0 km
				D = 5.87 Az = 35 (ISC)
7.	eP	A	08 03 38.5	<u>Hokkaido, Japan Region</u> 41.83 N 142.21 E
				H = 07 51 44.8 h = 75 km MB = 5.0
				D = 78.30 Az = 330.4 (NEIC)
				PV A 1.0s 11.8nm M = 4.8

Day	Phase		h m s	Remarks
7.	LmH	B	15 13.1	Northern Easter I., Cordillera
	LmV	B	13.1	4.61 S 105.42 W
				H = 14 12 28.4 h = normal
				MB = 5.0 MS = 5.6 (NEIC)
				D = 110.4
				LmH B 19s 0.85 µm M = 5.3
				LmV B 20 0.89 µm 5.3
7.	e	A	16 14 27.5	
8.	ePn	A	02 13 37	Swabian Jura, Federal Rep. of Germany
	ePg	A	13 46.5	48.3 N 9.1 E
	eSn	A	14 11	H = 02 12 53 h c. 20 km (BCIS)
	iSg	A	14 24.5	D = 2.84
8.	eP	AB	03 14 11	Southern Iran 27.05 N 54.48 E
	ePP	C	15 48	H = 03 06 36.4 h = 36 km MB = 5.6
	eiS	AB	20 03	D = 40.09 Az = 317.7 (NEIC)
	eiSS	B	23 22	PV A 1.5s 100.0nm M = 5.4
	LmH	B	33.4	LmH B 17 7.5 µm 5.6
	LmV	B	35.9	LmV B 16 6.7 µm 5.7
8.	eP	A	03 32 01.5	Southern Iran 26.90 N 54.46 E
				H = 03 24 26.6 h = normal MB = 4.9
				D = 40.19 Az = 317.9 (NEIC)
8.	LmV	B	12 43.8	Rat Islands, Aleutian Is.
	LmH	B	44.0	51.47 N 179.11 E
				H = 11 54 12.20 h = 2 km MB = 4.9 MS = 4.9
				Cannikin collapse (NEIC)
				D = 77.71
				LmH B 18s 0.71 µm M = 5.1
				LmV B 16 0.82 µm 5.2
8.	-iP	A	22 56 33.8	Mindanao, Philippine Islands
	e	A	59 15.5	5.77 N 126.51 E
	ePP	A	23 00 53	H = 22 43 00.6 h = 164 km MB = 5.7
	LmH	B	39.5	D = 100.93 Az = 324.0 (NEIC)
	LmV	C	45.0	PV A 1.2s 81.2nm M = 6.2
				LmV C 20 0.85 µm

Day	Phase		h m s	Remarks
9.	eP	A	00 24 35	Southern Iran 26.99 N 54.52 E H = 00 16 58.2 h = normal MB = 4.8 (NEIC) D = 40.2 PV A 1.8s 33.8nm M = 4.8
9.	eP	A	12 45 28	Ryukyu Islands 27.40 N 129.36 E
	e	A	47 22	H = 12 32 51.7 h = normal MB = 5.0
	LmH	C	13 28.0	D = 84.87 Az = 325.5 (NEIC)
	LmV	C	28.0	PV A 0.8s 11.5nm M = 5.1
				LmV C 16 0.93 μm 5.3
10.	LmH	C	07 21.8	New Ireland Region 3.80 S 151.63 E
	LmV	C	25.5	H = 06 12 15.7 h = 32 km MB = 4.9 MS = 4.9 (NEIC) D = 122.6 LmH C 22s 0.55 μm M = 5.2 LmV C 20 0.64 μm 5.3
10.	ePKIKP	A	10 41 38	New Britain Region 5.28 S 152.64 E
	LmH	B	11 35.5	H = 10 22 42.4 h = 49 km MB = 5.2
	LmV	B	41.6	D = 124.29 Az = 331.0 (NEIC) PKIKPV A 1.6s 22.0nm LmH B 18 0.5 μm M = 5.2 LmV B 20 0.9 μm 5.4
10.	eP	A	15 33 18	Iceland 64.06 N 22.41 W H = 15 28 21.2 h = normal MB = 4.4 D = 22.39 Az = 110.7 (NEIC) PV A 1.2s 12.2nm M = 4.2
11.	eP	A	04 51 58	Burma 21.43 N 93.87 E
	epP	A	52 13.5	H = 04 40 56.8 h = 48 km MB = 4.8 D = 68.89 Az = 317.5 (NEIC) h = 61 km PV A 1.7s 30.3nm M = 5.0 pPV A 1.4 16.3nm
11.	eP	A	10 31 49.5	Hokkaido, Japan Region 42.00 N 142.54 E
	epP	A	32 08	H = 10 19 55.9 h = 71 km MB = 5.3

Day	Phase		h m s	Remarks
ont.				
11.	esP	A	10 32 16	D = 78.28 Az = 330.6 (NEIC) h = 73 km
				PV A 1.3s 13.1nm M = 4.7
11.	LmH	C	12 27.3	<u>West New Guinea Region</u> 2.78 S 136.54 E
	LmV	C	27.3	H = 11 16 53.3 h = 17 km
				MB = 5.7 MS = 5.1 (NEIC)
				D = 113.7
				LmH C 20s 1.32 μm M = 5.5
				LmV C 20 2.65 μm 5.8
12.	ePKP	A	05 20 19	<u>Fiji Islands Region</u> 17.96 S 176.81 W
				H = 05 01 17.0 h = 360 km MB = 4.9
				D = 146.70 Az = 350.2 (NEIC)
				PKPV A 1.6s 33.0nm
12.	ePKP	A	06 01 54	<u>Fiji Islands Region</u> 18.43 S 177.76 W
				H = 05 43 15.0 h = 583 km MB = 4.7
				D = 147.00 Az = 349.0 (NEIC)
12.	eP	A	12 35 00.5	<u>Dodecanese Islands</u> 36.54 N 27.10 E
	LmH	C	42.0	H = 12 30 49.9 h = 23 km MB = 4.7
	LmV	C	43.5	D = 17.97 Az = 326.6 (NEIC)
				PV A 1.9s 41.7nm M = 4.2
13.	eP1	AB	15 56 02	<u>Ethiopia</u> 10.97 N 39.68 E
	eiP2	A	56 06	H = 15 47 41.5 h = 24 km MB = 5.3
	e(PcP)	A	57 44.5	D = 45.77 Az = 335.3 (NEIC)
	eS	C	16 02 55	P1V A 1.8s 40.5nm M = 5.1
	eSS	C	06 20	P2V A 1.5 72.9nm 5.4
	LmH	B	19.2	PcPV A 1.4 34.8nm
	LmV	B	19.5	LmH B 16 0.6 μm 4.6
				LmV B 12 0.8 μm 5.0
14.	eX	A	14 55 51.5	XV A 0.9s 9.7nm
14.	LmV	B	15 54.8	LmH B 16s 0.31 μm
	LmH	B	55.5	LmV B 16 0.47 μm

Day	Phase		h m s	Remarks
15.	ePKIKP	A	00 41 42	New Ireland Region 5.10 S 153.38 E
	LmH	B	01 39.8	H = 00 22 47.3 h = 59 km MB = 4.9
	LmV	B	39.8	D = 124.49 Az = 331.4 (NEIC)
				LmH B 19s 0.6/um M = 5.3
				LmV B 18 0.5/um 5.2
15.	ePg	A	11 13 54	Czechoslovakia 49°25.3' N 13°14.1' E
	eSg	A	14 16	Explosion yield 4.9 t (PRU)
				D = 1.6
15.	eP	A	12 46 18	Taiwan 23.70 N 121.58 E
				H = 12 33 53.4 h = 55 km MB = 5.0
				D = 83.74 Az = 323.0 (NEIC)
15.	+PKP	A	17 40 29.5	Tonga Islands 15.95 S 174.94 W
				H = 17 21 18.4 h = 230 km MB = 4.3
				D = 144.99 Az = 352.7 (NEIC)
				PKPV A 1.2s 20.3nm
15.	e	A	21 18 57.5	Belgium 50.9 N 6.0 E
	eSg	A	19 05	H = 21 17 10 (BCIS)
				D = 3.57
16.	eP	A	01 34 10.5	Sea of Okhotsk 49.6 N 148.0 E
	epP	A	36 14	H = 01 23 36 h = 585 km MB = 5.1
				D = 73.4 h = 617 km (ANUSSR)
16.	eP	A	19 18 26.5	Greece - Bulgaria Border Region
	e	A	19 33	41.72 N 26.40 E
				H = 19 15 11.9 h = 0 km
				D = 13.56 Az = 316 (ISC)
				PV A 2.0s 42.7nm
17.	eP	A	12 27 30	Jan Mayen Island Region 69.07 N 16.26 W
	eS	B	31 40	H = 12 22 29.7 h = normal MB = 4.7
	LmH	B	38.2	D = 22.80 Az = 129.8 (NEIC)
	LmV	B	38.2	PV A 2.0s 59.8nm M = 4.7
				LmH B 13.5 2.9/um 4.9
				LmV B 15 2.6/um 4.9

Day	Phase		h m s	Remarks
18.	+iPKP	A	05 51 51	Tonga Islands 16.10 S 174.87 W
	-i	A	51 55.5	H = 05 32 46.5 h = 290 km MB = 4.9
				D = 145.14 Az = 352.8 (NEIC)
				PKPV A 1.4s 55.8nm
18.	eP	A	07 39 10.5	Southeastern Uzbek SSR 38.35 N 66.77 E
	e	B	40 39	H = 07 31 32.8 h = 30 km MB = 5.3
	ePP	BE	40 48	D = 40.32 Az = 306.1 (NEIC)
	ePcP	B	41 14	PV A 1.4s 32.6nm M = 4.9
	eS	B	45 27	LmH B 16 1.8 μm 5.0
	LmH	B	59.7	LmV B 14 2.3 μm 5.3
	LmV	B	59.7	
18.	ePKHKP	A	19 47(18)	Kermadec Islands Region
	ePKP2	A	47 42	27.53 S 176.32 W
	LmH	B	21 08.2	H = 19 27 18.0 h = normal MB = 5.1
	LmV	B	08.6	D = 156.13 Az = 347.5 (NEIC)
				LmH B 16s 0.8 μm M = 5.5
				LmV B 16 0.8 μm 5.6
19.	eP	A	01 07 50	Kirgiz SSR 41.87 N 72.43 E
	LmH	B	23.6	H = 01 00 01.0 h = normal MB = 4.9
	LmV	B	26.7	D = 41.87 Az = 303.6 (NEIC)
				PV A 1.0s 19.7nm M = 4.8
				LmH B 14 1.0 μm 4.8
				LmV B 12 0.8 μm 4.9
19.	eP	A	01 25 24.5	Iceland Region 63.85 N 22.96 W
				H = 01 20 22.5 h = 6 km MB = 4.3
				D = 22.55 Az = 109.5 (NEIC)
				PV 'A 1.3s 17.5nm M = 4.4
19.	LmH	B	01 38.4	LmH B 12s 0.7 μm
	LmV	B	40.0	LmV B 12 0.8 μm
19.	eP	A	03 02 47.5	Iceland Region 63.72 N 22.92 W
	e	A	02 53.5	H = 02 57 48.8 h = normal MB = 4.8
	eS	C	07 00	D = 22.48 Az = 109.3 (NEIC)

Day	Phase		h m s	Remarks
cont. 19.	LmH	B	03 12.7	PV A 1.4s 41.8nm M = 4.7
	LmV	B	14.4	LmH B 12.5 1.3/um 4.6
				LmV B 12.5 1.6/um 4.8
19.	iPn	A	03 04 29	Swabian Jura, Federal Rep. of Germany
	iPg	A	04 38.5	48.37 N 8.99 E
	iSn	A	05 03.5	H = 03 03 44.6 h = 25 km
	iSg	A	05 14	D = 2.85 Az = 35.9 (NEIC)
19.	eP	A	06 02 03	Iceland Region 63.78 N 22.56 W
	e	A	02 09.5	H = 05 57 02.6 h = 16 km MB = 4.8
	LmH	B	11.9	D = 22.35 Az = 109.9 (NEIC)
	LmV	B	13.6	PV A 1.6s 27.5nm M = 4.5
				LmH B 12 0.6/um 4.3
				LmV B 12 0.8/um 4.5
19.	iPg	A	09 06 28	Czechoslovakia 50°31.3' N 13°43.8' E
	iSg	A	06 47	explosion yield 13 t (PRU)
				D = 1.35
20.	+iPKIKP	AB	07 46 46.5	South of Fiji Islands 23.40 S 179.92 W
	+iPKHKP	AB	46 53.8	H = 07 28 01.0 h = 551 km MB = 6.0
	iPKP2	AB	47 04.8	D = 151.36 Az = 344.6 (NEIC)
	ipPKIKP	AB	48 55	h = 565 km
	iX	A	49 00	PKIKPV A 1.5s 156.0nm
	iY	A	49 05	PKHKPV A 1.2 435.0nm
	i	A	49 19	PKP2V A 1.1 383.0nm
	ePP	B	50 32	XV A 1.4 167.0nm
	e	B	53 35	YV A 1.4 502.0nm
	ePSKS	B	08 01 00	
	e(SPP)	C	02 52	
	eSS	B	09 14	
	eSSP	B	10 20	
	esSS	C	12 44	
	e	C	14 10	
	eSSS	C	14 50	

Day	Phase		h m s	Remarks
20.	iPn	A	11 09 34.4	<u>Czechoslovakia</u> 50°35.2' N 14°03.2' E
	iSg	A	09 56	explosion yield 26.6 t (PRU)
				D = 1.55
20.	ePn	A	13 09 29	explosion
	ePg	A	09 35	D c. 2.8
	eSg	A	10 14	
20.	eP	A	17 00 16	<u>South of Honshu, Japan</u> 33.15 N 140.46 E
				H = 16 47 47.4 h = 82 km MB = 4.8
				D = 85.15 Az = 330.2 (NEIC)
20.	+iP	A	21 36 25.8	<u>Vancouver Island Region</u> 48.76 N 129.52 W
	eS	B	46 10	H = 21 24 42.6 h = normal MB = 5.5
	LmH	B	22 10.3	D = 75.52 Az = 24.4 (NEIC)
	LmV	B	12.1	PV A 1.3s 48.1nm M = 5.3
				SH B 10 1.27 μm 6.0
				LmH B 18 3.9 μm 5.8
				LmV B 16 4.0 μm 5.9
21.	ePKIKP	AB	06 16 11.5	<u>Santa Cruz Islands</u> 11.82 S 166.53 E
	eiX	A	16 21	H = 05 57 11.9 h = 115 km MB = 6.4
	eiY	A	16 28	D = 136.18 Az = 337.1 (NEIC)
	ePP	B	19 00	PKIKPV A 1.3s 15.3nm
	eSS	C	37 00	XV A 1.6 165.0nm
	eSSS	C	42 05	YV A 3 (2370)nm off scale
	LmH	B	07 18.2	PPV B 7 12.30 μm M = 7.1
	LmV	B	18.5	LmH B 20 29.3 μm
				LmV B 20.5 34.0 μm
21.	ePKIKP	A	06 39 10.5	<u>Santa Cruz Islands</u> 11.90 S 166.61 E
	e	A	39 21.5	H = 06 19 59.5 h = 110 km MB = 5.8
				D = 136.28 Az = 337.1 (NEIC)
				PKIKPV A 1.5s 40.2nm
22.	eP	AB	00 57 57	<u>Near Islands, Aleutian Is.</u>
	eS	B	01 07 40	52.27 N 174.32 E
	ePS	B	08 25	H = 00 46 11.1 h = 43 km MB = 5.6
	i	B	08 35	D = 76.42 Az = 348.8 (NEIC)

Day	Phase		h m s	Remarks
cont. 22.	eSS	C	01 12 40	PV A 1.4s 116.2nm M = 5.8
	LmV	B	30.3	LmH B 17 4.6 µm 5.9
	LmH	B	34.7	LmV B 20 3.7 µm 5.7
22.	ePP	A	08 38 10	<u>Northern Celebes</u> 0.2 N 123.8 E
	LmH	B	09 16.5	H = 08 19 56 h = 110 km MB = 5.9
	LmV	B	26.0	D = 103.7 (ANUSSR)
				PPV A 1.7s 51.5nm M = 5.9
22.	LmH	B	11 22.0	LmH B 12s 0.96 µm
22.	ePn	A	12 42 22	<u>Yugoslavia</u> 43.77 N 18.58 E
	eiPg	A	42 35	H = 12 40 20.7 h = 38 km
	eSn	A	43 52	D = 8.36 Az = 328 (ISC)
	ei	A	44 23.5	
22.	eP	A	19 31 09.5	<u>Dodecanese Islands</u> 35.32 N 27.75 E
	LmH	B	38.4	H = 19 26 45.3 h = 35 km MB = 4.8
	LmV	B	39.6	D = 19.27 Az = 327.6 (NEIC)
				PV A 1.3s 48.0nm M = 4.6
				LmH B 12 1.1 µm 4.4
				LmV B 12 0.7 µm 4.3
22.	eP	A	22 48 17	<u>North Atlantic Ridge</u> 30.15 N 42.67 W
	eS	B	55 00	H = 22 40 02.7 h = normal MB = 5.2
	LmH	B	23 04.2	D = 44.98 Az = 47.0 (NEIC)
	LmV	.B	04.4	PV A 1.6s 41.2nm M = 5.1
				SH B 14 1.35 µm 5.5
				LmH B 18 3.1 µm 5.3
				LmV B 18 4.0 µm 5.5
24.	eiP	A	01 09 56.5	<u>Kurile Islands</u> 46.68 N 152.55 E
	e	A	10 06.5	H = 00 58 06.1 h = 54 km MB = 5.3
	e	A	10 09	D = 77.36 Az = 335.7 (NEIC)
				PV A 1.0s 43.3nm M = 5.4
24.	eP	A	08 31 32.5	<u>Tadzhik-Sinkiang Border Region</u>
				38.69 N 73.35 E
				H = 08 23 24.6 h = normal MB = 5.1
				D = 44.27 Az = 306.6 (NEIC)

Day	Phase		h m s	Remarks
24.	ePKP	AB	18 25 04	Tonga Islands Region 17.84 S 172.92 W
	LmH	B	19 39.0	H = 18 05 23.5 h = normal MB = 5.1
	LmV	B	41.5	D = 147.07 Az = 354.7 (NEIC)
				PKPV A 1.9s 45.5nm
				LmH B 18 0.3/um M = 5.0
				LmV B 18 0.3/um 5.1
24.	+iP	AB	19 46 51.0	Off East Coast of Kamchatka
	esP	B	47 35	52.90 N 159.19 E
	ePP	B	49 40	H = 19 35 29.1 h = 106 km MB = 6.3 (NEIC)
	iS	A	56 13.5	D = 73.1 h = 120 km
	eiS	B	56 16	PV A 2.0s 1795.0nm M = 6.5
	eisS	B	56 56	PV B 6 43.4/um 7.0
	i	B	57 36	SH B 10 49.6/um 7.3
	e	B	20 00 55	P'P'V A 1.4 46.5nm
	esSS	B	01 45	LmH B 24 119.0/um
	e(SSS)	B	05 00	LmV B 17 71.3/um
	eP'P'	A	14 28	
	LmH	B	18.1	
	LmV	B	25.2	
25.	iPn	A	01 32 37	Austria 47.08 N 14.48 E
	eSn	A	33 23	H = 01 31 34.0 h = 0 km
	iSg	A	33 40	D = 4.04 Az = 333 (ISC)
25.	+iP	A	11 10 23.2	Alaska Peninsula 56.40 N 160.68 W
	epP	A	11 01	H = 10 59 06.4 h = 136 km MB = 5.3
	e	A	11 41.5	D = 73.13 Az = 5.1 (NEIC) h = 160 km
				PV A 1.1s 180.0nm M = 5.7
25.	eP	A	23 51 56	Vancouver Island Region
	e	A	52 00	48.78 N 129.38 W
	e	A	52 32	H = 23 40 12.1 h = normal MB = 5.1
	LmH	B	24 25.8	D = 75.46 Az = 24.5 (NEIC)
	LmV	B	26.0	PV A 0.9s 13.6nm M = 4.9
				LmH B 17 0.9/um 5.1
				LmV B 17 1.0/um 5.2

Day	Phase		h m s	Remarks
26.	eP	A	03 08 00.5	Near East Coast of Kamchatka
				53.79 N 160.45 E
				H = 02 56 34.2 h = normal MB = 4.6
				D = 72.70 Az = 339.8 (NEIC)
				PV A 1.5s 25.1nm M = 5.0
26.	LmV	C	15 53.1	Santa Cruz Islands 12.77 S 166.40 E
	LmH	C	16 04.0	H = 14 40 24.1 h = 36 km MB = 5.2 (NEIC)
				D = 137.1
				LmH C 20s 0.29 µm M = 5.0
				LmV C 25 0.61 µm 5.2
26.	eP	A	16 26 55	Turkey 36.07 N 29.07 E
	LmH	B	35.5	H = 16 22 30.0 h = normal MB = 4.4
	LmV	B	35.5	D = 19.25 Az = 324.6 (NEIC)
				PV A 1.6s 22.0nm M = 4.2
26.	ePg	A	22 59 23	Federal Republic of Germany
	eSg	A	59 56.5	51.6 N 6.8 E
				H = 22 58 24 h = 0 km
				D = 3.17 Az = 105 (ISC)
26.	eP	A	23 14 01.5	Eastern Greenland 79.44 N 17.77 W
	e	A	14 07.5	H = 23 07 47.4 h = 19 km MB = 5.2
	LmH	B	25.3	D = 30.66 Az = 142.2 (NEIC)
	LmV	B	26.0	PV A 1.2s 30.5nm M = 5.0
				LmH B 20 0.6 µm 4.2
				LmV B 20 0.8 µm 4.5
27.	eP	A	03 58 00	Aegean Sea 39.78 N 25.49 E
	-i	A	58 06	H = 03 54 27.5 h = 22 km MB = 4.6
	LmH	C	04 02.6	D = 14.59 Az = 322.7 (NEIC)
	LmV	C	03.8	PV A 1.3s 8.7nm M = 4.1
				pPV A 1.3 28.4nm
				LmH C 18 1.42 µm 4.1
				LmV C 16 1.15 µm

Day	Phase		h m s	Remarks
27.	LmH	C	11 56.0	Near Coast of Central Chile
	LmV	C	56.3	31.75 S 71.49 W
				H = 10 52 43.7 h = 64 km MB = 5.3 (NEIC)
				D = 109.7
				LmH C 20s 0.81 µm
				LmV C 20 1.48 µm
27.	+eP	A	13 57 36	Ryukyu Islands 29.19 N 130.11 E
	ePcP	A	57 40	H = 13 45 11.4 h = 56 km MB = 5.5
	epP	A	57 50	D = 83.77 Az = 325.7 (NEIC) h = 52 km
	eS	C	14 07 50	PV A 1.7s 121.1nm M = 5.6
	eSS	C	13.5	LmH B 17.5 8.2 µm 6.2
	LmH	B	38.9	LmV B 17 8.5 µm 6.3
	LmV	B	39.0	
27.	ePKP	A	18 19 41	Tonga Islands 16.79 S 173.06 W
	e	A	19 53.5	H = 18 00 01.9 h = normal MB = 4.8
				D = 146.02 Az = 354.7 (NEIC)
				PKPV A 1.3s 34.9nm
28.	ePKIKP	A	11 05 34	South Pacific Cordillera
	e	A	05 40	55.33 S 128.69 W
	ePKHKP	A	05 42	H = 10 45 41.5 h = normal MB = 4.7
				D = 155.89 Az = 84.6 (NEIC)
				PKIKPV A 1.3s 19.7nm
28.	ePKP	A	11 30 16.5	Chile-Argentina Border Region
	ePP	A	30 33.5	29.84 S 69.49 W
	epPP	A	31 15	H = 11 12 00.9 h = 105 km MB = 5.9
	LmH	B	12 11.0	D = 107.23 Az = 41.2 (NEIC)
	LmV	B	11.0	PKPV A 1.5s 40.2nm
				PPV A 1.8 60.8nm M = 5.8
				pPPV A 2.0 85.5nm
28.	eP	A	13 06 14	Iceland Region 62.90 N 25.4 W
	e	A	06 25.5	H = 13 01 04.8 h = 0 km MB = 4.7
	eX	A	09 14	D = 23.29 Az = 104 (ISC)
				PV A 1.5s 25.1nm
				XV A 2.0 51.3nm

Day	Phase		h m s	Remarks
28.	eP	A	13 09 14	Iceland Region 62.9 N 25.0 W H = 13 04 04 h = 0 km MB = 4.8 D = 23 14 Az = 105 (ISC) PV A 2.0s 51.3nm
28.	LmH	C	13 17.0	LmV C 16s 1.72/um
	LmV	C	17.0	
28.	e	A	15 18 56	Probably to : West of Macquarie Island 60.29 S 153.49 E
	e	A	19 02.5	H = 14 58 23.80 h = normal MB = 5.4 (NEIC)
	LmH	C	16 36.0	D = 156.7
	LmV	C	36.8	LmH C 23s 0.98/um M = 5.5 LmV C 22 1.56/um 5.8
28.	eP	A	17 59 47	Kurile Islands 49.48 N 155.51 E H = 17 48 09.7 h = 85 km MB = 5.0 D = 75.58 Az = 337.2 (NEIC) PV A 1.2s 56.9nm M = 5.3
29.	ePKHKP	A	03 57 23.6	Tonga Islands 21.46 S 174.28 W H = 03 37 32.5 h = normal MB = 4.8 D = 150.50 Az = 352.4 (NEIC) PKHKPV A 1.0s 13.8nm
29.	+iP	A	06 10 46.8	Eastern Kazakh SSR 49.76 N 78.13 E
	ePn	A	12 19	H = 06 02 57.1 h = 0 km MB = 5.5 D = 41.28 Az = 297.7 (NEIC) PV A 0.8s 65.4nm M = 5.4 Underground explosion
29.	eP	A	18 52 11	Southern Italy 40.40 N 15.80 E
	LmH	B	56.0	H = 18 49 36.4 h = 13 km MB = 4.7
	LmV	B	56.9	D = 10.65 Az = 345.5 (NEIC) PV A 0.5s 11.5nm M = 5.5 LmH B 11 1.5/um 4.1 LmV B 7.5 1.3/um

Day	Phase		h m s	Remarks
29.	e(P)	A	19 25 30.5	Poland 50.35 N 18.93 E H = 19 22 57.6 (WAR) D = 4.7 PV A 1.5s 20.1nm
29.	ePP	A	20 31 45	Near Coast of Peru 11.24 S 77.75 W
	eSKS	C	38 16	H = 20 14 12.2 h = 54 km MB = 5.3 (NEIC)
	ePS	C	40 50	D = 98.2
	eSS	C	46 15	LmH B 18s 1.4 /um M = 5.5
	LmV	B	21 12.2	LmV B 18 1.6 /um 5.6
	LmH	B	12.8	
30.	eSg	A	05 28 48	Bay of Biscay 46.8 N 2.1 W H = 05 23 18 (BCIS) D = 9.87
30.	eP	A	06 55 53	Rat Islands, Aleutian Is.
	LmH	C	07 33.6	51.11 N 179.54 E
	LmV	C	36.0	H = 06 43 56.5 h = 44 km MB = 5.0 D = 78.10 Az = 352.2 (NEIC) LmH C 19s 0.64 /um M = 5.0 LmV C 18 0.55 /um 5.0

Day	Phase		h m s	Remarks
1.	eP	A	04 14 51	<u>Andaman Islands Region</u> 13.85 N 96.05 E
	e	A	14 59	H = 04 03 06.6 h = 47 km MB = 4.8
				D = 75.91 Az = 319.2 (NEIC)
				PV A 1.4s 14.0nm M = 4.7
2.	eP1	A	17 30 26.5	<u>Kurile Islands Region</u> 44.83 N 153.34 E
	eiP2	AB	30 28	H = 17 18 21.9 h = 24 km MB = 6.2
	+iP3	A	30 30	D = 79.27 Az = 336.3 (NEIC)
	+iX	A	30 46.5	P2V A 1.2s 81.2nm M = 5.6
	ePP	BC	33 25	P2V B 10 8.0µm 6.7
	eiS	B	40 20	P3V A 1.5 337.0nm 6.2
	e(SKS)	B	40 35	XV A 1.5 422.0nm
	eSS	C	45 10	PPV B 7.5 2.7µm 6.5
	eP'P'1	A	57 14.5	SH B 9.5 6.4µm 6.6
	eP'P'2	A	57 21.5	P'P'2V A 1.8 33.8nm
	LmH	B	18 04.9	LmH B 22 22.4µm 6.5
	LmV	B	14.8	LmV B 16 16.4µm 6.5
2.	iP	A	23 18 10	<u>Unimak Island Region</u> 53.70 N 163.65 W
	e	A	18 19	H = 23 06 24.9 h = normal MB = 4.7
				D = 75.95 Az = 3.1 (NEIC)
				PV A 1.2s 28.4nm M = 5.1
3.	eSn	A	04 16 03	<u>Yugoslavia</u> 45.99 N 15.73 E
	eiSg	A	16 35	H = 04 13 39.8 h = 0 km
				D = 5.41 Az = 331 (ISC)
3.	LmH	C	15 41.5	<u>Near Coast of Northern Chile</u>
				22.8 S 70.6 W
				H = 14 46 01.3 h = 37 km MB = 5.1 (NEIC)
				D = 102.5
				LmH C 22s 0.4µm M = 4.9
3.	ePg	A	16 17 04	<u>Federal Rep. of Germany</u> 49.96 N 8.4 E
	i	A	17 08.5	H = 16 16 23.1 h = 0 km
	iSn	A	17 26	D = 2.18 Az = 70 (ISC)
	iSg	A	17 32.5	

Day	Phase		h m s	Remarks
3.	eP	A	19 24 59	Kurile Islands 47.12 N 153.09 E
				H = 19 13 07.7 h = normal MB = 4.8
				D = 77.11 Az = 336.0 (NEIC)
4.	ePKIKP	AB	02 44 45	Solomon Islands 6.00 S 154.56 E
	eX	A	46 08.5	H = 02 25 51.6 h = 78 km MB = 6.0
	ePP	B	46 40	D = 125.83 Az = 331.8 (NEIC)
	ePKS	B	48 02	PKIKPV A 1.3s 26.2nm
	ePS	B	56 40	XV A 2.6 156.0nm
	ePcPPKP	A	58 22	PPV B 18 2.2/um M = 5.9
	ePPS	B	58 25	LmH B 21 9.4/um
	eSS	B	03 04 00	LmV B 19 5.4/um
	LmH	B	31.4	
	LmV	B	40.4	
4.	eP	A	08 48 09	Nepal 27.90 N 87.87 E
				H = 08 38 00.7 h = 32 km MB = 5.0
				D = 60.48 Az = 314.7 (NEIC)
				PV A 1.1s 14.1nm M = 5.0
4.	eP	A	16 09 22	Mindanao, Philippine Islands
	LmV	B	57.6	8.09 N 126.61 E
	LmH	B	57.9	H = 15 55 45.1 h = 54 km MB = 5.5
				D = 99.12 Az = 324.2 (NEIC)
				PV A 1.4s 14.0nm M = 5.3
				LmH B 18 0.4/um
				LmV B 20 1.0/um
4.	eP	A	23 19 35	Crete 35.00 N 23.23 E
				H = 23 15 27.5 h = 47 km MB = 4.4
				D = 17.77 Az = 335.2 (NEIC)
5.	eP1	A	06 01 50	Vancouver Island Region
	eP2	A	01 54	49.63 N 129.45 W
	iS	B	11 30	H = 05 50 05.8 h = 5 km MB = 5.6 MS=6.0
	ePS	B	11 56	D = 74.71 Az = 24.5 (NEIC)
	eSS	B	16 17	P1V A 2.0s 34.2nm M = 5.0

Day	Phase		h m s	Remarks
cont. 5.	LmV	B	06 35.9	P2V A 2.5s 123.0nm M = 5.5
	LmH	B	36.5	SH B 12 2.1 μm 6.0
				LmH B 17 5.3 μm 5.9
				LmV B 17 7.0 μm 6.1
5.	eP1	A	06 24 31	Vancouver Island Region
	eP2	A	24 36	49.73 N 129.08 W
				H = 06 12 51.1 h = 16 km MB = 5.1
				D = 74.51 Az = 24.7 (NEIC)
				P1V A 1.5s 12.6nm M = 4.7
				P2V A 1.5 35.2nm 5.2
6.	-iP	A	00 21 27.5	Kamchatka 54.37 N 158.55 E
				H = 00 10 22.9 h = 171 km MB = 5.1
				D = 71.75 Az = 338.5 (NEIC)
				PV A 0.9s 37.0nm M = 5.1
6.	ePKIKP	A	02 26 11.5	New Hebrides Islands 17.54 S 167.86 E
				H = 02 06 43.5 h = 27 km MB = 4.7
				D = 141.91 Az = 335.4 (NEIC)
				PKIKPV A 0.6s 9.6nm
6.	ePKIKP	A	02 28 37	New Hebrides Islands 17.60 S 167.81 E
	LmH	C	03 31.3	H = 02 09 10.0 h = 27 km MB = 4.6
	LmV	B	34.5	D = 141.94 Az = 335.4 (NEIC)
				PKIKPV A 2.5s 76.8nm
				LmH C 22 0.60 μm M = 5.2
				LmV B 14 0.55 μm 5.5
6.	LmH	C	14 04.3	LmH C 19s 0.65 μm
	LmV	C	12.4	LmV C 18 0.65 μm
7.	ePP	A	03 44 38	Bouvet Island Region 54.38 S 5.91 E
	eS	C	52 35	H = 03 26 19.9 h = normal
	ePS	C	53 55	MB = 5.7 MS = 6.1 (NEIC)
	eSS	C	59 40	D = 104.7
	LmV	B	04 28.2	LmH B 17.5s 4.2 μm M = 6.0
	LmH	B	28.5	LmV B 18 3.9 μm 6.0

Day	Phase		h m s	Remarks
7.	eiP	A	06 37 07	North of Halmahera 3.70 N 128.02 E
	epP	A	37 22.5	H = 06 23 11.8 h = 69 km MB = 6.0
	e	A	41 13	D = 103.48 Az = 324.1 (NEIC)
	ePP	A	41 25	h = 55 km.
	ePS	C	50.5	PV A 2.0s 59.9nm M = 6.1
	eSS	C	57 00	PPV A 3.0 368.2nm 6.5
	LmV	C	07 20.0	LmH C 17 2.1 µm
	LmH	B	22.3	LmV B 36 4.4 µm
7.	eP1	A	12 11 42	East of Labrador 55.08 N 54.35 W
	eP2	A	11 45	H = 12 04 19.7 h = normal MB = 5.4
	ePP	A	13 15	D = 38.74 Az = 68.3 (NEIC)
	eS	C	17 38	P1V A 0.8s 13.5nm M = 4.8
	LmH	B	26.5	P2V A 1.6 46.7nm 5.1
	LmV	B	26.5	PPV A 2.4 152.0nm 5.4
				SH C 22 1.4 µm 5.4
				LmH B 19 1.6 µm 4.9
				LmV B 19 1.8 µm 5.0
7.	eP	A	15 05 38.5	Turkey 39.11 N 40.11 E
				H = 15 00 34.6 h = 71 km MB = 4.3
				D = 23.10 Az = 309 (ISC)
				PV A 1.3s 10.9nm M = 4.2
8.	ePg	A	04 55 33.5	Northern Italy 45.1 N 11.3 E
	eSn	A	56 16	H = 04 53 41 h = 0 km (ISC)
	eSg	A	56 52	D = 5.69
8.	ePKIKP	A	06 29 06	South of Fiji Islands. 25.37 S 177.32 W
	ePKHKP	A	29 14.5	H = 06 09 32.6 h = 156 km MB = 5.1
	eiPKP2	A	29 28	D = 153.84 Az = 347.0 (NEIC)
				PKHKPV A 1.2s 18.3nm
				PKP2V A 1.2 38.6nm
8.	eP	A	08 50 03	Vancouver Island Region
	e	A	50 20	49.13 N 128.52 W
	eS	C	59 46	H = 08 38 24.0 h = normal MB = 5.2
	LmH	B	09 23.9	D = 74.90 Az = 25.0 (NEIC)

Day	Phase		h m s	Remarks
cont. 8.	LmV	B	09 27.0	LmH B 17s 2.8/um M = 5.6
				LmV B 15 3.0/um 5.8
8.	e	A	16 19 30	Near Coast of Northern Chile
	ePP	B	23 33	22.91 S 70.76 W
	eSKS	B	30 02	H = 16 05 23.2 h = 18 km MB = 5.6
	eS	C	31 20	D = 102.74 Az = 40.3 (NEIC)
	eiPS	B	32 45	LmH B 22s 6.2/um M = 6.1
	ePPS	B	33 42	LmV B 22 6.7/um 6.1
	eSS	C	38 24	
	LmH	B	17 00.8	
	LmV	B	00.8	
8.	e(P)	A	19 12 59	Fox Islands, Aleutian Is.
				51.01 N 170.95 W
				H = 19 00 53.8 h = normal MB = 4.1
				D = 78.70 Az = 358.3 (NEIC)
9.	eP	A	01 50 16	Southern Iran 27.24 N 56.40 E
	Pm	A	50 22	H = 01 42 30.7 h = 15 km MB = 5.3
	eS	B	56 32	D = 41.11 Az = 317.0 (NEIC)
	LmH	B	02 07.9	PV A 1.4s 20.9nm M = 4.7
	LmV	B	10.3	PmV A 2.0 85.5nm 5.1
				LmH B 19 9.5/um 5.7
				LmV B 18 4.8/um 5.5
9.	eP	A	12 38 34.5	Carlsberg Ridge 2.44 S 67.98 E
				H = 12 27 16.4 h = normal MB = 5.4
				D = 71.35 Az = 326.0 (NEIC)
				PV A 1.1s 24.2nm M = 5.2
9.	PKP	AB	15 20 39.5	New Hebrides Islands Region
	LmH	B	16 22.5	21.74 S 173.24 E
	LmV	B	24.0	H = 15 00 59.1 h = 41 km MB = 5.6
				D = 147.75 Az = 337.9 (NEIC)
				PKPV A 2.4s 200.5nm
				LmH B 19 1.7/um M = 5.8
				LmV B 23 3.1/um 6.0

Day	Phase		h m s	Remarks
9.	LmH	B	20 11.5	New Britain Region 5.80 S 152.08 E
	LmV	B	12.0	H = 18 55 27.2 h = 47 km MB = 5.2 (NEIC)
				D = 124.5
9.	eP1	A	22 51 49.5	Northeast of Taiwan 25.59 N 124.43 E
	eP2	A	51 52	H = 22 39 34.6 h = 145 km MB = 5.3 (NEIC)
	LmH	C	23 38.5	D = 83.78
	LmV	B	39.0	P1V A 1.8s 27.0nm M = 4.8
				P2V A 1.5 37.7nm 5.0
10.	e(Sb)	A	02 56 18	Poland 50.37 N 18.87 E
	eSg	A	56 30	H = 02 53 58.5 M = 3.0 (WAR)
				D = 4.64 Az = 276 (ISC)
10.	ePKIKP	A	05 15 41.5	New Britain Region 6.58 S 150.28 E
	e	A	15 50.5	H = 04 56 41.9 h = 14 km MB = 5.2
				D = 124.24 Az = 329.4 (NEIC)
10.	e	A	08 52 03	New Hebrides Islands Region
				21.09 S 173.62 E
				H = 08 32 14.2 h = 50 km MB = 4.9
				D = 147.28 Az = 338.7 (NEIC)
11.	ePP	A	03 48 15.5	Halmahera 1.75 N 127.27 E
				H = 03 29 57.5 h = 114 km MB = 4.2 (NEIC)
				D = 104.60
				PPV A 1.8s 43.9nm M = 5.8
11.	ePKIKP	A	07 44 16	Solomon Islands 6.07 S 154.51 E
	e	A	44 40	H = 07 25 19.1 h = 65 km MB = 5.9
	ePP	B	46 14	D = 125.87 Az = 331.7 (NEIC)
	e	A	46 23	PPV B 15s 1.7/um M = 5.9
	e	B	47 28	LmH B 21 7.7/um
	ePS	B	56 00	LmV B 21 9.8/um
	ePPS	C	57 38	
	eSS	C	08 03 40	
	eSSS	C	08 20	
	LmH	B	42.0	
	LmV	B	42.2	

Day	Phase		h m s	Remarks
11.	LmH	C	22 22.3	Near Coast of Northern Chile
	LmV	B	23	23.02 S 70.60 W
				H = 21 27 10.4 h = 38 km
				MB = 5.2 MS = 4.9 (NEIC)
				D = 102.73
				LmH C 25s 0.65/um M = 5.1
				LmV B 22 0.90/um 5.3
11.	eP	A	22 43 52	Oaxaca, Mexico 16.45 N 94.09 W
				H = 22 31 15.4 h = 95 km MB = 5.0
				D = 87.04 Az = 37.9 (NEIC)
12.	eP	A	13 50 09	Kirgiz-Sinkiang Border Region
				41.38 N 79.24 E
				H = 13 41 39.8 h = normal MB = 4.7
				D = 46.36 Az = 305.6 (NEIC)
				PV A 1.2s 18.3nm M = 4.9
12.	eP	A	22 35 47	Tadzhik-Sinkiang Border Region
	LmH	C	56.0	39.48 N 73.23 E
	LmV	C	56.0	H = 22 27 41.1 h = normal MB = 4.8 (NEIC)
				D = 43.73
				LmH C 20s 0.4/um M = 4.3
				LmV C 20 1.0/um 4.8
13.	eP	A	16 31 09	Western Brazil 8.89 S 71.64 W
				H = 16 19 00.1 h = 614 km MB = 5.3
				D = 92.55 Az = 39.3 (NEIC)
				PV A 2.0s 51.3nm M = 5.2
14.	eP	A	02 41 09.5	Near Islands, Aleutian Is.
				52.79 N 172.04 E
				H = 02 29 26.8 h = 38 km MB = 4.9
				D = 75.63 Az = 347.3 (NEIC)
14.	eP	A	09 49 35	Mascarene Islands Region 17.30 S 66.47 E
				H = 09 37 09.1 h = normal MB = 4.7
				D = 82.98 Az = 328.4 (NEIC)
				PV A 1.4s 16.3nm M = 4.9

Day	Phase		h m s	Remarks
15.	eP	A	03 08 49	Hokkaido, Japan Region 41.86 N 142.31 E
				H = 02 56 52.9 h = 70 km MB = 4.4
				D = 78.32 Az = 330.5 (NEIC)
				traces
15.	eP	A	05 11 04.5	Komandorsky Islands Region
				53.71 N 169.46 E
				H = 04 59 29.2 h = 35 km MB = 5.1
				D = 74.36 Az = 345.6 (NEIC)
				PV A 1.1s 16.1nm M = 4.9
15.	iPKP2	. A	07 47 30	Kermadec Islands Region 31.18 S 179.83 E
				H = 07 27 36.5 h = 355 km MB = 4.8 (NEIC)
				PKP2V A 1.5s 62.8nm D = 158.76
15.	ePn	A	08 01 31.5	Czechoslovakia 50°41' N 14°39.5' E
	ePg	A	01 34	explosion yield 12.3 t (PRU)
	eSn	A	01 57	D = 1.91
	eSg	A	02 00	
15.	+iP1	AB	08 41 12.3	Near East Coast of Kamchatka
	iP2	A	41 15	56.00 N 163.26 E
	iP3	Á	41 24	H = 08 29 55.3 h = normal MB =6.1 MS=7.8
	eX	A	41 55	D = 71.15 Az = 341.4 (NEIC)
	eS	B	50 32	P1V A 1.3s 153.0nm M = 5.9
	eiPS	B	51 05	PV B 9.5 13.1/um 6.9
	eSS	B	55 06	P2V A 1.8 872.0nm 6.5
	eSSS	B	58 19	P3V A 1.9 2082.0nm 6.8
	eP'P'	A	09 09 09	XV A 2.6 5460.0nm
	LmH	B	19.0	LmH B 17 454.0/um 7.8
	LmV	B	19.6	LmV B 16 482.0/um 7.9
				N-component off scale
15.	eP	A	09 01 39.5	Off East Coast of Kamchatka
				55.78 N 163.58 E
				H = 08 50 21.8 h = normal MB = 4.5
				D = 71.41 Az = 341.6 (NEIC)

Day	Phase		h m s	Remarks
15.	e	A	09 11 51.5	North Pazific Ocean 42.5 N 173.8 W H = 08 58 55.2 (HFS 1) D = 87.12 Az = 357 (ISC)
15.	eP	A	09 22 25.5	Off East Coast of Kamchatka 55.79 N 163.4 E H = 09 11 02.5 h = 0 km D = 71.36 Az = 341 (ISC)
15.	eP	A	09 31 24.5	Eastern Siberia 65.8 N 165.6 E H = 09 20 59 h = 0 km D = 62.22 Az = 342
15.	+iP	A	09 33 54	Off East Coast of Kamchatka 55.79 N 163.68 E H = 09 22 35.8 h = normal MB = 5.2 D = 71.42 Az = 341.7 (NEIC) PV A 1.2s 52.9nm M = 5.5
15.	eP	A	10 05 05	Near East Coast of Kamchatka 56.0 N 162.8 E H = 09 53 49 D = 71.04 Az = 341 (ISC)
15.	eP eX	A A	10 17 14 17 17.5	Komandorsky Islands Region 56.11 N 164.07 E H = 10 05 54.2 h = normal MB = 4.9 D = 71.18 Az = 341.9 (NEIC) XV A 0.9s 13.6nm
15.	eP e	A A	10 38 37 38 43	Near East Coast of Kamchatka 55.89 N 162.93 E H = 10 27 19.8 h = normal MB = 4.8 D = 71.19 Az = 341.2 (NEIC) PV A 1.4s 23.3nm M = 5.0

Day	Phase		h m s	Remarks
15.	eP	A	11 13 18.5	<u>Off East Coast of Kamchatka</u> 55.63 N 163.88 E H = 11 02 00.0 h = 40 km MB = 4.7 D = 71.60 Az = 341.8 (NEIC)
15.	eP	A	11 20 19	<u>Off East Coast of Kamchatka</u> 55.73 N 163.74 E H = 11 08 59.4 h = 25 km MB = 4.5 D = 71.48 Az = 341.7 (NEIC)
15.	eP	A	11 22 07.5	<u>Off East Coast of Kamchatka</u> 55.68 N 163.94 E H = 11 10 48.0 h = normal MB = 4.7 D = 71.57 Az = 341.8 (NEIC) PV A 1.2s 12.2nm M = 4.8
15.	eP	A	11 22 26.5	<u>Komandorsky Islands Region</u> 55.67 N 163.99 E H = 11 11 07.0 h = normal MB = 5.2 D = 71.59 Az = 341.9 (NEIC) PV A 2.0s 94.1nm M = 5.5
15.	eP	A	11 44 30	<u>Komandorsky Islands Region</u> 55.70 N 164.07 E H = 11 33 10.6 h = 35 km MB = 4.7 D = 71.57 Az = 341.9 (NEIC)
15.	eP	A	12 13 06	<u>Komandorsky Islands Region</u> 55.67 N 164.06 E H = 12 01 46.4 h = normal MB = 5.1 D = 71.60 Az = 341.9 (NEIC) PV A 1.5s 40.2nm M = 5.2
15.	eP	A	12 44 30	<u>Off East Coast of Kamchatka</u> 55.81 N 163.51 E H = 12 33 12.4 h = normal MB = 4.7 D = 71.37 Az = 341.5 (NEIC) PV A 1.2s 16.3nm M = 4.9

Day	Phase		h m s	Remarks
15.	eP	A	13 03 00	<u>Near East Coast of Kamchatka</u> 55.80 N 162.80 E H = 12 51 42.9 h = 35 km MB = 4.9 D = 71.25 Az = 341.1 (NEIC) PV A 1.1s 8.1nm M = 4.7
15.	eP	A	13 19 02	<u>Komandorsky Islands Region</u> 55.71 N 164.11 E H = 13 07 42.6 h = 32 km MB = 5.3 D = 71.57 Az = 341.9 (NEIC) PV A 1.6s 60.5nm M = 5.4
15.	eP e	A A	13 29 35 29 39	<u>Near East Coast of Kamchatka</u> 56.08 N 163.02 E H = 13 18 19.9 h = 35 km MB = 5.0 D = 71.02 Az = 341.2 (NEIC) PV A 1.6s 24.7nm M = 5.0
15.	ePKP e	A A	13 41 11.5 41 28	<u>Loyalty Islands Region</u> 22.09 S 169.9 E H = 13 21 34.5 D = 146.83 Az = 335 (ISC) PV A 1.1s 20.2nm
15.	eP	A	15 18 08	<u>Off East Coast of Kamchatka</u> 55.82 N 163.60 E H = 15 06 50.2 h = normal MB = 5.1 D = 71.38 Az = 341.6 (NEIC) PV A 1.4s 44.2nm M = 5.3
15.	eP	A	15 31 52	<u>Iran</u> 30.25 N 50.57 E H = 15 24 57.4 h = 42 km MB = 4.9 D = 35.42 Az = 316.3 (NEIC)
15.	eP	A	17 11 15	<u>Komandorsky Islands Region</u> 55.68 N 163.97 E H = 16 59 54.2 h = 23 km MB = 4.7 D = 71.58 Az = 341.9 (NEIC) PV A 1.1s 12.1nm M = 4.9

Day	Phase		h m s	Remarks
15.	ePKP	A	19 28 55.5	<u>Tonga Islands</u> 15.64 S 173.22 W H = 19 09 20.4 h = normal MB = 4.9 D = 144.87 Az = 354.6 (NEIC)
15.	ePKP	A	23 47 45	<u>Tonga Islands</u> 16.39 S 174.50 W H = 23 28 04.2 h = normal MB = 4.8 D = 145.48 Az = 353.1 (NEIC) PKPV A 1.6s 30.2nm
16.	eP	A	00 14 24.5	<u>Komandorsky Islands Region</u> 55.73 N 163.98 E H = 00 03 06.3 h = 41 km MB = 4.8 D = 71.53 Az = 341.9 (NEIC) PV A 1.5s 22.6nm M = 4.9
16.	eP epP	A A	02 49 13.5 49 21	<u>Off East Coast of Kamchatka</u> 55.79 N 163.66 E H = 02 37 54.1 h = 25 km MB = 4.8 D = 71.42 Az = 341.6 (NEIC) h = 28 km PV A 1.1s 16.1nm M = 5.0 pPV A 1.2 18.3nm
16.	eiP	A	08 25 00	<u>Near East Coast of Kamchatka</u> 55.99 N 162.95 E H = 08 13 43.7 h = 35 km MB = 5.0 D = 71.10 Az = 341.2 (NEIC) PV A 1.0s 29.5nm M = 5.3
16.	eP ePP LmH LmV	A A C C	12 37 05.5 39 44 13 08.8 11.2	<u>Off East Coast of Kamchatka</u> 55.89 N 163.56 E H = 12 25 48.0 h = normal MB = 5.6 D = 71.30 Az = 341.6 (NEIC) PV A 1.9s 182.0nm M = 5.8 PPV A 2.4 124.3nm 5.7 LmH C 20 3.0 μm 5.5 LmV C 18 2.8 μm 5.6

Day	Phase		h m s	Remarks
16.	eiP	A	12 40 49.5	Off East Coast of Kamchatka 55.84 N 163.63 E H = 12 29 31.4 h = normal MB = 5.2 D = 71.36 Az = 341.6 (NEIC) PV A 1.3s 34.9nm M = 5.2
16.	eP	A	13 22 29	Off East Coast of Kamchatka 55.87 N 163.54 E H = 13 11 10.0 h = 24 km MB = 4.8 D = 71.32 Az = 341.6 (NEIC) PV A 1.2s 16.3nm M = 5.0
16.	eP	A	14 39 36.5	Near East Coast of Kamchatka 55.97 N 163.39 E H = 14 28 19.0 h = normal MB = 4.8 D = 71.19 Az = 341.5 (NEIC) PV A 1.1s 16.1nm M = 5.0
16.	eP	A	15 18 58.5	Near East Coast of Kamchatka
	e	A	19 02	55.96 N 162.85 E
				H = 15 07 39.4 h = 25 km MB = 4.8
				D = 71.11 Az = 341.1 (NEIC)
16.	+eP	A	18 41 29	Svalbard Region 77.86 N 17.85 E
	e	A	41 49.5	H = 18 35 45.5 h = normal MB = 5.0
	eS	C	46 30	D = 27.42 Az = 188.6 (NEIC)
	LmH	B	51.1	PV A 1.2s 34.6nm M = 4.9
	LmV	B	54.4	LmH B 18 2.5/um 4.8
				LmV B 16 2.3/um 5.0
17.	eP	A	00 29 14	Komandorsky Islands Region 54.7 N 165.0 E H = 00 17 40 h = 0 km MB = 4.6 D = 72.64 Az = 343 (ISC) PV A 1.3s 10.9nm
17.	eP	A	01 14 05	Off East Coast of Kamchatka 55.86 N 163.54 E

Day	Phase		h m s	Remarks
cont. 17.	e	A	01 14 12	H = 01 02 47.2 h = normal MB = 4.3 D = 71.32 Az = 341.6 (NEIC)
17.	eP	A	02 10 16	<u>Crete</u> 34.96 N 23.97 E
	e	A	10 22	H = 02 06 05.2 h = 33 km MB = 4.7
	LmH	C	18.7	D = 18.07 Az = 333.9 (NEIC)
	LmV	C	18.7	LmV C 15s 1.3 µm M = 4.5
17.	eP	A	04 18 49.5	<u>Near East Coast of Kamchatka</u>
	ipP	A	18 56	55.87 N 162.88 E
	LmH	C	52.6	H = 04 07 31.1 h = 27 km MB = 5.0
	LmV	C	53.0	D = 71.20 Az = 341.1 (NEIC) h = 24 km PV A 1.1s 20.2nm M = 5.1 pPV A 1.0 29.6nm
17.	e	A	05 49 31.5	<u>Off East Coast of Kamchatka</u> 55.89 N 163.52 E H = 05 38 04.9 h = 22 km MB = 4.6 D = 71.30 Az = 341.5 (NEIC)
17.	ePn	A	10 48 23.5	D = 4.2
	ePg	A	48 42	
	eSn	A	49 13	
	eSg	A	49 36	
17.	e	A	13 04 11	<u>Off East Coast of Kamchatka</u>
	e	A	04 22	55.70 N 163.66 E H = 12 52 39.4 h = 35 km MB = 4.2 (NEIC) D = 71.50
17.	eP	A	18 11 54	<u>Northern Sumatra</u> 1.65 N 99.80 E
	epP	A	12 41	H = 17 59 20.7 h = 170 km MB = 5.0 D = 87.58 Az = 320.4 (NEIC) h = 178 km

Day	Phase		h m s	Remarks
17.	eP	A	18 36 45.5	<u>Off East Coast of Kamchatka</u>
	epP	A	36 57.5	55.52 N 163.91 E
	LmH	B	19 08.8	H = 18 25 24.6 h = normal MB = 4.9
	LmV	B	13.0	D = 71.72 Az = 341.8 (NEIC)
				h = 45 km
				PV A 0.8s 11.5nm M = 4.9
				pPV A 1.2 26.4nm
				LmH B 16 2.3µm 5.5
				LmV B 16 0.9µm 5.2
17.	eP	A	19 02 42	<u>Komandorsky Islands Region</u>
				54.7 N 165.8 E
				H = 18 51 08 h = 0 km
				D = 72.78 Az = 343 (ISC)
17.	+iP	AB	19 17 27.5	<u>Komandorsky Islands Region</u>
	eipP	A	17 37.5	55.52 N 163.95 E
	eisP	AB	17 40.5	H = 19 06 07.1 h = normal MB = 5.5
	eS	B	26 45	D = 71.72 Az = 341.9 (NEIC)
	e(SKS)	B	27 24	h = 37 km
	LmH	B	49.5	PV A 1.5s 85.5nm M = 5.5
	LmV	B	53	pPV A 1.3 52.4nm
				sPV A 1.6 187.0nm
				LmH B 17 11.4µm 6.2
				LmV B 16 4.7µm 5.9
18.	eP	A	02 37 17	<u>Southern Greece</u> 36.70 N 22.86 E
	eX	A	37 20	H = 02 33 24.5 h = 7 km MB = 4.4
				D = 16.12 Az = 333.4 (NEIC)
				PV A 1.0s 13.8nm M = 4.0
				XV A 0.8 19.2nm
18.	ePn	A	06 04 04	<u>Central Italy</u> 43.39 N 12.75 E
	e	A	04 51	H = 06 02 11.5 h = 46 km
	eSn	A	05 28	D = 7.37 Az = 354 (ISC)
	eSg	A	06 25	

Day	Phase		h m s	Remarks
18.	e	A	06 28 32	Near East Coast of Kamchatka
	e	A	28 35	55.30 N 162.95 E
	LmH	C	07 00.0	H = 06 17 03.5 h = normal MB = 4.8 MS=4.9
				D = 71.75 Az = 341.2 (NEIC)
				LmV C 20s 0.6/um M = 4.9
18.	e	A	06 55 25	Off East Coast of Kamchatka
	LmV	C	07 29.5	55.31 N 162.99 E
	LmH	B	30.0	H = 06 43 55.0 h = normal MB = 4.9
				D = 71.75 Az = 341.3 (NEIC)
				LmV B 22s 1.1/um M = 5.1
18.	e	A	06 58 40	Near East Coast of Kamchatka
	e	A	58 45	55.36 N 162.74 E
	e	A	58 51	H = 06 47 10.3 h = normal MB = 5.0
				D = 71.65 Az = 341.1 (NEIC)
18.	e	A	07 27 37.5	Off East Coast of Kamchatka
	e	A	27 42	55.35 N 162.98 E
	e	A	27 50	H = 07 16 08.3 h = normal MB = 4.6 MS=4.8
	LmH	C	08 03.5	D = 71.72 Az = 341.3 (NEIC)
	LmV	C	02.0	LmH C 18s 0.6/um M = 4.9
				LmV C 20 0.7/um 5.0
18.	ePg	A	13 34 03	Federal Republik of Germany - France Border Region 48°55' N 8°12' E
	eSg	A	34 41	H = 13 33 10.0 (BCIS)
				D = 2.81 Az = 51 (ISC)
18.	eX	A	17 05 39	Off East Coast of Kamchatka
	e	A	05 56.5	55.98 N 162.61 E
				H = 16 54 21.5 h = 35 km MB = 4.4
				D = 71.04 Az = 340.9 (NEIC)
				XV A 1.1s 12.1nm
18.	+eP	A	22 09 11.5	Off East Coast of Kamchatka
	LmH	B	41.2	55.58 N 163.27 E

Day	Phase		h m s	Remarks
cont. 18.	LmV	B	22 46.5	H = 21 57 52.6 h = normal MB = 5.5
				D = 71.54 Az = 341.4 (NEIC)
				PV A 1.5s 85.4nm M = 5.6
				LmH B 16 1.5 µm 5.3
				LmV B 14 1.1 µm 5.3
18.	eP	A	22 33 26.5	East of Lake Baikal 56.58 N 113.97 E
				H = 22 23 52.7 h = normal MB = 4.8
				D = 55.47 Az = 311.0 (NEIC)
				h = 26 km
				PV A 1.2s 28.5nm M = 5.2
				pPV A 1.3 21.8nm
19.	ePKP	A	06 18 24.5	New Hebrides Islands 14.84 S 167.39 E
	ePP	A	21 18.5	H = 05 59 18.2 h = 147 km MB = 5.1
	e	A	22 00	D = 139.26 Az = 336.4 (NEIC)
	e	A	22 10	
19.	eP	A	08 01 44.5	Off East Coast of Kamchatka
	LmH	B	40.9	55.95 N 163.07 E
	LmV	B	44.1	H = 07 50 27.9 h = normal MB = 5.0
				D = 71.16 Az = 341.2 (NEIC)
				PV A 1.2s 24.4nm M = 5.1
				LmH B 17.5 2.6 µm 5.6
				LmV B 13 5.6 µm 5.6
19.	iP	A	15 43 39.5	Off East Coast of Kamchatka
	epP	A	43 51	55.24 N 163.03 E
	e	A	43 57.5	H = 15 32 18.3 h = normal MB=5.1 MS=5.1
	e	A	44 10	D = 71.83 Az = 341.3 (NEIC)
	LmH	B	16 14.8	h = 44 km
	LmV	B	23.4	PV A 1.6s 44.0nm M = 5.2
				LmH B 16 1.3 µm 5.3
				LmV B 15 1.3 µm 5.3

Day	Phase		h m s	Remarks
19.	eX	A	15 55 56	Off East Coast of Kamchatka
				55.2 N 163.6 E
				H = 15 44 22
				D = 72.00 Az = 342 (ISC)
				XV A 1.8s 37.2nm
20.	eP	A	00 35 10.5	Near East Coast of Kamchatka
				55.95 N 163.83 E
				H = 00 23 53.7 h = 34 km MB = 4.9
				D = 71.29 Az = 341.7 (NEIC)
20.	eP	A	01 34 59.5	Eastern Caucasus 41.21 N 48.34 E
	e	A	35 02	H = 01 29 18.6 h = normal MB = 5.0
	e	A	35 16	D = 26.99 Az = 303.0 (NEIC)
	e	A	35 35	PV A 1.4s 18.6nm M = 4.5
	LmH	B	46.0	LmH B 14.5 6.2 µm 5.3
	LmV	B	48.3	LmV B 15 4.4 µm 5.3
20.	eP	A	01 46 46.5	Eastern Caucasus 41.14 N 48.39 E
	LmH	B	56.8	H = 01 41 04.9 h = normal MB = 5.2
	LmV	B	02 00.2	D = 27.06 Az = 303.1 (NEIC)
				PV A 1.4s 34.8nm M = 4.8
				LmH B 22 7.5 µm 5.2
				LmV B 12 3.8 µm 5.3
20.	e	A	07 58 58	Eastern Caucasus 41.23 N 48.33 E
				H = 07 53 11.4 h = normal MB = 4.8
				D = 26.97 Az = 302.9 (NEIC)
20.	eP	A	16 34 48	Off East Coast of Kamchatka
	epP	A	34 57	55.80 N 163.64 E
				H = 16 23 25.9 h = normal MB = 4.5
				D = 71.40 Az = 341.6 (NEIC)
				h = 42 km
20.	ePn	A	18 33 42	Central Italy 43.17 N 12.87 E
	ePg	A	34 22	H = 18 31 53.6 h = 43 km
	eSn	A	35 02	D = 7.53 Az = 354 (ISC)

Day	Phase		h m s	Remarks
20.	e	A	21 53 32	Near East Coast of Kamchatka 55.53 N 162.94 E H = 21 42 00.4 h = normal MB = 4.4 D = 71.53 Az = 341.2 (NEIC)
20.	eP	A	23 35 18	Southern Iran 28.35 N 57.23 E H = 23 27 38.5 h = normal MB = 4.9 D = 40.82 Az = 315.9 (NEIC) PV A 1.1s 12.1nm M = 4.5
21.	eP e(pP)	A A	10 03 12.5 03 17	Northwestern Kashmir 35.48 N 74.16 E H = 09 54 42.3 h = 25 km MB = 5.2 D = 46.75 Az = 309.1 (NEIC) pPV A 1.4s 32.5nm
21.	e	A	17 30 36	Near East Coast of Kamchatka 55.98 N 163.23 E H = 17 18 54.8 h = normal MB = 4.3 D = 71.16 Az = 341.3 (NEIC)
21.	eP epP	A A	19 36 04 36 40	Northern Sumatra 1.88 N 99.22 E H = 19 23 34.1 h = 142 km MB = 5.0 D = 87.04 Az = 320.4 (NEIC) h = 144 km
22.	eP	A	00 51 49	Northern Peru 4.00 S 76.24 W H = 00 38 53.0 h = 117 km MB = 5.2 D = 91.70 Az = 39.5 (NEIC)
22.	+iP1 +iP2 +iP3 eiPP	A A A A	07 05 12.6 05 14.5 05 18.3 05 46.5	Western Kazakh SSR 47.87 N 48.22 E H = 06 59 56.3 h = 0 km MB = 6.0 D = 23.90 Az = 290.4 (NEIC) P1V A 0.8s 53.8nm M = 5.2 P2V A 1.1 318.0nm 5.8 P3V A 1.0 342.0nm 5.8

Day	Phase		h m s	Remarks
22.	eX	A	11 14 02	Off East Coast of Kamchatka 55.92 N 163.73 E H = 11 02 36.1 h = normal MB = 4.9 D = 71.31 Az = 341.7 (NEIC) XV A 1.4s 14.0nm
22.	eP	A	13 35 03	Alaska Peninsula 57.98 N 154.53 W H = 13 23 49.4 h = 69 km MB = 3.7 D = 71.14 Az = 9.3 (NEIC)
22.	eP	A	20 30 18	Komandorsky Islands Region 54.69 N 164.50 E H = 20 18 52.0 h = normal MB = 4.4 D = 72.62 Az = 342.3 (NEIC) PV A 0.6s 7.7nm M = 4.9
23.	e	A	00 20 16.5	Windward Islands 14.60 N 60.91 W H = 00 08 50.1 h = 16 km MB = 4.8 D = 67.78 Az = 41.0 (NEIC)
23.	eP	A	01 23 04.5	Near East Coast of Kamchatka 55.96 N 163.70 E H = 01 11 43.7 h = 13 km MB = 4.8 D = 71.26 Az = 341.7 (NEIC)
23.	eSg	A	08 46 57	Yugoslavia 45.9 N 15.6 E H = 08 44 00 (BCIS) D = 5.47
23.	ePn	A	19 20 32	Northern Italy 46.5 N 12.7 E
	ePg	A	20 52.5	H = 19 19 29 h = 0 km
	eSn	A	21 23	D = 4.22 Az = 350 (ISC)
	eSg	A	21 42.5	
23.	ePn	A	20 22 07	Austria 47.43 N 11.08 E
	ePg	A	22 21	H = 20 21 17.1 h = 0 km
	iSn	A	22 46.5	D = 3.24 Az = 6 (ISC)
	eSg	A	23 02	

Day	Phase		h m s	Remarks
24.	eP	A	21 24 45	Taiwan Region 21.77 N 121.00 E H = 21 12 11.0 h = normal MB = 5.0 D = 84.95 Az = 322.9 (NEIC) PV A 1.7s 30.3nm M = 5.2
24.	ePKIKP	A	21 31 56.5	Tonga Islands 19.14 S 175.69 W
	ePKHKP	A	32 00	H = 21 12 42.5 h = 253 km MB = 5.0
	epPKIKP	A	33 00.5	D = 148.03 Az = 351.2 (NEIC)
	e	A	33 06	h = 250 km PKHKPV A 0.8s 15.4nm
24.	ePKIKP	A	22 40 40	New Ireland Region 4.95 S 153.29 E
	epPKIKP	A	40 50.5	H = 22 21 42.8 h = 50 km MB = 5.0
	LmH	C	23 20.5	D = 124.31 Az = 331.4 (NEIC)
	LmV	C	(40)	h = 37 km LmH C 35s 1.0 µm M = 5.2
24.	eX	A	24 05 10	Atlantic-Indian Rise 36.22 S 52.51 E
	LmH	C	42.0	H = 23 51 46.3 h = normal MB = 5.1
	LmV	C	49.7	D = 93.69 Az = 335.3 (NEIC) XV A 1.4s 13.9nm LmH C 30 0.9 µm M = 5.1 LmV C 18 2.0 µm 5.6
25.	eP	A	06 48 32	Unimak Island Region 53.90 N 163.61 W H = 06 36 47.1 h = normal MB = 4.3 D = 75.75 Az = 3.1 (NEIC) PV A 1.2s 12.2nm M = 4.8
25.	eP	A	09 02 17	Off East Coast of Kamchatka 55.75 N 163.66 E H = 08 50 45.3 h = normal MB = 4.4 MS = 4.3 (NEIC) D = 71.46
25.	ePKP	A	12 55 56	Tonga Islands 15.60 S 173.06 W H = 12 36 20.7 h = normal MB = 4.9 D = 144.84 Az = 354.8 (NEIC) PKPV A 1.4s 16.3nm

Day	Phase		h m s	Remarks
26.	ePKP	A	04 40 41	Halmahera 1.93 N 128.03 E
	ePP	A	45 05	H = 04 26 35.5 h = normal MB = 5.6
				D = 104.91 Az = 323.8 (NEIC)
				PPV A 1.9s 53.0nm M = 5.8
26.	eP1	A	13 31 03.5	Andreanof Islands, Aleutian Is.
	eP2	A	31 06	50.57 N 175.14 W
	e	A	31 16.5	H = 13 19 02.0 h = normal MB = 5.2
	e	A	31 26.5	D = 78.99 Az = 355.6 (NEIC)
				P1V A 1.1s 28.2nm M = 5.2
				P2V A 1.2 38.6nm 5.3
26.	eP	A	14 32 48	Kurile Islands 43.51 N 147.91 E
	epP	A	32 56.5	H = 14 20 46.5 h = 27 km MB = 5.2
	esP	A	33 00.5	D = 78.80 Az = 333.4 (NEIC)
				h = 33 km
				PV A 1.3s 15.3nm M = 4.9
26.	ePKIKP	A	16 09 54	Solomon Islands 6.28 S 154.84 E
	e	A	10 10	H = 15 50 57.1 h = 56 km MB = 5.6
	e	A	10 24	D = 126.21 Az = 331.8 (NEIC)
	e	A	10 30	PKIKPV A 1.1s 14.1nm
	LmH	C	17 09.0	LmH C 20 0.70/um
	LmV	C	09.2	LmV C 18 1.55/um
27.	eP	A	00 30 10	Sakhalin Island 46.51 N 142.18 E
				H = 00 18 34.2 h = 39 km MB = 4.5
				D = 74.24 Az = 329.8 (NEIC)
27.	ePKIKP	A	11 20 14.5	Tonga Islands 19.86 S 175.92 W
	-iPKHKP	AB	20 19.5	H = 11 00 56.8 h = 222 km MB = 5.4
	ePKP2	A	20 25	D = 148.70 Az = 350.8 (NEIC)
	epPKIKP	A	21 14	h = 236 km
	ipPKHKP	AB	21 17	PKIKPV A 1.8s 54.1nm
	ipPKP2	A	21 21	PKHKPV A 1.7 248.2nm
	eSS	C	42 40	pPKHKPV A 1.4 153.7nm
	e	C	42 55	pPKP2V A 1.3 109.3nm
	LmH	C	12(05)	

Day	Phase		h m s	Remarks
28.	ePKP	A	06 25 35	Tonga Islands 15.25 S 175.37 W
	e	A	25 54.5	H = 06 05 59.9 h = 35 km MB=5.1 MS=5.3
	LmH	B	07 34.3	D = 144.25 Az = 352.4 (NEIC)
	LmV	B	39.7	LmH B 18s 0.6 µm M = 5.4
				LmV B 18 0.8 µm 5.5
28.	eP	A	12 57 55	Nicobar Islands Region 9.06 N 94.19 E
	e	A	57 58	H = 12 45 56.7 h = normal MB = 5.1
				D = 78.35 Az = 319.9 (NEIC)
28.	eP	A	19 47 15	Off East Coast of Kamchatka
				55.78 N 163.88 E
				H = 19 35 55.5 h = normal MB = 5.0
				D = 71.46 Az = 341.8 (NEIC)
				PV A 1.4s 23.2nm M = 5.0
29.	eiP	A	09 53 14.5	Komandorsky Islands Region
	epP	A	53 26	55.23 N 164.53 E
				H = 09 41 51.7 h = 35 km MB = 5.0
				D = 72.10 Az = 342.3 (NEIC)
				h = 39 km
				PV A 1.0s 29.5nm M = 5.2
				pPV A 0.9 15.6nm
29.	eP	A	21 19 49	Southern Iran 29.55 N 52.85 E
				H = 21 12 37.3 h = 18 km MB = 4.9
				D = 37.29 Az = 316.2 (NEIC)
				PV A 1.2s 10.2nm M = 4.5
29.	+eP	A	22 37 52	Burma-India Border Region
	eX	A	37 56.5	25.12 N 94.67 E
	eY	AB	38 09.5	H = 22 27 02.0 h = normal MB = 5.5
	ePS	B	47 05	D = 66.71 Az = 316.5 (NEIC)
	eSS	B	54 35	PV A 1.0s 35.4nm M = 5.4
	e(P'P')	A	23 07 04.5	XV A 1.3 48.0nm
	e	A	07 19	YV B 1.7 172.8nm
	LmH	B	07.8	LmH B 12 0.8 µm 5.1
	LmV	B	12.5	LmV B 10 0.6 µm 5.2

Day	Phase		h m s	Remarks
29.	eP	A	23 30 39	Greece 39.32 N 21.26 E
	eS	A	32 53	H = 23 27 30.2 h = 18 km MB = 3.7
	LmH	B	35.3	D = 13.21 Az = 332.2 (NEIC)
	LmV	B	36.4	LmH B 14s 0.9/um M = 3.9
				LmV B 7.5 1.0/um
30.	eP	A	02 30 22	Dodekanese Islands 35.44 N 27.82 E
				H = 02 25 53.2 h = 0 km
				D = 19.20 Az = 327 (ISC)
30.	+eP	A	05 12 47.5	Near West Coast of Colombia
				5.64 N 77.75 W
				H = 05 00 12.9 h = 43 km MB = 4.9
				D = 85.28 Az = 39.7 (NEIC)
				PV A 1.3s 17.5nm M = 5.1
30.	+iP	AB	06 28 46.5	Eastern Kazakh SSR 49.75 N 78.13 E
	ePn	A	30 19.5	H = 06 20 57.7 h = 0 km MB = 5.8
	LmH	B	46.0	D = 41.28 Az = 297.7 (NEIC)
	LmV	B	46.8	PV A 0.7s 176.1nm M = 5.9
				PnV A 1.2 36.6nm
				LmV B 10 0.8/um 5.0
				underground explosion
30.	ePKIKP	A	06 57 30	Tonga Islands 19.39 S 173.96 W
	ePKHKP	A	57 33	H = 06 37 44.1 h = normal MB = 4.6
	ePKP2	A	57(39.5)	D = 148.50 Az = 353.2 (NEIC)
				PKIKPV A 1.5s 20.1nm
30.	LmH	C	11 30.0	Eastern China 30.86 N 122.62 E
	LmV	C	36.0	H = 10 47 05.1 h = normal MB = 4.7 (NEIC)
				D = 78.62
30.	+ePKIKP	A	15 58 23.5	New Britain Region 4.73 S 151.87 E
	e	A	59 55	H = 15 39 37.4 h = 109 km MB = 6.0
	ePP	AB	16 00 09	D = 123.44 Az = 330.8 (NEIC)
	e	C	00 46	PKIKPV A 1.8s 81.0nm
	e	C	10 32	PPV A 1.7 54.5nm M = 5.2

Day	Phase		h m s	Remarks
cont. 30.	eSS	C	16 16 50	LmH B 23s 1.7 µm
	LmH	B	43.2	LmV C 22 1.15 µm
	LmV	C	51.2	
30.	eP	A	19 18 32	Fox Islands, Aleutian Is.
	LmV	C	57.0	53.04 N 167.59 W
	LmH	C	20 01.5	H = 19 06 42.3 h = 27 km MB = 4.9
				D = 76.69 Az = 0.5 (NEIC)
				PV A 1.5s 40.2nm M = 5.2
31.	ePKP	A	01 07 08.5	Tonga Islands 15.32 S 173.67 W
				H = 00 47 35.1 h = 38 km MB = 5.0
				D = 144.51 Az = 354.2 (NEIC)
				PKPV A 0.8s 23.1nm
31.	ePn	A	09 08 46.5	Austria 47.17 N 16.52 E
	ePg	A	09 00	H = 09 07 36.1 h = normal
	eSn	A	09 33.5	D = 4.75 Az = 318.9 (NEIC)
	eiSg	A	09 51	
	LmH	B	10.0	
31.	e(PKHKP)	A	14 57 15	West of Macquarie Island
	ePKP2	A	57 26.5	56.68 S 147.83 E
	LmH	C	16 10.0	H = 14 37 13.20 h = normal MB = 5.1 (NEIC)
	LmV	C	10.0	D = 153.73
31.	eP	A	20 03 04	Andreanof Islands, Aleutian Is.
				51.91 N 179.93 W
				H = 19 51 18.8 h = 99 km MB = 5.4
				D = 77.36 Az = 352.5 (NEIC)
				PV A 1.5s 30.2nm M = 4.9
31.	eP	A	22 08 27	Komandorsky Islands Region
				56.0 N 166.3 E
				H = 21 57 01 h = 0 km MB = 4.9
				D = 71.70 Az = 343 (ISC)

Day	Phase		h m s	Remarks
31.	e	A	22 54 45	Probably to : Mindanao Philippine Islands 6.09 N 123.26 E H = 22 37 25.0 h = 608 km MB = 5.5 (NEIC) D = 98.76

Bei Fragen zur Produktsicherheit wenden Sie sich bitte an:
If you have any questions regarding product safety,
please contact:

Walter de Gruyter GmbH
Genthiner Straße 13
10785 Berlin
productsafety@degruyterbrill.com